ECOLOGY AND RESOURCE MANAGEMENT

A Quantitative Approach

KENNETH E. F. WATT   *Institute of Ecology*
University of California, Davis

Ecology and

A QUANTITATIVE APPROACH   # Resource

Management

New York   San Francisco   St. Louis

McGRAW-HILL BOOK COMPANY   Toronto   London   Sydney

ECOLOGY AND RESOURCE MANAGEMENT

Printed in the United States of America

Library of Congress catalog card number: 67-22973

234567891011   CZ   7543210698

68573

This book is dedicated to

Dr. Malcolm L. Prebble

Assistant Deputy Minister (Research)

Canada Department of Forestry

who for many years has labored to foster an environment in which large-scale, long-term basic ecological research on resources could flourish. His organization, and many others, have made it abundantly clear that the only sure road to management of renewable resources is via a deep understanding of ecological principles.

PREFACE

Some zoology and biology departments in universities have courses intended to show the relationship among animals, plants, and the world of man. These courses, given titles such as "Economic Vertebrate Zoology" or "Management of Natural Resources," are one of the few means by which it is possible to get a broad perspective of the field of natural resource management—and, unfortunately, such courses are rare. Most students gain their insight into this field from more specialized courses dealing with specific aspects of natural resources, such as fisheries, forestry, agriculture, tropical agriculture, range management, epidemiology, population ecology, economic and tropical entomology, and wildlife management. Consequently, the field of resource management has tended to fragment, and there is inadequate exchange of principles, theories, methods, and ideas about data processing and interpretation techniques among the various professional groups of resource managers. This book is intended to help remedy this situation by providing a text for courses that cut across disciplinary lines and consider all resources in a comparative approach. Since resource management is of necessity very much concerned with statistics, mathematics, and computer methods, the book considers these subjects at some length, and hence may be of use in courses on biomathematics or biostatistics. Also, it is my fervent wish that the book will be read by laymen who will then demand that local, state, national, and international agencies undertake the deep and clear rethinking of issues, policies, and legislation that is required in this area of increasingly desperate problems.

At the Davis campus of the University of California this book is used in undergraduate courses in resource management as well as for graduate courses in biomathematics. The graduate schedule also has programming workshops in problems related to those discussed in the text. The biomathematics course has been taken by students, faculty, and staff in the departments of agronomy, animal husbandry, biochemistry, entomology, genetics, geology, landscape horticulture, mathematics, nematology, and zoology, and also in Agricultural Extension, the Computer Center, and the National Center for Primate Biology. The second half of this book is designed to be of interest to people in a wide variety of applied biological disciplines.

A word is in order about the level of mathematical preparation assumed by the text. Calculus and at least a first course in statistics are essential prerequisites for understanding many of the chapters. While additional training beyond this level would certainly be helpful, it is not absolutely necessary because most of the other mathematical topics in this book are developed from first principles, using only calculus and an elementary knowledge of statistics.

The aim in this book is to reach an audience of biologists, rather than mathematicians. Therefore there is no attempt at mathematical elegance, existence theorems, or exhaustive theoretical treatment of mathematical methods. Only as much mathematical treatment is presented as experience has shown to be useful. No more space is devoted to derivation of formulas than is necessary to teach the biologist how to develop similar formulas for himself.

No man is an island unto himself; particularly in ecology, which weaves together skeins from many sources, each worker owes part of his intellectual development to many other people. I have been particularly fortunate in this regard. While a graduate student at the University of Chicago, I gained much from Dr. Dennis Strawbridge, and from the inspired teaching of Professors Thomas Park, Alfred Emerson, and Sewall Wright. Subsequently I have profited much from contacts with Professor Frederick Fry, and Drs. Frank Morris, Malcolm Prebble, Eugene Munroe, and C. S. Holling. Drs. Richard Bellman and Robert Kalaba were most kind and helpful in giving me the benefit of their insight into novel methods in applied mathematics. Drs. Holling, Munroe, Douglas Robson, Scott Urquart, and Burwell Gooch read the entire manuscript and offered a great many helpful suggestions. Dr. William Longhurst read Sec. 12.5 to its benefit. To these men and many others, I am deeply indebted.

Kenneth E. F. Watt

CONTENTS

THE *Problem* OF RESOURCE MANAGEMENT

THE *Theory* OF RESOURCE MANAGEMENT

THE *Principles* OF RESOURCE MANAGEMENT

ONE | THE *Problem* OF RESOURCE MANAGEMENT

1 | NATURE OF THE PROBLEM

This book has been written to satisfy two objectives. First, I wish to present a general theory of resource management that will be useful in dealing with all kinds of resource management problems, from oceanic fisheries to rodent control, to management of forests, rangelands, savanna, or orchards. Indeed, it is my major intent to show that there is a developing science of resource management per se, even though there is so little contact between the various resource management professions that this fact is often veiled. However, all the resource management fields are related to each other by dependence on *a common science*: ecology; by *a common problem*: the optimization, or "extremum problem" of mathematicians; and by the need for a *common set of tools*: scientific sampling, statistical and mathematical analysis, the intellectual procedures of operations research and systems analysis, and computers.

My second objective is to explain a set of techniques that are ushering in an entirely new era in the management of complex resource optimization problems. I am particularly interested in showing the great pertinency of certain new methods of applied mathematics and computer simulation for solving problems of enormous complexity in this field. To attain these objectives, a number of somewhat novel procedures are adopted in this book.

Throughout, there is great emphasis on a comparative approach. Darwin has taught all biologists that an enormously useful scientific method is to gain deeper insight into a principle by comparing the ways in which differ-

3

ences in various organisms affect the mode of operation of the principle. It is not possible to gain extensive insight into all the principles or procedures of resource management by looking at any one type of resource or even a small number of types. For example, because several commercial fish populations of the North Sea have been monitored carefully for many years, and because the two world wars relaxed fishing pressure and caused vast population dispersals and changes, these resources are among the best for gaining insight into population processes. Also, because of the great difficulty in obtaining information directly from mobile fish stocks, the most sophisticated theory extant for population estimation is based on catch per unit effort and was developed by fishery workers. But, because of the great difficulty of learning about community organization in the sea, terrestrial situations, as in East Africa, have done more to stimulate theory in this area. Deer and pheasant hunting stimulated the development of theories for population estimation based on change in composition of stock, since the bulk of the hunting was directed against males. The sophisticated modern theory of predator-prey and host-parasite interaction originated largely in entomology, as did theories on the effect of density on reproductive rate; modern thinking on the effects of herbivore community structure on succession in plant-animal ecosystems is largely the creation of range and wildlife managers and animal husbandrymen.

The point is that some fields of resource management are strong in some aspects of theory and technique, and others are strong in other aspects. Each field of resource management has much to learn from the others and this process of communication can be facilitated by creating a common body of theory and methods applicable to all fields, and contributed to by all fields. In this book each major problem in resource management will be illustrated using that type of resource by which the problem is best understood, and for which the documentation is most complete. For example, lampreys in the Great Lakes have been chosen to illustrate the population dynamics of a pest simply because there has been a tremendous amount of recent work on this problem. Similarly, to illustrate large-scale biological spatiotemporal waves, we discuss predator-prey phenomena among fur-bearers in the boreal forest and among forest-insect defoliators, because such waves have been most deeply studied in these cases. However, when we consider the resolution of conflicts of interest, it turns out that the resource most carefully studied by econometricians is water. The central theme running through this book therefore is not a particular resource or class of resources, but rather the strategy of managing resources of all kinds.

Throughout the book, it is pointed out exactly how mathematical, statistical, and computer procedures can be applied to the solution of large-scale biological problems. However, many readers will find the approach to mathematics is unusual in several respects.

Large-scale biological systems are dynamic phenomena of great complexity, embracing lag effects, cumulative effects, thresholds, interactions between variables, large number of variables, and nonlinear causal relations. Because of this complexity, we must concede at the outset that even mathematical models based on oversimplified and hence unrealistic assumptions would be insoluble using traditional paper-and-pencil methods. Therefore the emphasis is shifted from ingenuity of mathematical manipulation to realism of description, with computer simulation being substituted for paper-and-pencil solution. This does not lead to less complication in the mathematics. Rather it leads to complexity of a new type. The requirement now is for depth of insight into the character of the processes dealt with, in order that computer simulation studies can in fact mimic nature. An enormous amount of detail associated with such depth of insight can now be handled because of the great arithmetic speed and size of memory in the new computers. We propose that computer experimentation be used as a complement to actual experimentation, because costs per unit computation are dropping sufficiently low to make this an economic tool for research on systems where many variables are involved.

In essence, our approach is that management of a natural resource, and also the sampling processes by which we learn about the resource, can be thought of as a game. The object of research is to gain insight into the structure of the game and determine the procedure that optimizes the outcome when the game is played. A natural and logical way to gain this insight is to construct games, either in physical form, played with dyed chenille balls and baby bathtubs, for example, or electronic games played in the computer. The structure of the game is based on an elaborate program to measure the real system to be simulated, followed by development of a mathematical model mimicking the real system. Such games are then played repeatedly to determine the optimal strategy. In effect, we simulate nature.

A difficulty that arises immediately is that, because of the great complexity of nature, innumerable games may be necessary to determine the parameter values that optimize the resource situation in the game. This leads us to yet another subject, techniques for speeding up the search by trial and error for optimum management strategies. Several such techniques are presented in Chap. 13, on systems optimization.

Problems of manipulating animal populations, either to maximize the production of useful species or to minimize the survival of pest species, are all fundamentally similar. This statement holds true if the population to be maximized is bees, whales, deer, fish, or cattle, and if the population to be minimized is mosquitoes, grasshoppers, lampreys, rodents, or spruce budworms. The fundamental similarities are due to the fundamentally similar nature of the problems themselves. Note that optimization does not necessarily imply a search for maximum or minimum values of a variable. The value of an endemic-disease reservoir in maintaining resistance in a population whose treatment resources are inadequate to control outbreaks is an example of the optimum level of a harmful organism being above the practically attainable minimum. In addition to maximum, minimum, and optimum problems there is also a stability problem, in which the goal may well be not to attain a particular mean or extreme value but rather to damp wild fluctuations. This would be desirable, for instance, in changing an outbreak pest species into a stable species subsisting at a subeconomic level.

Resource management problems have two classes of features in common: *a common set of processes*, such as dispersal, predation, competition, parasitism, disease, the effects of weather on animal processes, and the effects of various factors on the structure and dynamics of communities; and *a common set of mathematical properties*, lag effects, cumulative effects, thresholds, interactions between variables, large number of variables, and nonlinear causal relations. Also, the problems are similar in that a particular group of mathematical, statistical, and computer techniques can be gainfully applied to all problems of this general type.

The availability of a common set of mathematical approaches is due to the fact that it makes no difference from a mathematical standpoint whether we wish to maximize or minimize some variable; in each case, we are dealing with what the mathematicians call an extremum problem. Because of the great interest in systems optimization in recent years in the military, business, industrial, and economic areas, mathematicians have provided us with an impressive arsenal of weapons to be used in solving optimization, or extremum, problems. Unfortunately these mathematical methods, which are by now routine in most other areas of science and economic life, have seen little application in resource management. This book aims to remedy this situation by presenting an integrated discussion of biology and mathematical methods. At every step of the way we shall attempt to point out as explicitly as possible why a particular mathematical method is needed and appropriate, and exactly how it is applied to the particular sets of biological data under consideration.

The main body of the text is in two sections. Parts One, Two, and Three have as their central theme the principles of ecology relevant to resource management, comprehensive theories of management derived from the principles, and case studies showing the application of the principles. In Part Four, we consider in turn the sequence of operations involved in scientific management of a resource: the measurement, analysis, description, simulation, and optimization of complex biological systems. Systems *measurement* is concerned with obtaining accurate and precise estimates of the dependent and independent variables. Systems *analysis* is determining which variables of those measured are most important in regulating the system. Systems *description* is concerned with taking those variables shown to be important and incorporating them into a systems model. Systems *simulation* is exploring the behavior of the systems model to evaluate the consequences of various strategies and policies for managing the system. Systems *optimization* is concerned with approaches for the systematic determination of optimal strategies. Of course, it is necessary to distinguish variables before beginning to measure them. Therefore a stage of survey and qualitative analysis is extremely important before we begin measurement followed by quantitative analysis. Also there is an important later stage of verification in which the results of simulation and optimization are checked in practice. The techniques of preliminary survey and final verification are typically so influenced by the nature of a particular problem, however, that we will not deal with them in this book.

2 | MAN'S POSITION AND FUTURE PROSPECTS IN THE WORLD

There is much discussion in newspapers and magazines to the effect that the world's population of human beings is rising dangerously fast, indeed to such an extent that the supply of natural resources will soon be depleted, with a resultant sharp drop in standards of living on an international scale. Is there any truth in these assertions, or are they merely uninformed attempts at scare sensationalism? In order to provide the reader with motivation for studying the principles and procedures in this book, it behooves us to define the exact nature of humanity's predicament with considerable care.

2.1 GROWTH OF THE WORLD'S HUMAN POPULATION

One means of determining how big the human population will be at various times in the future is to find the population growth equation that most accurately describes growth to the present, then use the equation to extrapolate into the future. To explain the basis of the equation, it is necessary to describe the logical rationale underlying mathematical descriptions of population growth.

If a population is introduced into an environment where it is not already found, population growth occurs initially at a rate dependent only on the size of the population present at any instant in time. Where N represents the population size, t is time, b is the instantaneous birth rate, d is the instantaneous death rate, and r, the intrinsic rate of natural increase, is defined as $b - d$, we have

$$\frac{dN}{dt} = (b - d)N \qquad \text{or} \qquad \frac{dN}{dt} = rN \qquad (2.1)$$

However, after a population has been in a new environment for some time, and the density of the population has increased to a high level, competition for food and other resources becomes severe, and this is reflected in dropping rates of birth and survival. We say that the population is being limited by environmental resistance. For any given population of animals or plants, and any given environment, there is a certain maximum number of members of the population which the environment is able to support. We shall call this maximum number K. As the population grows more and more dense, the growth rate becomes smaller and smaller as N approaches K. Hence, beginning with the first small group of immigrants, and following the population to the time at which N equals K, the form of population growth will be described reasonably well by the differential equation

$$\frac{dN}{dt} = rN(K - N) \qquad (2.2)$$

Allee and his colleagues (1949) give several examples showing the universal applicability of (2.2). A population growing in accord with (2.2) will have its maximum rate of growth when $d^2N/dt^2 = 0$, or when $N = K/2$. However, when a population is just beginning to expand in a new environment, N is very small relative to K, and, to a very close approximation, changes in the term $K - N$ occur so slowly that we can treat (2.2) as

$$\frac{dN}{dt} = rKN \qquad (2.3)$$

We can determine how important the term $K - N$ has become for any population as follows. Equation (2.3) integrates to yield

$$\ln N_t = \ln N_0 + rKt \qquad \text{or} \qquad N_t = N_0 e^{rKt} \qquad (2.4)$$

We can determine if (2.4) describes the growth of a particular population by plotting N_t, the number at a particular time for that population, against t, on semilog graph paper. If $K - N$ has become important for the population in question, the plotted points will not fall on a straight line, but will bend more and more downward from the left to the right side of the graph. The reader may determine for himself if (2.4) describes human population growth, by using the data on world total human population found in the tables referred to by Foerster, Mora, and Amiot (1960), supplemented with more recent data from the tables in the Demographic Yearbooks published by the United Nations Statistical Office. The reader will make the amazing discovery that his plotted values of N_t against year show that the world total human population is growing according to some law different from either the logistic (2.2) or the exponential (2.1), because exponential growth is the fastest a logistic population will ever show. Thus we have uncovered the surprising fact that the human population is growing in accord with a law different from that for any other plant or animal population ever observed, either in the laboratory or in nature! Kleiber (1961, table 19.1) has made this same observation.

Foerster et al. (1960) have proposed a new law to describe the world population growth of humans, which they derived as follows. Because of advances in medicine, and the speed with which these advances can be communicated from one country to another, death rates are dropping faster than birth rates, and hence r is not constant, but steadily rising. Furthermore, analysis of demographic data shows that the rate of rise in r is itself a function of N, expressed by the relation

$$r = aN^{1/k} \tag{2.5}$$

The support for this particular formula is particularly striking in present data for countries such as India where a rapidly dropping death rate and slowly dropping birth rate together are defined by (2.5). Substituting for r in (2.1) yields the new growth law

$$\frac{dN}{dt} = (aN^{1/k})\, N = aN^{1+1/k}$$

or

$$\int_{t_1}^{t} \frac{dN}{N^{1+1/k}} = a\int_{t_1}^{t} dt$$

from which we obtain

$$\frac{1}{-(1/k)N_t^{1/k}} - \frac{1}{-(1/k)N_1^{1/k}} = a(t - t_1)$$

or

$$\frac{k}{N_t^{1/k}} = a(t_1 - t) + \frac{k}{N_1^{1/k}} = \frac{aN_1^{1/k}(t_1 - t) + k}{N_1^{1/k}}$$

From this, we see that

$$N_t = N_1 \left[\frac{k}{aN_1^{1/k}(t_1 - t) + k} \right]^k \tag{2.6}$$

Multiplying the numerator and denominator of the expression in square brackets by $N_1^{-1/k}/a$, and adding and subtracting t_1 in the numerator, gives

$$N_t = N_1 \left[\frac{(k/a)N_1^{-1/k} + t_1 - t_1}{(k/a)N_1^{-1/k} + t_1 - t} \right]^k \qquad (2.7)$$

Foerster and his associates made the perceptive observation that the expression

$$\frac{k}{a} N_1^{-1/k} + t_1$$

is a measure of time, and the values are all constants describing the system. Therefore the time expressed is itself a constant, t_E. Thus (2.7) reduces to

$$N_t = N_1 \left(\frac{t_E - t_1}{t_E - t} \right)^k \qquad (2.8)$$

in which t_E is the end time for the system. As the difference $t_E - t$ approaches zero, N_t expands very rapidly, approaching infinity at $t = t_E$. For this reason, Foerster et al. have labeled the time remaining, or $t_E - t$, as "time to doom." Note that in (2.8), N_1, t_E, t_1, and k are all constants, so that defining two new symbols

$$K = N_1(t_E - t_1)^k \qquad \text{and} \qquad T = t_E - t$$

simplifies (2.8) still further to

$$N_t = KT^{-k} \qquad (2.9)$$

or
$$\ln N_t = \ln K - k \ln T \qquad (2.10)$$

We can determine if this equation describes the growth of the world human population by plotting population size against year on log-log graph paper, with years plotted backward (that is, with the earliest year plotted at the right-hand side of the graph and the latest at the left). Foerster and his colleagues (1960) used least squares analysis to obtain the values.

$$K = 1.79 \times 10^2$$
$$k = 0.990$$

and
$$t_E = 2026.87 \text{ A.D.}$$

The reader may determine for himself if these parameter values are still realistic by calculating N_t for each year from

$$N_t = K(t_E - t)^{-k} \qquad (2.11)$$

and plotting observed and calculated values of N against year on semilog graph paper. When my students perform this exercise, they discover to their horror that the world population is growing faster than the rate predicted by (2.11) in 1960! If this equation is correct, the world population of humans will become infinite, and therefore squeeze itself to death, as Foerster et al. describe our doom, some time prior to 2026 A.D.!

However, as might be suspected, there is a flaw in the reasoning underlying (2.11). Returning to (2.1) and (2.5), we see that

$$r = b - d \quad \text{and also} \quad r = aN^{1/k}$$

hence
$$b - d = aN^{1/k} \tag{2.12}$$

Clearly, there is an absolute lower limit below which the death rate cannot drop, and an absolute upper limit above which the birth rate cannot rise. Therefore $aN^{1/k}$ cannot increase forever, but ultimately will reach an absolute upper limit, which will occur when the difference $b - d$ is $b_{max} - d_{min}$. The difficulty is that because of changing medical procedures, we do not know exactly how low d can sink before it reaches d_{min}, and because of changing social values, we do not know how high b_{max} will be. All we do know is that $aN^{1/k}$ can rise considerably above its present level, and that (2.11) will therefore apply for some time into the future, during which the average international standard of living will drop to a really appalling level.

2.2 THE ULTIMATE CAPACITY OF THE EARTH

How, then, can we determine exactly how large the human population can become, and exactly how great the pressure on natural resources will be? The answer can be obtained by making use of the following pieces of information, all essentially fixed values, and all readily determined using methods acceptable to experts:

1. The amount of solar energy that falls on a unit area of land per annum
2. The number of units of land on the earth's surface suitable for growing crops, and the number of units of water in all the oceans, lakes, rivers, and streams
3. The percentage of the radiant energy from the sun that appears as chemical energy in each possible type of crop
4. The number of units of energy required to fuel an average human being for 1 year

If we know (1) and (3), we can compute the amount of energy captured on a unit of the earth's surface per annum for each type of crop. This result can be combined with (2) to calculate for each crop the total amount of energy that could possibly be captured on the earth's surface per annum. Then from (4), we can use division to obtain the total human population that can be supported by each crop, or by various combinations of crops.

Our four sets of information are as follows. They should be regarded only as very rough estimates, for reasons that will be explained in each case.

1. The flux density of the sun is different at different points on the surface of the earth, and because of long-term climatic cycles, will be very different during successive years at any particular location. However, as a very approximate average of all international measurements, we are taking the flux as about 1.6×10^{10} kilocalories per hectare per year, the California flux density. (A kilocalorie is 1,000 times the amount of heat required to raise the temperature of 1 gram of water from 14.5 to 15.5 degrees centigrade. A hectare is 10^4 square meters.) (Kleiber, 1961.)

2. The total available arable land in the world is about 4.56×10^9 hectares, which includes all land now under cultivation and all additional land cultivable in principle. (Pawley, 1963, and Christian, 1964.)

3. Conversion efficiencies of the various crops appear in Table 2.1.

4. Human beings of 70 kilograms of body weight (about 154 pounds for a man) have a basal metabolism of 1,700 kilocalories per diem. The Asian

TABLE 2.1

The capacity of the earth to support human beings* (Kleiber, 1961)

Crop	Percentage of the radiant energy from the sun appearing as chemical energy in crop	Approximate area required to grow enough food to feed one man 10^6 kcal/year, in sq m	World population of human beings that could be supported by terrestrial resources alone
Algae	12.5	5	9.1×10^{12}
Potatoes	0.10	600	7.6×10^{10}
Grain	0.05	1,200	3.8×10^{10}
Prunes	0.04	1,500	3.0×10^{10}
Milk	0.04	1,500	3.0×10^{10}
Pork	0.015	4,000	1.1×10^{10}
Eggs	0.002	30,000	1.5×10^9

*Current population is 3.0×10^9.

and Far Eastern mean intake per day in 1958 was 2,070 kilocalories per diem. A generous international average allowance per diem per human being is about 3,000 kilocalories, or roughly 10^6 kilocalories per annum. (Kleiber, 1961.)

TABLE 2.2

Trends in world production of major foods, in millions of metric tons (metric ton = 2,205 pounds)*

Commodity	Average 1948– 1952	1954– 1955	1955– 1956	1956– 1957	1957– 1958	1958– 1959	1959– 1960	1960– 1961	1961– 1962	1962– 1963	1963– 1964	1964– 1965	1965– 1966 (Prelim- inary)
Wheat	155.5	171.4	183.5	201.3	197.5	228.2	229.0	219.9	210.3	235.6	217.2	250.6	238.0
Rice	70.6	77.8	83.0	87.5	81.4	90.3	95.6	100.6	100.3	102.0	111.3	112.9	105.3
Milk	259.6	288.2	297.2	310.0	321.2	329.8	335.8	342.8	348.0	354.7	354.7	358.8	371.9
Meat	36.2	42.6	44.6	46.9	47.8	48.6	50.0	50.5	52.2	54.9	65.3	65.1	66.6
Eggs	8.8	10.3	10.6	11.0	11.4	11.7	12.2	12.5	13.0	13.3	13.3	13.8	14.2
Seafood†	21.2	27.0	28.3	29.9	30.9	32.2	35.7	38.0	41.8	44.9	47.4	51.6	52.9

*Data on seafood from table II-11, all others from annex table 2A in *The State of Food and Agriculture*, 1963, and annex tables 3A and 4 in *The State of Food and Agriculture*, 1966, Food and Agriculture Organization of the United Nations, Rome.
†Figures for seafood include mainland China; other figures do not. "Seafood" as used here includes fish, crustaceans, and mollusks.

For example, for algae, the energy trapped per annum in a hectare is $1.6 \times 10^{10} \times 0.5$ kilocalories. Therefore the approximate area required to grow enough algae in 1 year to support one person for that year is

$$\frac{10^6}{1.6 \times 10^{10} \times 0.5}$$

or, very roughly, 1 square meter.

It will be clear from inspection of Table 2.1 that a principal factor determining the world human population is the number of steps involved in converting incident solar radiation into food. Thus algae use the sun directly, and efficiency is high. Pigs, which eat plant material, are an extra step removed from incident radiation, and therefore use the land surface much less efficiently. The whole question of energy-conversion efficiencies in trophic communities will be explored at some length in Sec. 3.2.

The international picture with respect to utilization of various resources is changing from time to time and is also different in different places. Table 2.2 shows that the relative importance of food from the oceans is rising rapidly. Table 2.3 demonstrates that diets are different in different parts of the world and, for most of the world's inhabitants, protein intake is diminishing. It is also noteworthy that in most parts of the world, human beings are

TABLE 2.3

The relative importance of animal and vegetable protein in different parts of the world (food/person)*

Area	Time	Total protein (gms/day)†	Animal protein (gms/day)
North America	Prewar	86	51
	1961–1962	93	66
Western Europe	Prewar	85	36
	1961–1962	83	39
Mexico	Prewar	53	18
	1961–1962	68	20
South Asia	Prewar	52	8
	1961–1962	50	7
Far East, including mainland China	Prewar	61	7
	1961–1962	56	8

*From annex table 13B, *The State of Food and Agriculture*, 1963, Food and Agriculture Organization of the United Nations, Rome.
†100 grams = 3.527 ounces.

pressing too hard on resources to allow the luxury of growing plants to feed animals, which are then eaten. In South Asia and the Far East it is necessary to eat the plants directly, thus eliminating the lowered efficiency of the extra step in the trophic pyramid. Table 2.4 shows that on an international level, mankind's per capita intake of food peaked in 1959 to 1960, and has not risen since, despite the fact that much of the world's population is suffering from chronic malnutrition (Table 2.3).

Confronted with this alarming picture, most readers will be quick to ask if there may not be salvation hidden in some relatively untapped resource, such as the seas. A few simple calculations show how much help we can reasonably expect from this direction. Cod have about 70 kilocalories per 100 grams; salmon, 92; sardines canned in oil, 266; canned dry shrimps, 120; canned tuna, 255 (Wooster and Blanck, 1949). A safe assumption is that we can get 100 kilocalories per 100 grams of marine food of the present type. The 1964 world catch of marine food, including all countries, was about 46 million metric tons (46×10^9 kilograms). This present world catch of marine food represents approximately $46 \times 10^9 \times 10^3$ kilocalories. This will support

$$\frac{46 \times 10^9 \times 10^3}{10^6}$$

or 4.6×10^7 people per annum. Suppose the marine catch for the world could be made 20 times as great as at present. Then marine resources alone could support 9×10^8 people, or about one-third the present world population. This rather disheartening picture would only change greatly if we

TABLE 2.4

The race with fate:
*World per capita food production index numbers**

Time	Index number	Time	Index number
Prewar average	95	1959–1960	107
Average 1948–1953	95	1960–1961	107
1953–1954	100	1961–1962	106
1954–1955	99	1962–1963	107
1955–1956	100	1963–1964	108
1956–1957	103	1964–1965	109
1957–1958	102	1965–1966	107
1958–1959	106	1966–1967	106 (est.)

*From table II-2, *The State of Food and Agriculture* 1963, 1966, Food and Agriculture Organization of the United Nations, Rome.

switch to extensive harvesting of sea plants instead of sea animals. The difficulty with this possibility is that much of the world's sea plant material is low-density populations of algae; we might expend more energy sieving this out of vast quantities of sea water than we would obtain energy from the process, and energy, not money, will be the coin of the realm for human populations within a few decades.

In conclusion, the prospects for the future are rather bleak. It will become progressively more important to manage natural resources with great skill and foresight, and to use the most sophisticated tools at our disposal.

REFERENCES

Allee, W. C., A. E. Emerson, O. Park, T. Park, and K. P. Schmidt: "Principles of Animal Ecology," W. B. Saunders Company, Philadelphia, 1949.

Christian, C. S.: The Use and Abuse of Land and Water, in S. Mudd (ed.), "The Population Crisis and the Use of World Resources," pp. 387–406, Indiana University Press, Bloomington, Ind., 1964.

Foerster, H. V., P. M. Mora, and L. W. Amiot: Doomsday: Friday, 13 November, A.D. 2026, *Science*, **132**:1291–1295 (1960).

Kleiber, M.: "The Fire of Life," John Wiley & Sons, Inc., New York, 1961.

Pawley, W. H.: "Possibilities of Increasing World Food Production," Food and Agriculture Organization of the United Nations, Rome, 1963.

Wooster, H. A., and F. C. Blanck: "Nutritional Data," H. J. Heinz Co., Pittsburgh, 1949.

TWO | THE *Theory* OF RESOURCE MANAGEMENT

3 THE PRINCIPLES OF ECOLOGY

3.1 POPULATION ECOLOGY

Whenever man is actively exploiting a particular species population of animals, he must be concerned with the factors regulating numbers and average weight in that population, because these factors determine the yield, or harvest to man. Typically, the variable we wish to maximize is the biomass yield, that is, the product of the number harvested and their average weight. We must be careful to distinguish between biomass yield and two other measures: the *standing crop* and the *productivity*. Standing crop is the biomass present in the population at any time we happen to measure it, and biomass productivity is the rate of weight produced by a population. Studies on community energetics (Chap. 3.2) show that productivity is really a measure of energy flux per unit area per unit time. However, in most studies on the dynamics of exploited populations, it is more operationally feasible to define biomass productivity as the difference between the biomass left in the population after harvesting at time t and the biomass present in the population just before harvesting at some subsequent time $t + 1$. The maximum yield that can be removed from a population repeatedly is equivalent to the productivity. Typically, man removes a biomass yield that is considerably less than the productivity; the difference, productivity minus yield, is lost to natural mortality, or competition, as we shall see in this chapter. The object of rational management is to gain enough insight into the population ecology of the

21

exploited species so that we can keep productivity as high as possible. We attempt to adjust the amount and type of effort expended on the harvesting procedure so that the entire productivity can be cropped as biomass yield. As we shall see in subsequent chapters, high productivity, and hence high yield, can only be maintained if we exercise considerable care with respect to the allocation of harvesting effort against the two sexes and various age groups of the resource. One more term is required to show the relation between biomass yield and standing crop. The rate of exploitation is defined by the relation:

$$\text{Biomass yield} = \text{rate of exploitation} \times \text{standing crop}$$

There are five causal pathways by which a population's productivity can be regulated. A factor or group of factors can affect the: natality, mortality, individual growth rates, population dispersal, or partition coefficient between useful and nonuseful components of an organism. *Natality*, or the birthrate, may be influenced through fecundity, the production of eggs or sperm, or fertility, the proportion of the eggs that develop into living offspring. *Survival* is equal to 1.000 minus the proportion dying, or the proportion surviving at each of the sequence of stadia, or stages of life. The *individual growth rate* governs the biomass productivity. *Dispersal* may increase or decrease productivity, depending on the terrain. For example, dispersal can destroy a population by forcing it outward to where living sites are unfavorable, or alternatively, it can enhance population growth, by movement to previously underutilized resources. Partition coefficient is important since the important food-producing proportion of a plant is only one of its many elements and can alter widely without any alteration in the overall biomass productivity of the plant.

Several factors can operate through each of the causal pathways regulating productivity. These may be classified as: the animal itself; other animals of the same species; other animals of other species, including man; disease; food; weather; and site factors (soil, plants, topography).

The most important point to be brought out in the following discussion is that some of these factors are density-dependent, while others are density-independent. This distinction has been made for a long time by population ecologists (for history see Allee et al., 1949, p. 331) and refers to the fact that some variables operate against a population with an intensity that depends on population density. Such factors are called density-dependent; variables that have a constant effect without regard to density are called density-independent. Note, however, that such things as weather and site factors vary in their individual effects in proportion to the density of the population.

The animal itself is an important factor in population productivity because the reproductive rate, mortality rate, growth rate, and dispersal tendency are all age-dependent. Therefore a change in the age composition of a population in response to harvesting will have an important effect on all the attributes of that population. In Table 3.1 we see that in buffalo, as in most mammals, the reproductive rate is highest in females of intermediate age, and lower in very young and very old individuals. However, in fish, fecundity may increase gradually to an asymptote at the highest age measured (Table 3.2). The point made by these two tables is that while many attributes of animals change with age, the way in which a particular attribute changes with age may be quite different in different kinds of animals. This remark applies to natality, mortality, growth, and dispersal. Banding studies with birds have provided one of the best sources of data on the way in which mortality changes with age. The studies show a considerable variation from one

TABLE 3.1

The effect of age on reproductive rate of bison of Hay Camp herds, Wood Buffalo National Park, Northern Alberta, and District of Mackenzie (Fuller, 1962)

Age	Number in sample	Percentage pregnant
2 years	160	36
3 years	92	52
4 years	120	37
Young adult	95	40
Adult	192	28
Aged	74	8

TABLE 3.2

The effect of age on reproductive rate in herring (Hickling, 1940)

Age	Mean fecundity
3	14,620
4	17,679
5	20,482
6	23,102
7	25,580
8	27,931

type of bird to another. For example, Lack (1951) found that lapwings had a remarkably constant mortality rate from the first to the ninth year of life (between 29 and 40 percent per annum); but Richdale (1949) found a 58 percent mortality in yearling penguins, which declined to 10 percent per annum when the penguins were 4 to 10 years of age.

Table 3.3 shows the effect of age on the growth rate for market samples of plaice. These data are typical of age-growth curves in many animals: Growth rate spurts during adolescence, and then declines slowly. The rate of decline can be very erratic in middle and old age. Table 3.4 illustrates the way in which dispersal changes with age in smallmouth bass. These data, obtained from tagging studies where numbered tags identified an individual fish permanently, show a pattern found in many types of animals: a gradually increasing tendency to disperse as age increases.

TABLE 3.3

The effect of age on growth rate in plaice,
measured in market samples from 1929–1938*

Age	Percentage increase in weight, next year of life	Age	Percentage increase in weight, next year of life
2	15.4	11	2.7
3	22.1	12	12.7
4	13.6	13	8.1
5	12.9	14	9.2
6	12.9	15	5.5
7	12.7	16	6.4
8	12.2	17	− 3.0
9	11.4	18	9.4
10	12.8	19	− 4.2

*Calculated from table 16.2, Beverton and Holt, 1957.

Other animals of the same species can also affect natality, mortality, growth, and dispersal. They can interfere with copulation, and compete for oviposition or nesting sites, and for food and places to live. However, in density-dependent processes such as these, the optimum population density is not necessarily the lowest density; typically it is at some intermediate density. Allee noted this phenomenon early in the history of population ecology (Allee, 1931; Allee et al., 1949, Chap. 23). Reproduction rate will be highest at intermediate densities for the following reasons. When density is

TABLE 3.4

The effect of age on dispersal of South Bay smallmouth bass (Watt, 1959)

Age	Mean minimum distance traveled, in miles per day
2	0.014
3	0.050
4	0.075
5	0.088
6	0.105
7	0.142

very low, prospective mates will have difficulty finding each other. This phenomenon is occurring now in blue whales (Chap. 5.2). When density is very low or very high, the percentage of females in copula can decline. Mac-Lagan and Dunn (1935) observed this decline experimentally in *Sitophilus oryzae*, the rice weevil. As densities go to very high levels, oviposition rates decline from an optimum level because of competition for reproduction sites and interference with the mating process (Chap. 11.4). Survival is also higher at intermediate densities in many organisms. Lack (1948) found that survival rates in Swiss starlings in their first year of life were optimal at intermediate densities (Table 3.5).

Competition between species can also affect the attributes of a population. It is important to note that this competition can occur in a subtle and com-

TABLE 3.5

The effect of population density on survival of Swiss starlings ringed in nest and recovered at least 3 months after leaving nest (Lack, 1948)

Clutch size	Number of young ringed	Percentage recovered
2	328	1.8
3	1,278	2.0
4	3,956	2.1
5	6,175	2.1
6	3,156	1.7
7	651	1.5
8	120	0.8
9, 10	28	0.0

plicated way, and often without any direct aggression being observed between individuals which are, however, competing intensely in a real and measurable fashion. Johannes and Larkin (1961) describe an interesting example of competition between redside shiners (*Richardsonius balteatus*) and rainbow trout (*Salmo gairdneri*) that illustrates these points. The nature of the inter-specific relationship varies with the age of individuals in both species: Shiners prey on trout fry, shiners and juvenile trout compete for resources, and large trout prey on shiners. Shiners entered Paul Lake, British Columbia, between 1945 and 1950, and between 1946 and 1957 there was a marked decrease in the growth and survival of Paul Lake trout under 20 centimeters, fork length. The competition between shiners and juvenile trout happened in an interesting way: The shiners overgrazed *Gammarus* by pursuing them deeper into the weeds than the trout did. This forced the shiners and the juvenile trout to switch to other food. Thus there was no interspecific aggressive behavior. An especially noteworthy element in this example is that proof of competition would not have been observed unless scientists had been studying Paul Lake while the overgrazing of *Gammarus* and the consequent switch to less satis-factory food was in progress.

Predation and parasitism are other forms of interspecific relationships that can have marked effects on a population; they will be discussed at length in later chapters. It is sufficient to say here that rapid annihilation of a popula-tion by a parasite or predator can only be expected where the parasite or predator has immigrated into an area new to it, and the inhabitants have not evolved behavioral mechanisms allowing them to cope with the pressures of the new enemy. (Perhaps the most spectacular recent example is that of the marine lampreys in the Great Lakes.) A predator and prey that have evolved together exist in a state of balance, in which much of the predation or parasitism pressure is directed against sick, old, or superfluous young individuals, as Fuller (1962) found in wolf attacks against buffalo. The state of balance must persist, or all the prey or hosts will be destroyed and the predators or parasites will become extinct themselves because of starvation.

Man has been the most destructive predator, since he suffers no severe penalty if he exterminates a prey species, but simply shifts his predation pressure elsewhere. A particularly rapacious example of man's effect on a species is given in Table 3.6, which documents the slaughter of fur seals of the North Pacific islands in the last decade of the nineteenth century. Proof of the great impact on the stock is that even though the number of vessels dropped to 38 in 1897 from 87 the previous year, in response to a declining catch per unit effort, the catch per unit effort remained low.

Man may have a far more destructive effect on many species uninten-

TABLE 3.6

Catastrophic overharvesting by man:
Decimation of North Pacific stocks of pelagic fur seals*

Year	Number of vessels	Catch	Catch per ship
1894	37	55,686	1.53×10^3
1895	59	56,291	0.95×10^3
1896	87	43,917	0.51×10^3
1897	38	24,321	0.64×10^3

*From Jordan (1898), part 1, pp. 174–175.

tionally, by polluting their environment, than by overharvesting. The side effects of insecticides have received so much discussion elsewhere that it is not necessary to dwell on this matter here, but another type of phenomenon that receives almost no attention is noteworthy. The following account is drawn from the monograph by Tuck (1960). The murre (genus *Uria*) is probably the most abundant Northern Hemisphere sea bird. They weigh about 2 pounds as adults, and with their black heads and markings and white bellies might be mistaken for small penguins (which however are confined to the Southern Hemisphere). Murres are important to man as food, since the crop is a million birds and 2 million eggs a year in the Northern Hemisphere. But, they are far more important for recycling nitrates and phosphates in the ocean. The total world population of murres is about 50 million birds. These consume about 100 million pounds of plankton and fish weekly. Since murres feed on a large number of bottom fish in the summer, much of the nitrates and phosphates returned to the top layers of the sea in their excretions might otherwise be lost by drifting to the bottom of the sea. There is little circulation in the arctic marine regions, and for this reason recycling by murres may be very important.

We will now consider how this ecologically important mechanism may be destroyed by man through thoughtlessness. Ocean vessels usually flush out waste and dirty oil along the coast of Newfoundland. This oil destroys the insulating air pockets of the murres, so the birds ultimately die of exposure. The oil also increases the birds' specific gravity, causing them to be unable to fly or dive, so they drown. Many birds probably drown without being observed, but at least 1,000 murres are destroyed per day along the southeast coast of Newfoundland. For much of the coast, no attempt has been made to accurately count the mortality from oil pollution, so the total figure may be several million birds a year.

It is part of our everyday experience that disease can have a sudden and enormous effect on animal or plant populations. Any observant person will have noticed evidence of mass mortality caused by pathogens to grasshoppers, caterpillars, or trees. The same phenomenon can occur in the ocean. Tibbo and Graham (1963) reported that at least half the mature herring stocks in the western Gulf of St. Lawrence were wiped out during 1954 to 1956 by a fungus infection. This was reflected in the decrease in the catch of herring larvae per sampling tow in the area (Table 3.7). Note that disease is a density-dependent mortality factor (Chaps. 6.3 and 11.8), and because of its tremendous mortality effect in certain cases, may so diminish the intensity of the struggle for existence that the population is changed for some time after the epidemic.

TABLE 3.7

The effect of disease on abundance of herring larvae in the western Gulf of St. Lawrence (Tibbo and Graham, 1963)

Year	Catch per sampling tow
1952	1,897.8
1953	855.9
1954	127.4
1955	407.8
1956	31.6
1957	32.5
1958	9.1
1959	1.5

In the case of the herring stock in the Gulf of St. Lawrence, after the epidemic, the mean age of the fish in the stock dropped from 7 to 6 years, and the growth rate increased. It is important to note that disease will strike wounded, weakened, or crowded members of a population first and hardest; a diseased individual is also more likely to succumb to other mortality factors. For example, Fuller (1962) noted that bison are much more likely to be attacked by wolves if they have tuberculosis. This is an example of an interaction effect: Occurrence of one type of event increases the likelihood of occurrence of another type of event. In Sec. 11.2 the process of determining how best to mathematically express the operation of interactions, given adequate data, is described.

The availability of food also has a great effect on a population. Of importance are the degree of contagiousness, or spatial clumping of the distribution of food, and the way in which the clumps are arranged (large numbers of small clumps or small numbers of large clumps). Such matters are considered in great detail, experiments are described, and mathematical models proposed in the important treatise on feeding by Ivlev (1961). Holling (1965, 1966) discusses feeding behavior and physiology in great detail as it relates to feeding by predators. These matters will be considered in later chapters; here we will confine ourselves to an interesting example showing the great impact that increased food can have on the growth of animals. Cooper and Steven (1948) discussed the benefits to be gained by fertilizing lakes, inlets, fjords, and arms of the sea. (Many of the world's coastlines, as in Scotland, Norway, British Columbia, and the states of Washington, Virginia, and North Carolina are very much indented by arms of the ocean.) Loch Craiglin, Argyllshire, was fertilized. The bottom fauna population increased so that its dry weight was several times that of the prefertilization level. Plaice laid on 2 years' growth in 1 year, and flounders increased in weight up to 18 times per annum that of the species in average natural conditions. Cooper and Steven suggested damming up arms of the sea to attain two goals at once: conversion of tidal power to electricity, and marine fish farming. The cost of the dam could not be justified by either goal above but might be met by both together.

Weather has a tremendous effect on many animal populations, and this effect can be achieved in many ways. If an entire area warms up or cools down, the reproduction, growth, and survival of organisms in that area will be affected, provided the temperature change is sufficient. For example, Fry and Watt (1957) found, for a group of unconnected bodies of water at the southeast corner of Manitoulin Island, a simple linear relationship between the contribution of year classes of smallmouth bass to the sport fisheries, and the summed monthly deviations of temperatures from long-term means for the period June to October in the year the fish were spawned. (A year class is the group of animals of a particular species born in the same year; being able to assign animals correctly to year classes depends on being able to ascertain their ages from annual rings in scales, otoliths, bones, teeth, or horns. This can be done for many species.) Subsequently Watt (1959) found that in the third year of life, temperature accounted for 76 percent of the basses' variance in weight growth from year to year, and 95 percent of the year-to-year variance in recruitment. A sudden temperature drop caused a tremendous mortality of warmwater fish of all ages on the coast of Texas in January, 1940 (Gunter, 1940), so temperature is important for the survival of adult fish as well as of fry and eggs. The flow of hot or cold masses of air, or streams of air or water through other masses normally at a different temperature, can also affect fish mortality. Carruthers and various associates (1937, 1951, 1952) have shown the relation between winds and brood strength of North Sea fisheries, and Sette (1943) and Stevenson (1962) have shown the importance of ocean currents for the survival of young Atlantic mackerel and Pacific herring, respectively. Stevenson showed that young herring washed out to sea by currents do not survive, and the proportion of the young stock not so dispersed determines the strength of a year class.

The foregoing remarks have as a corollary that environmental factors (called "site factors" by foresters) can affect organisms in three ways. First, if organisms stay where they are, and the environment changes, it is probable that natality or mortality and the growth rate will be affected accordingly. Second, if animals reach a certain point in their life at which they must move to a favorable site but do not, the probability of success for the population is lowered. Third, if organisms move, or are passively transported by wind or water currents to a new site, their probability of survival will be raised or lowered depending on the suitability of the new site. The environment can become unfavorable to animals in many ways: A decrease in dissolved oxygen as a result of pollution will eliminate trout, drying up swamps will eliminate hippopotami and ducks, clearing forested land will eliminate the natural habitat of the big cats and certain kinds of warblers.

It is important to note that different factors will probably be restrictive for the same species in different parts of its range. That is, near the central part of the range of a species, climate will probably be optimal, and weather will rarely be an important regulator. Under such optimal weather, the population will be regulated by density-dependent factors, such as competition for food, disease, parasitism, and predation. At the edge of its range, the weather will only be good enough to support population increase in, say, 4 years out of every 10, and the population will never become dense enough for density-dependent factors to regulate it. Precisely this state of affairs was discovered in the smallmouth bass previously mentioned. In the north end of Lake Huron, these fish approach the northern edge of their range. In a 9-year period in which the bass were studied carefully, reproduction was successful in only 3 years; in 2 of the 9 years, reproduction was only about 2 percent as successful as in the most successful year (Watt, 1959, table 24). Therefore the only universally applicable rule of population regulation is: the particular factor regulating a species population at a particular place depends on the suitability of the climate at that place relative to the physiological tolerance limits of the species in question.

A second difficulty arises because whenever we observe a striking effect, a plethora of factors are possible causes simply because an enormous number of factors are always operative in any natural situation. For this reason we run a great risk of deciding that the wrong factor caused the observed effect unless our explanation is based on a carefully designed sampling scheme conducted in enough places, and for an adequate length of time, with a mathematically and statistically sound analysis of the resultant data. The following example illustrates the nature of the problems which can confront the resource manager. Between 1941 and 1947, the lake trout and whitefish fisheries of Georgian Bay, Lake Huron, suffered a decline. The catch of trout dropped from 1,501,600 to 368,100 pounds, and the catch of whitefish dropped from 748,000 to 87,300 pounds (Frankland, 1955). Six theories to account for this decline were available, depending on who was sought for an explanation. A favorite theory from anglers was that overfishing by commercial fishermen caused the drop. Commercial fishermen were inclined to believe that the damage was due to smelt eating the eggs of the larger fish, or to predation by lampreys (both smelt and lampreys had recently invaded Georgian Bay and were building up to enormous population densities in the period 1941 to 1947). Other theories attributed the decline to destruction of the spawning beds by pollution, weather, and endogenous population factors. But, beginning in 1948, while trout catches remained low, whitefish catches increased spectacularly, and by 1953 the catch was 6,166,200 pounds, the largest on record to that time (Frankland, 1955). Unless one had a long run of data extending over, say, a 15- or 20-year period, he might assume that a relationship observed between two variables over a short period of time implied some causal connection, when in fact the relationship was fortuitous (smelt just happened to be building up when whitefish were declining). Another logical difficulty in interpreting such situations can arise because two variables are both changing under the influence of a third, which is the real cause. For example, fish species A might be declining and B expanding not because B was eating A, but because the water was warming up $2°F$ every 50 years, and A was a coldwater species and B a warmwater one. It takes sophisticated statistical and experimental procedures to discover the truth in such situations.

A third difficulty in understanding resource management problems arises because growth, survival, and reproductive rate may all be increased by harvesting a population, providing it is not harvested too hard. Harvesting may actually *increase* the productivity of a population (this is an example of homeostasis).

Interaction also compounds the difficulty of interpreting data. For example, deer are more likely to starve if their population density is high, if the lowest foliage on tall trees is too high for them to reach, and if snow covers low browse. Here we have an interaction of population density, snow depth, and height of lowest tree foliage. Bison are much more likely to be killed by wolves if they are young, aged, or have tuberculosis. Predation by wolves is thus related to an interaction of age, health, prey density, and predator density.

There are numerous pitfalls in interpreting population data which can only be circumvented by good statistical design in the research program. For example, suppose that growth and survival rates are measured at three different sites. Suppose that despite the fact that densities are different at all three sites, growth and survival rates are the same. Does this mean that density has no effect on growth and survival? No, because density might have had an effect that was masked by an environmental gradient. The solution is to use an experimental layout or sample survey design in which different sites and different densities are used as two of the factors in a factorial design, that is, to try to sort out the effects of density and site factors by taking measurements and samples at several different density locales for each type of site.

This discussion leads to a consideration of the requirements for a proof that natality, growth, or survival of a species are regulated by a particular factor or factors. First there must be adequate replication in space and time to explore the effects of various site and weather combinations. That is, at each site weather is constantly changing, and to be able to sort out the effects of weather changes through time from those of various sites at different locations in space, we need an adequate number of years of data for each site, and an adequate number of sites measured in each season. Of course, the same site will be measured in each of the seasons, so that at the end of a sequence of years, the statistical analysis can be performed on a set of data corresponding to the output from a factorially designed experiment. This comment may seem unnecessary to some readers, but it is made because many scientists have a habit of letting curiosity overcome their common sense, with the result that some study plots are discontinued and others initiated partway through a research program. How many site times weather combinations are necessary? This depends on the number of independent variables being measured. One would never try to define the shape of a curve on a plane with less than six experimentally determined points. Ideally, we would like 6^n site times weather sets of data if there are n independent variables. However, this number is impractical if n is large, and it must be cut down by means of Youden squares, confounding, or other tricks of experimental design as explained in texts on experimental designs. There must be adequate replication at each sampling locus to distinguish real effects from the effects of chance (Sec. 9.1), and there must be a sufficiently elaborate experimental design to allow statistical separation of the effects of the various factors operating at each site. Note that the particular types of designs required depend on whether or not there are environmental gradients.

We shall now use a mathematical model, and graphical representation, to explore the implications of the facts presented. Consider any age group of animals in a population. Let $_jN_i$ denote the number, N, of animals of age-group j in year-class i. Thus, $_3N_{1946}$ refers to the 3-year-olds born in 1946 which are still alive in 1949. The fraction of these animals surviving 1 year is given by $_jS_i$. Thus

$$_{j+1}N_i = {_jN_i}(_jS_i)$$

If the per capita reproductive rate is $_jr_i$, the total contribution to the stock 1 year later from the $_jN_i$ is given by

$$_jN_i(_jS_i) + {_jN_i}(_jr_i)(_0S_{i+j})$$

Now suppose we consider the number of eggs (or living young) produced by animals of a particular age j^*, in year $j^* + i$. This number of eggs will be

$$_{j*}N_i(_{j*}r_i)$$

We will refer to this number of eggs as N_p, the number in the parental generation. Now consider the number of eggs produced by the survivors of N_p when the survivors are j^* years old. This number of eggs will be referred to as N_o, the number in the offspring generation. We have

$$N_o = \left[_{j*}N_i(_{j*}r_i) \right] \left(\prod_{j=0}^{j*-1} {_jS_{i+j*+j-1}} \right) {_{j*}r_{i+2j*}}$$

For example, N_p might refer to the number of eggs produced by 6-year-olds in 1946, then N_o must refer to the number of eggs produced by 6-year-olds in 1952.

As indicated previously in this chapter, the r and S values are all under the influence of a great many factors, and if we developed an analogous formula for biomass, the growth rates would likewise be under the influence of a great many factors. The whole point of research on population ecology in resource management is to find out how to make $N_o = N_p$, and to optimize N_p by manipulating the value of N_p, and the r and S values. We shall now consider the general theory of how this maximization can be achieved.

We might explore the relationship between N_o and N_p by considering either plots of N_o versus N_p, or N_o/N_p versus N_p. The latter leads to statistical difficulties because where $j^* = 1$ (as for most insects, for example), we are left with the problem of analyzing the relationship between N_{t+1}/N_t and N_t. This formulation produces a spurious artifact hyperbola, and a misleading "density-dependent relation," even where there is no density dependence at all (Watt, 1964a). Consider a random number series. If we divide each number in the series by the preceding number, and then plot the resulting ratios against the denominators, we must obtain a hyperbola, around which, of course, there is a great deal of scatter. For a given N_t, since we are dealing with a random number series, any number in the series, except the first, could in principle be the numerator of the ratio N_{t+1}/N_t. If E denotes "average, for all possible samples" (that is, the "expected value") then for N_t^*, a fixed N_t,

$$E(N_{t+1}/N_t^*) = E(N_{t+1})/N_t^* = \frac{\sum\limits_{t=2}^{n} N_t/(n-1)}{N_t^*}$$

The numerator of this ratio is a constant, and hence the expected form for the plot of N_{t+1}/N_t versus N_t will be a hyperbola. For this reason it is much better to develop our theoretical discussion around the plot of N_{t+1} versus N_t, rather than N_{t+1}/N_t versus N_t. The more general graph of N_o versus N_p will appear somewhat as in Fig. 3.1 for any species of organism. At $N_p = 0$, N_o must equal 0. As N_p rises, N_o rises at an increasing rate because of the increased probability that reproduction will occur as population density rises. However, as N_p continues to rise beyond an optimum value, the rate dN_o/dN_p decreases because of intense competition. Note that the position taken by this curve with respect to a 45° line through the origin is critical for describing the temporal trend of any species population. Where the curve lies above the 45° line, the population is increasing; where it lies below, the population density decreases. For all populations, the curve must cross the 45° line at two points: A and B, the equilibrium points. The curve must drop below the 45° line below A because as population densities become very low, the probability of successful reproduction occurring must approach zero for all species. For example, where there are two sexes, the probability of prospective mates encountering each other drops to zero as the density drops. The curve must drop below the 45° line above B because of intensifying density-dependent mortality as population density rises. Note that A is an instable equilibrium point because populations slightly greater than A tend to increase; those slightly less tend to decrease. On the other hand, B is a stable equilibrium point because populations slightly greater than B tend to drop toward B; those slightly less than B tend to rise to meet it.

Two features of this curve define the pattern of the temporal trend for any species population. The particular shape of the curve determines the

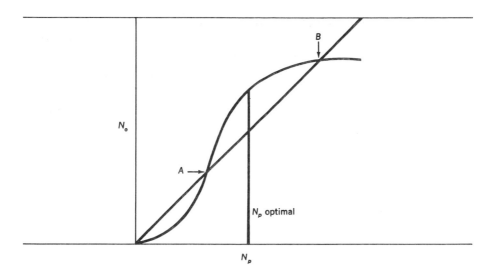

FIGURE 3.1

Relation between numbers in the offspring population and numbers in the parental population. (See text for discussion.)

density at which equilibrium occurs. The amount of scatter about the curve when actual data for any species is plotted shows the relative importance of density-dependent and density-independent factors in regulating abundance of the species. Where there is a great deal of scatter, non-density-dependent factors such as climate are of primary importance in regulating abundance, and a great deal of variability in numbers from one year to the other is to be expected. Where there is little scatter, density-dependent factors are of primary importance in regulating the species, and there is relatively little variation in numbers from one year to another.

The ideas presented in the preceding two paragraphs have been the subject of much discussion by ecologists for several decades. Nicholson and Bailey (1935) and Morris (1959) have discussed plots of N_{t+1} versus N_t, particularly as they related to host-parasite systems. Ricker (1954a,b) has discussed the relation between the shape of curves, as in Fig. 3.1, and the resulting patterns of population fluctuation. Utida (1941), Odum and Allee (1954), Slobodkin (1955), Klomp (1962a), Watt (1963), and Takahashi (1964) have all perceived the importance of the equilibrium points. Radovich (1962) noted that scatter about such a curve expressed the effect of weather, or other non-density-dependent factors on populations of sardines.

In Secs. 4.1 and 4.2, we shall explore the implications of Fig. 3.1 for maximization of productivity and pest control, respectively. However, before concluding this chapter, it behooves us to mention a problem that the student will discover for himself as he studies the literature of population ecology. A controversy about the relative importance of weather and density-dependent factors in regulating animal populations has concerned ecological literature for more than 3 decades. The interested reader may gain insight into this through the following publications and their lists of references: Andrewartha (1957), Andrewartha and Birch (1954), Klomp (1962b, 1964), Milne (1957a,b), Nicholson (1954a,b, 1958), H. S. Smith (1935), Thompson (1939, 1956), Varley (1959). The truth in this controversy, as in so many, is contained in part in all of the arguments. Density-dependent factors are the ultimate regulators of density for all populations, because if all other factors fail to regulate a species, starvation certainly will not fail. However, there are differences in the degree to which density-dependent factors are actually controlling various populations, and in the cases discussed by Radovich (1962), Ricker (1954b), and Watt (1959), non-density-dependent factors seem to have the primary role in many instances. Climate is important either at the edge of a species range, or in a situation where climate is typically instable because of erratic winds or ocean currents. As we shall show in subsequent chapters, the relative effect of climate and density-dependent factors is of the utmost importance in development of management policies for a particular resource.

3.2 COMMUNITY ENERGETICS

When we examine the details of food cycles, we see that the nonliving and living elements in nature are bound together in a system through which energy cascades and matter cycles: the ecosystem. Plants use incident solar energy and minerals to manufacture living tissue later eaten by herbivores, which in turn are eaten by carnivores. As the living tissue dies at each stage, it may be broken down into mineral material which then reenters the cycle. It is important to note that while matter can cycle through the ecosystem repeatedly, energy passes through only once (for a given unit of incoming radiation), and there is wastage at each level of the food pyramid (plants, herbivores, primary and secondary carnivores). This is why we say that energy cascades, whereas matter cycles through the system. Lotka (1925) and MacFadyen (1948) made this important distinction. Because an individual unit of matter may be reutilized by a trophic pyramid, and a quantum of energy may be used only once, productivity must be defined in terms of energy flux per unit area per unit time. The caloric equivalents for matter are determined by burning material in an oxygen-bomb calorimeter. When matter is used as the basis for computing productivity, the energy flux in a

36 *The theory of resource management*

trophic pyramid will be underestimated, and the more the flux per unit time, the greater the percentage underestimate for a given sampling interval.

Our interest in this section is in discovering the various factors determining the efficiency of utilization of incident radiation, in order to get clues as to how to optimize resource management policy. The main features of a typical trophic pyramid are brought out in Table 3.8. Note that there is tremendous energy wastage at each step because plants do not capture much of the incident solar radiation; and only a tiny proportion of the living tissue produced at each step is eaten by the next step. Few food pyramids have more than five levels because of the tremendous wastage at each step. Note

TABLE 3.8

Energy flow* in a terrestrial community (Golley, 1960)

Trophic level i	Flux-incident solar radiation, gross photosynthetic production, or consumption λ_i	Respiration R_i	R_i/λ_i	Efficiency λ_i/λ_{i-1}
Sunlight	47.1×10^8			
Vegetation	58.3×10^6	8.76×10^6	0.150	0.012
Microtus (Field mouse—herbivore)	250×10^3	170×10^3	0.680	0.004
Mustela (Least weasel—carnivore)	5,824	5,434	0.933	0.023

*All measurements in calories per hectare per year.

that respiration loss is an increasing proportion of the total flux as we go from plants to herbivores to carnivores. The price an organism pays for increasing mobility (implying freedom to search considerable distances for food, and a lack of dependence on the spot where it is born) is in increased need for energy. Finally, note that efficiency is greatest in the terminal stage of the food pyramid, the largest animals. This efficiency is bought with the high energy expenditure of predators. That is, increased respiration implies increased activity and a better chance of locating prey. Man's current philosophy of harvesting the oceans is a product of the foregoing principles. We let the fish gather up the plankton for us, saving us the energy expenditure that would be required if we gathered up the plankton. All trophic pyramid caloric measurements obtained to date reveal the same picture presented by the data in Table 3.8 [see, for example, the tables in Lindeman (1942) on Cedar Bog Lake and Lake Mendota].

Ecosystems go through an evolutionary sequence during which their productivity changes. Lindeman (1942) has explained the reasons for the changes. Consider a deep lake with a limited number of dissolved nutrients. Such a lake is called oligotrophic because of the paucity of food. Much of the solar energy hitting the surface of the lake is dissipated in heat because of a low concentration of photosynthesizing plants per unit volume of water. Gradually, the productivity of such a lake increases with increasing influx of nutrients from the surrounding drainage basin, and the lake accumulates living organisms faster than they can be removed by respiration, predation, and bacterial decomposition. The lake becomes shallower at an increasing rate, because of the aggregation of organic debris at the bottom, and is said to be "senescent" because of oxygen lack. Note that we are describing a sequence of events that could be prevented by intense harvesting. Finally, the entire area is filled with rooted plants, and the lake fills up very rapidly with organic debris, becoming a bog. At this point, the productivity of the lake drops to a very low level because the habitat is not suitable for aquatic or terrestrial vegetation. Gradually, as the bog dries, it becomes a bog forest, and finally, a climax forest. The transition from bog to climax forest is accompanied by a gradual increase in productivity.

Some recent work by Ovington, Heitkamp, and Lawrence (1963) indicates the productivity of various terrestrial plant associations (see Table 3.9). From these data we can draw several conclusions. The most striking point is that the greater the number of layers of vegetation, the greater the productivity will be, all other things being equal. More layers of vegetation increase the probability that a given quantum of incident solar radiation will hit—and be used by—a photosynthesizing surface before striking the ground and being dissipated as heat. Maize productivity is high because of fertilizer, indicating the tremendous dependence of productivity on the quantity of nutrient material available. Plant productivity is also enormously dependent on the presence of woody perennials with strong enough stems (trunks) to support a multistoried canopy or a thick top canopy. Woody perennials provide more efficient utilization of cubic space for growing room or, more likely, the development of their much more extensive root systems permits greater utilization of mineral and organic nutrients and even water from the soil. This is supported by the fact that fertilizer, not increased utilization of solar energy, is responsible for the high productivity of maize.

The general conclusion is that man can make best use of incident solar

TABLE 3.9

Annual primary net productivity in oven-dry weight, kilograms per hectare*

Vegetation layer	Prairie	Savanna	Oakwood	Maize
Herbaceous	920	1,886	182	9,456
Shrub	10	41	389	0
Tree:				
Yearling shoots	0	2,833	4,046	0
Older shoots	0	503	3,575	0
Total, aerial parts	930	5,263	8,192	9,456

*Ovington et al. (1963), data for central Minnesota.

energy by cropping the trophic pyramid as close to the plant layer as possible, if he does not crop the plant layer itself. This policy maximizes short-term productivity by minimizing wastage of energy, but it may impair long-term productivity by creating great instability, as we shall see in Sec. 3.3.

An even more important conclusion is suggested by these facts, but not proven by them. Since productivity changes with succession, and is different in different places, this suggests that the ecosystem which occurs naturally is the one that maximizes the energy flux, given the soil type at that time and place. Many experiments and measurements will be required to test the validity of this notion, and of course the concept is subject to qualification, as when fertilizer, plant breeding, etc., are used. Nevertheless, if true, the concept has important implications for land management policies (Sec. 4.3). Another, though perhaps more brutal, hypothesis is that the stable ecosystem is that which is successful in inhibiting other ecosystems, regardless of whether the inhibited system might be more productive. Effective shading, poisoning, repellents, deconditioning of environment, and even crowding by cumulative woody or other persistent plants can effectively suppress competitors with higher apparent productivity rates.

3.3   COMMUNITY ORGANIZATION AND STABILITY

The stability and productivity of species populations are determined in part by a number of factors that can be included under the general heading of *community organization*. These factors fall into three groups. First, we may speak of community organization as the pattern of the trophic web in a community. Different types of pattern are illustrated in Fig. 3.2. Second,

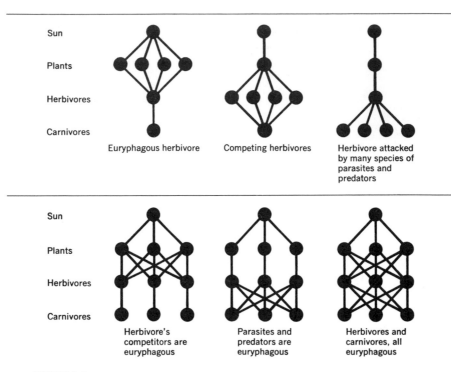

Sun

Plants

Herbivores

Carnivores

Euryphagous herbivore   Competing herbivores   Herbivore attacked
                                                 by many species of
                                                 parasites and
                                                 predators

Sun

Plants

Herbivores

Carnivores

Herbivore's   Parasites and   Herbivores and
competitors are   predators are   carnivores, all
euryphagous   euryphagous   euryphagous

FIGURE 3.2

Various patterns of three-level trophic-web organization. Each dot represents a species, and lines between dots indicate that the species in the lower level eats the species on the level above.

we may think of community organization as the spatial distribution pattern of the organisms in the community (for example, are the animals solitary or gregarious; are the plants isolated or growing together in large tracts of the same species?). Third, we may consider community organization as the organization of species in supraspecific phylogenetic units.

In this chapter we will explain the practical significance of these factors in the management of natural resources. Then we will explain the body of biological theory concerned, outline the data bearing on the theory, and explain how the present theories of community organization can be applied to the decision-making process in the resource management field.

Man is constantly changing the structure of the organization of communities of plants and animals, or creating communities with unusual structure. For example, whenever we annihilate a carnivore species in a geographic location, we simplify the structure of the community. If we cultivate wheat, apples, corn, or oranges, and weed out other plants, we are attempting to create a monoculture, or artificially simplified community, and this has implications for animal populations in the area. If we raise one, two, or three species of fish in a pond or lake after removing the other species with rotenone, we have again produced an artificially simple community, with the attendant biological consequences. Biological control of insect pests involves adding one or more entomophagous organisms to a community, and thus increases the complexity of the trophic organization pattern. Forest management practices based on the one-species, one-age stand also change the degree of organization in the trophic web. Furthermore, all the practices I have mentioned may involve changes in the spatial distribution of plants and animals, and in the organization of supraspecific biological units. The whole matter of trophic-web organization and community stability is complicated because a change in the organization at one trophic level may have one type of effect on that level, and another or even opposite effect on another level.

Elton (1958, pp. 145–150) develops six arguments to show that complexity of the trophic-web organization in a community causes greater species population stability through time than would be the case where the trophic-web organization is of a simpler pattern. His first and second arguments are that mathematical work in population dynamics and laboratory experiments with populations show that very simple trophic webs, with one predator and one prey species, are highly unstable; they oscillate violently, reflecting the quick annihilation of both species unless there are refuges for the prey to hide in, or a huge experimental universe. His third argument, verified by the historical record, is that the simpler communities on islands have been more readily colonizable by invading species than have the more complex communities of large continental areas. Since invasion and subsequent rapid population buildup by a new species leads to population instability, such cases support Elton's argument. His fourth argument is that invasions and outbreaks typically occur not in natural communities, but in cultivated or planted land. Man has simplified these latter communities in three different ways. Monoculture cuts down the variety of flora, and since the crop is often a foreign plant species, there has not been time for a naturally full fauna to develop. Furthermore, pest control simplifies the herbivore trophic level intentionally, and the carnivore level unintentionally. Elton's fifth argument is that pest outbreaks do not occur in tropical forests, where there is a tremendous variety of plant and animal species, and hence, a very complex pattern of community organization. The final argument is that pest control is an increasing problem in orchards where there is indiscriminant use of insecticides, which as one of their side effects kill the entomophagous species or mite predators that keep many pests in check.

What are the dynamics of the process by which complex communities prevent violent oscillations in the population density of a particular species? MacArthur (1955) theorized that the stability of a community is a function of the number of links between species in a trophic web. The essence of this argument is that the more trophic links there are, the more likelihood there is of compensatory mechanisms operating if one species becomes rare or abundant. If a species becomes very abundant, competitors may become less abundant, and predators and parasites may then shift their attack pressure to the superabundant species. Such a shift in diet is well known; it has been called the functional response by Solomon (1949), and has been studied experimentally by Holling (1965, 1966). Conversely, if a herbivore becomes rare, other herbivores can serve as alternate prey or hosts for its predators and parasites, provided they are polyphagous, of course. MacArthur's paper demonstrated the basic analogy between the topological characteristics of food webs and information theory. That is, if a predator species eats n species of prey, and the probability of energy flowing from a particular prey species to the predator species is equal for all prey species, then this probability p_i for the ith prey species is $1/n$. MacArthur argues that the stability of this section of a food web is given (by information theory) as

$$S = -\sum_{i=1}^{n} p_i \log p_i \tag{3.1}$$

Since the summation is over n prey species, and p_i is defined as $1/n$, (3.1) reduces to

$$S = -n(1/n) \log(1/n) = \log n$$

(A discussion of the mathematical reasoning underlying this equation will be given in Sec. 11.9.) MacArthur noted that the shortage of a large number of alternate food species in the arctic would tend to produce precisely that wide amplitude in species population fluctuations for which the arctic is noted. Until recently, MacArthur's (1955) short but succinct and cogent discussion of the relation between community organization and stability was completely in accord with the "conventional wisdom" of ecologists on this problem. For example, the comprehensive reviews by Elton (1958) and Pimentel (1961b) both present data and arguments from many sources showing that, typically, increased complexity of trophic-web structure leads to increased community stability.

Recently, however, a number of papers have appeared suggesting that this question of the relationship of stability to trophic-web structure is more complicated than it has hitherto appeared. We may illustrate the possible sources of complication by referring back to Fig. 3.2. A community trophic web can be of a complicated structure in several different senses. For example, an herbivore community may have several competing herbivore species, may be attacked by many species of carnivores, may have stenophagous (restricted diet) rather than euryphagous (broad diet) predators or competitors, or all levels of the trophic pyramid may contain euryphagous rather than monophagous (one-food species) or stenophagous organisms. The complication arises because complexity of structure in one level of the trophic web may produce stability at that level, but instability at another level. For example, consider what happens when there is great competition at the predator level. If one predator species increases in abundance because of extrinsic factors (perhaps a shift in weather increases optimality for the predator), the intensity of competition among all the predator species is increased, and tends to hold down the total predation pressure exerted against the herbivore level by all carnivore species. Then, if one or more species of herbivores build up in numbers suddenly because of extrinsic factors, they may "escape" from control by the carnivore trophic level because there is too much stability in the carnivore level to allow rapid buildup in numbers concurrent with the buildup of the rapidly increasing pest species. In this way, stability in one trophic level might conceivably cause or allow instability in another. There are a number of lines of evidence to support this a priori line of argument. First, it is a disturbing fact that many of the most historically important pest species (rodents, locusts, and grasshoppers, and forest-insect defoliators such as the spruce budworm) are attacked by an enormous number and variety of species. Grasshoppers, for example, are attacked by a variety of pathogens, mites, nematodes, spiders, wasps, tachinid and sarcophagid fly parasites, bee fly egg parasites, praying mantises, snakes, mammals, and birds—it is truly a wonder that any grasshoppers ever survive. The fact that they do, often with spectacular success, raises the question as to whether such pest species have a wide variety of attackers because they are so productive and unstable, or whether they are so unstable because they have such a wide variety of attackers.

Two important recent papers suggest that interspecies competition among attackers depresses the total effectiveness of all attacking species. Turnbull and Chant (1961, pp. 731–741) discussed the strategy of biological control of insect pests. They pointed out that a superior parasite, in terms of biotic potential and searching ability, may be replaced by an inferior parasite with which it competes if the larva of the inferior parasite survives better than that of the superior where eggs of both species have been deposited in the same host individual. Therefore, Turnbull and Chant argued, it is far better to release one preselected superior parasite than a number of different species in the hope that one will work well. One species that might have worked well by itself may never be able to in the face of competition by inferior parasites or predators. Aware of the controversy surrounding the Turnbull and Chant paper, Zwölfer (1963) reported some significant observations in support of their position. He discussed the dynamics of the parasite complexes of six species of *Lepidoptera*. In each of the six cases, the herbivore is attacked by one, two, or three specialized and well-synchronized parasite species. In two of the *Lepidoptera* species these superior parasites are subjected to heavy competition by other species of the carnivore trophic level. For the other four species of *Lepidoptera*, there are fewer species in the carnivore trophic level. The species-rich complexes of entomophagous organisms attacking the two species are not able to prevent their hosts from building up to high population levels; however, in the case of the other four species of *Lepidoptera*, the relatively simple complexes do exert control. This illustrates clearly how too much trophic-web complexity at one trophic level relative to that at another can produce stability at one level at the expense of instability at another. Zwölfer's argument is particularly effective because the parasite species effective against *Lepidoptera* species with few parasites were the same parasite species that were not effective against *Lepidoptera* when many species of parasites were present.

Nevertheless, this whole matter is still highly controversial. Holling (personal communication) has argued that a high degree of stability at the carnivore level would encourage stability at the herbivore level by damping the oscillations of increasing amplitude inherent in a predator-prey system with delayed negative feedback. The complexity of community interactions is too great to allow us to discriminate between alternative interpretations unless a systems approach is adopted of the type outlined in Part Four of this book. DeBach et al. (1964, pp. 124–128 and elsewhere) discuss the question of multiple- versus single-species importation of parasites and show that in some cases one might be best, and in some cases the other. In fact, as they point out, the results depend on a great many other factors, such as the sequence of life history events in a typical season, relative to the sequence of weather conditions; and the subdivision of the range of a pest into different climate types, in each of which the weather is optimal for a different species of entomophagous organism.

I have tried to discover, by empirical means, generalizations about community-web-structure factors affecting species population stability. I analyzed the pooled data of the Canadian Forest Insect Survey (Watt, 1964, 1965) for 552 species of forest *Lepidoptera* and for the tree species they eat.

An extensive program of computer analysis showed that:

1. Stability of individual species populations of *Lepidoptera* was greater, the greater the number of competitor species for the tree species eaten by each species of *Lepidoptera*.
2. Stability of the *Lepidoptera* was less, the larger the number of tree species fed on by any species of *Lepidoptera*.

The first finding is in accord with Zwölfer's position, whereas the second is directly contrary to MacArthur's. There are at least two possible interpretations for the apparent paradox between my finding and MacArthur's. First, Holling (personal communication) notes that two kinds of community complexity can be distinguished: interspecific and trophic. When a community is depauperate (as in the north temperate forest communities monitored by the Canadian Forest Insect Survey), generalized feeders are common, allowing many connections between a species on one trophic level and others on another trophic level. If we assume that interspecific competition occurs whenever more than one species shares a common resource, then in the case of the Forest Insect Survey data, interspecific competition would seem great. In fact, however, it can be argued that the trophic complexity is low since there are many fewer inter- and intratrophic connections than in a community with many specialized species. Perhaps the inverse correlation between interspecific and trophic complexity (represented by northern boreal forests and tropical rain forests, respectively) explains the conflict between the Forest Insect Survey data and MacArthur's hypothesis. The existence of such major theoretical dilemmas demonstrates the need for systems-oriented field work on communities, to determine what really is going on. Furthermore, in much of the literature on community organization and stability, two different kinds of community organization have been discussed as if they were the same group of factors: web structure and spatial distribution of plants and animals. In fact, these are two different sets of factors, involving two different sets of mechanisms, as we shall explain shortly.

It is important to note that the preceding remarks, based on terrestrial communities, must be reconsidered for aquatic communities. Larkin (1956) remarks that there is only vague demarcation of ecological zones in freshwater environments, and this results in a type of community-trophic-web structure in which there is much more breadth than height. For example, Johannes and Larkin (1961) found that both shiners and trout eat almost all the available species of food organisms in their home lake, including each other. At different times of the day, year, and season, fish may be found leading a pelagic, shoal, or bottom existence, with their food habits varying accordingly. For example, both shiners and trout have an enormous range of food organisms. The ability of both species to change their distribution and diets tends to reduce intensity of competition. After amphipods had been severely depleted in Paul Lake, British Columbia, both trout and shiners found substitutes for them in their diets. The new diet is not as satisfactory for young trout as the old, indicated by a reduced growth rate at early ages. Since shiners have been added into the community, large old trout now grow faster than they used to. Hence, competition can have opposite effects on two different age and size groups in the same species. This implies that adding species into an aquatic community produces a much greater increase in the complexity of the trophic-web structure than would be the case with a corresponding increase in the number of species in a terrestrial community.

In view of the far-reaching implications of such additions in the aquatic case, such moves should be thought through carefully before being undertaken. An apparent blessing could turn out to be a calamity in disguise. For example, adding a small fish into a reservoir to be food for a species of fish predator would be calamitous if the small fish had an insatiable appetite for the eggs of the larger. Gerald Marten has informed me, on the basis of skin-diving experience off the coast of Puerto Rico, that trophic-web organization can be just as complex, or more, in marine situations as in fresh water. This complexity in rich marine subtropical environments, such as around coral reefs, may be of a quite different type from that in north temperate lakes. The complexity of the community in fresh water may be an outcome of the relative paucity of organisms, and the generalized nature of their feeding habits. C. S. Holling (personal communication) has spent considerable time skin diving off the coast of Hawaii and is impressed with the evidence for adaptation to specialized feeding, in the form of mouthparts adapted specifically for crushing, grazing, grasping, or probing. The result is that any one species has relatively few connections with other food species. The high degree of specialization also produces many connections between and within trophic levels. The contrast between north temperate lakes and tropical marine coral reefs is thus analogous to that between northern boreal forests and tropical rain forests. In the former of each pair, interspecific competition pressure is great but trophic-web complexity is low; in the latter of each pair, the reverse is true.

The second group of factors important in community organization as it affects species stability concerns the spatial distribution of organisms. I have mentioned that analysis of the Canadian Forest Insect survey data showed that the greater the number of tree species eaten by a *Lepidoptera* species, the less stable that species was, though conventional ecological wisdom led us to expect the reverse. The reason for this paradox lies in the nature of dispersal mortality and ecological interfaces. If an animal species lives in a geographical location where the entire environment, or almost all the environment, is filled with suitable food for the species, then the probability of death from intentional or accidental dispersal is low because there will almost always be suitable food wherever dispersal takes the animal. The lower the probability of dispersal mortality is for any species, the more likely it is to be able to build up in numbers rapidly, all other things being equal. There are two ways in which the entire environment may be filled with food suitable to an organism. First, the species may feed on a plant species occurring in huge tracts of land, as in the case of the spruce budworm, which lives in vast stands of only two species (white spruce and balsam fir), or grasshoppers, which live in huge tracts planted out to wheat. Second, the species may eat such a wide variety of plant species that virtually its entire environment is filled with suitable food. Several colonial, tent-spinning *Lepidoptera* are in this position. Pimentel (1961b) has reviewed a number of instances from forestry and agriculture where it is clear that mixed-species associations of plants are less likely to suffer insect-pest outbreaks than are single-species plots, as in monoculture. There is clear evidence from forest entomology that pest outbreaks build up to higher densities in large contiguous stands than they do in isolated plots of the same tree species (see Morris et al., 1963, fig. 28.1). In addition to dispersal losses, insects in isolated stands are victims of the fact that interfaces between a dense forest stand and the surrounding cut-over or agricultural country are very rich in parasites and predators, vertebrate and invertebrate. The smaller the forest stand (which means the more isolated the stand), the closer, on the average, each insect will be to high densities of potential enemies. This is another reason why animal populations of a given species tend to be more unstable in monocultures occupying large tracts than in mixed-species associations of plants.

A second factor, which has been studied by Pimentel (1961a), is the spacing of the individual plants in a monoculture. He found that the proportion of herbivore taxa to carnivore taxa was greatest in the dense planting and least in the dispersed planting, but that the density of individual herbivores in dispersed plantings was more than five times that in the dense plantings. Thus, if one must have a monoculture, it is necessary to grow the plants as closely together as possible if the plants are to survive. I have noticed in Saskatchewan, Canada, that densely planted wheat may be without any grasshoppers, while sparsely planted wheat may be almost eliminated by hordes of grasshoppers. Close inspection of the densely planted crop shows that the plants create their own microclimate: Immediately below the tops of the plants, and from there down to the soil surface, the air space between the plants is dark, moist, and cool. This is not only suboptimal for grasshoppers, but may be optimal for certain kinds of fungi and bacteria that will wipe out an entire grasshopper population in laboratory stocks in a few days.

It is, of course, dangerous in the present state of knowledge to generalize; dense planting of monocultures may be the very worst thing, for example, in the case of a pest that thrives on the microclimate so created.

The third important factor of community organization concerns the organization of species into supraspecific phylogenetic units. Margalef (1958) introduced the notion of using the information content of an assemblage of organisms to express the degree of order in the assemblage. The most useful statistic, since it corrects for collection size, is

$$I = \frac{1}{N} \ln \frac{N!}{N_1! \, N_2! \, \cdots \, N_n!} \tag{3.2}$$

where I = information content, per individual, of assemblage
N = total number of individuals
N_i = number of individuals in ith species

We shall show in Sec. 11.9 that equations (3.1) and (3.2) are in fact the same equation in different guises, and in fact both measure the same thing: the amount of order present in a system. As Margalef has pointed out, the information content of any group of individuals is in fact a measure of precisely the same thing as the "indices of diversity" used by many ecological writers to describe the groupings of individuals in a community. It occurred to me, however, that information content might be more useful, not as a measure of community organization in an area, but of organization of species within supraspecific taxons. The reasoning behind this step can be explained by referring to two other authors. Mozley (1960) has developed a philosophy of pest ecology and control through a lifetime of work on pest mollusks, particularly the snail hosts of *Bilharzia* in Africa. Mozley has had extensive experience of vast territories in the old and new worlds, and his impressions are worthy of careful attention. He feels that pests are characteristically species that thrive where circumstances favor an opportunist species more than other species, and that this is likely to occur where there are marked fluctuations in living conditions. These fluctuating conditions are difficult for most species, and are likely to eliminate them. This gives the opportunist species a "clear field," as Mozley puts it, which is the same situation as in the invasion-prone, oversimplified communities discussed by Elton (1958). Mozley finds that pests have four characteristics:

1. They occupy new territory readily.
2. Their rate of reproduction is relatively high.
3. They tend to take over exclusive occupancy of disturbed areas (areas where extrinsic factors fluctuate to an unusual degree).
4. They are transient, or fugitive species in a landscape, seldom occupying a site exclusively for a long period of time.

The significance of these remarks for Margalef's use of information theory becomes clear when we consider them in the light of another article. Downes (1965) states, in connection with a discussion on insect adaptations to arctic conditions, that a relationship exists between genetic heterozygosity (physiological versatility to colonize a wide and randomly varying environment) and restriction of the opportunity for more precise adaptation and speciation. Adding up the remarks of Mozley and Downes, we note that the characteristics—versatility and adaptability—of arctic insects also happen to be the characteristics of many pests. Downes' point about restriction on speciation also suggests that a way of detecting species with the genetic characteristics for versatility, adaptability, and high reproductive rate would be to look for certain statistical properties in the genera containing such species. If any such relation could be found, it would be a valuable aid to help locate optimal biological control agents. I assume here that it would be best to attempt control of an unstable species with another unstable species. The implication is that species that are unusually heterozygous and adaptable tend to occur in genera with few species, or in which the most abundant species are very abundant relative to the least abundant species. This is precisely what information content measures. Consider equation (3.2). When N is 1,

$$I = \frac{1}{N} \ln \frac{N!}{N!} = 0$$

Information content takes the maximum possible value when each specimen in a collection within one genus is in a different species. The more uniform the number of specimens per species and the more species there are, the greater the information content per genus. The more nonuniform the number of specimens per species in any sample, and the fewer species there are, the less the information content. We should therefore expect to find pests, and the biological control agents most useful for controlling them, in genera with very low information content. In preliminary tests of these notions (Watt, 1964), the results turned out as expected. In data of the Canadian Forest Insect Survey, information content was highest in genera containing the most stable species. The most unstable species occurred in genera with low information content. We would expect the most adaptable, versatile species to be those with the largest number of tree-host species, and we would expect these species to occur in genera with unusually low information content per individual in a sample. This was found to be so. Computer analysis of the data in catalogs of parasite *Hymenoptera* also showed that species with a large number of insect-host species occurred in genera with low information content. That is, genetic plasticity reveals itself in versatility within the species (as, for example, a large number of host species for entomophagous insects) but restriction on speciation (expressed as a low information content per genus).

In conclusion, a great many types of factors affect the productivity and stability of resources. From the standpoint of community energetics, a short and simple trophic pyramid constitutes the best way to use the biosphere; community-organization theory shows us that such trophic-web simplicity would lead to a wildly fluctuating system. By trading stability for productivity, we could produce a system with high average-productivity values, but there would be such violent oscillations around the long-term mean that at the nadir points in fluctuations in productivity, massive famine would result. Therefore we conclude that it is urgently necessary to keep the world human population low enough so that we do not have to sacrifice stability for productivity, and thus reap the resultant harvest of violent fluctuations. Furthermore, we conclude that management practices must be carefully conceived in terms of basic ecological principles if productivity and stability are to be maintained at optimal levels. One of the aims of this book is to show that resources have rarely been managed scientifically, even in instances where it was generally considered that this was being done. We also wish to indicate the principles and procedures that must be employed in developing more rational resource management policies in the future.

REFERENCES

Allee, W. C.: "Animal Aggregations," The University of Chicago Press, Chicago, 1931.

————, A. E. Emerson, O. Park, T. Park, and K. P. Schmidt: "Principles of Animal Ecology," W. B. Saunders Company, Philadelphia, 1949.

Andrewartha, H. G.: The Use of Conceptual Models in Population Ecology, *Cold Spring Harbor Symp. Quant. Biol.*, **22**:219–236 (1957).

———— and L. C. Birch: "The Distribution and Abundance of Animals," The University of Chicago Press, Chicago, 1954.

Beverton, R. J. H., and S. J. Holt: On the Dynamics of Exploited Fish Populations, *Fish Investment* (London), ser. 2, no. 19, 1–533 (1957).

Carruthers, J. N., and W. C. Hodgson: Similar Fluctuations in the Herrings of the East Anglian Autumn Fishery and Certain Physical Conditions, Rapp. et Procés-Verb., *J. Conseil Perm. Intern. Exploration Mer*, **18**:354–358 (1937).

————, A. L. Lawford, V. F. C. Veley, and B. B. Parrish: Variations in Brood Strength in the North Sea Haddock, in the Light of Relevant Wind Conditions, *Nature*, **168**:317–319 (1951).

————, A. L. Lawford, and V. F. C. Veley: Winds and Fish Fortunes, *J. Conseil Perm. Intern. Exploration Mer*, **18**:354–358 (1952).

Cooper, L. H. N., and G. A. Steven: An Experiment in Marine Cultivation, *Nature*, **161**:631–633 (1948).

DeBach, P. (ed.): "Biological Control of Insect Pests and Weeds," Reinhold Publishing Corporation, New York, 1964.

Downes, J. A.: Adaptations of Insects in the Arctic, *Ann. Rev. Entomol.*, **10**:257–274 (1965).

Elton, C. A.: "The Ecology of Invasions by Animals and Plants," Methuen & Co., Ltd., London, 1958.

Frankland, E. D.: "The Canadian Commercial Fisheries of the Great Lakes," Markets and Economics Service, Department of Fisheries of Canada, 1955.

Fry, F. E. J., and K. E. F. Watt: Yields of Year Classes of the Small Mouth Bass Hatched in the Decade of 1940 in Manitoulin Waters, *Trans. Am. Fish. Soc.*, **85**:135–143 (1957).

Fuller, W. A.: The Biology and Management of the Bison of Wood Buffalo National Park, *Can. Dept. Northern Affairs Natl. Resources, Wildlife Mgt. Bull.*, Series 1, no. 16 (1962).

Golley, F. B.: Energy Dynamics of a Food Chain of an Old-Field Community, *Ecol. Monogr.*, **30**:187–206 (1960).

Gunter, G.: Death of Fishes Due to Cold on the Texas Coast, *Ecology*, **22**:203–208 (1940).

Hickling, C. F.: The Fecundity of the Herring of the Southern North Sea, *J. Marine Biol. Assoc. U.K.*, **24**:619–632 (1940).

Holling, C. S.: The Functional Response of Predators to Prey Density and Its Role in Mimicry and Population Regulation, *Mem. Entomol. Soc. Can.*, no. 45, 1965.

————: The Functional Response of Invertebrate Predators to Prey Density, *Mem. Entomol. Soc. Can.*, no. 48, 1966.

Ivlev, V. S.: "Experimental Ecology of the Feeding of Fishes," Yale University Press, New Haven, Conn., 1961.

Johannes, R. E., and P. A. Larkin: Competition for Food between Redside Shiners (*Richardsonius balteatus*) and Rainbow Trout (*Salmo gairdneri*) in Two British Columbia Lakes, *J. Fish. Res. Bd. Can.*, **18**:203–220 (1961).

Jordan, D. S.: "The Fur Seals and Fur-Seal Islands of the North Pacific Ocean, Parts I–IV," Government Printing Office, Washington, D.C., 1898.

Klomp, H.: Discussion, in E. D. LeCren and M. W. Holdgate (eds.), "The Exploitation of Natural Animal Populations," pp. 376–377, Blackwell Scientific Publications, Inc., Oxford, 1962a.

————: The Influence of Climate and Weather on the Mean Density Level, the Fluctuations and the Regulation of Animal Populations, *Arch. Neerl. Zool.*, **15**:68–109 (1962b).

————: Intraspecific Competition and the Regulation of Insect Numbers, *Ann. Rev. Entomol.*, **9**:17–40 (1964).

Lack, D.: Natural Selection and Family Size in the Starling, *Evolution*, **2**:95–110 (1948).

————: Population Ecology in Birds, A Review, *Proc. X Intern. Ornithol. Cong.*, **1950**:409–448 (1951).

Larkin, P. A.: Interspecific Competition and Population Control in Freshwater Fish, *J. Fish. Res. Bd. Can.*, **13**:327–342 (1956).

Lindeman, R. L.: The Trophic-Dynamic Aspect of Ecology, *Ecology*, **23**:399–418 (1942).

Lotka, A. J.: "Elements of Physical Biology," The Williams & Wilkens Company, Baltimore, 1925.

MacArthur, R.: Fluctuations of Animal Populations, and a Measure of Community Stability, *Ecology*, **36**:533–536 (1955).

MacFadyen, A.: The Meaning of Productivity in Biological Systems, *J. Anim. Ecol.*, **17**:75–80 (1948).

MacLagan, D. S., and E. Dunn: The Experimental Analysis of the Growth of an Insect Population, *Proc. Roy. Soc., Edinburgh*, **55**:126–139 (1935).

Margalef, D. R.: Information Theory in Ecology (in Spanish), *Mem. R. Acad., Barcelona*, **23**:373–440 (1947), republished in English in *Gen. Systems*, **3**:36–71 (1958).

Milne, A.: The Natural Control of Insect Populations, *Can. Entomol.*, **89**:193–213 (1957a).

————: Theories of Natural Control of Insect Populations, *Cold Spring Harbor Symp. Quant. Biol.*, **22**:253–271 (1957b).

Morris, R. F.: Single-Factor Analysis in Population Dynamics, *Ecology*, **40**:580–588 (1959).

————: The Dynamics of Epidemic Spruce Budworm Populations, *Mem. Entomol. Soc. Can.*, **31**:332 (1963).

Mozley, A.: "Consequences of Disturbance; The Pest Situation Examined," H. K. Lewis, London, 1960.

Nicholson, A. J.: Compensatory Reactions of Populations to Stresses and Their Evolutionary Significance, *Australian J. Zool.*, **2**:1–8 (1954a).

————: An Outline of the Dynamics of Animal Populations, *Australian J. Zool.*, **2** (1):9–65 (1954b).

————: Dynamics of Insect Populations, *Ann. Rev. Entomol.*, **3**:107–136 (1958).

———— and V. A. Bailey: The Balance of Animal Populations, *Proc. Zool. Soc., London*, **1935**:551–598 (1935).

Odum, H. F., and W. C. Allee: A Note on the Stable Point of Populations Showing Both Intraspecific Cooperation and Disoperation, *Ecology*, **35**:95–97 (1954).

Ovington, J. D., D. Heitkamp, and D. B. Lawrence: Plant Biomass and Productivity of Prairie, Savanna, Oakwood and Maize Field Ecosystems in Central Minnesota, *Ecology*, **44**:52–63 (1963).

Pimentel, D.: The Influence of Plant Spatial Patterns on Insect Populations, *Ann. Entomol. Soc. Am.*, **54**:61–69 (1961a).

————: Species Diversity and Insect Population Outbreaks, *Ann. Entomol. Soc. Am.*, **54**:76–86 (1961b).

Radovich, J.: Effects of Sardine Spawning Stock Size and Environment on Year-Class Production, *Calif. Fish Game*, **48**:123–140 (1962).

Richdale, L. E.: A Study of a Group of Penguins of Known Age, *Biol. Monog.*, (Dunedin, N.Z.), **1**:1–88 (1949).

Ricker, W. E.: Effects of Compensatory Mortality upon Population Abundance, *J. Wildlife Mgt.*, **18**:45–51 (1954a).

————: Stock and Recruitment, *J. Fish. Res. Bd. Can.*, **11**:559–623 (1954b).

Schwerdtfeger, F.: Uber die Ursachen des Massenwechsels der Insekten, *Z. Angew. Entomol.*, **28**:254–303 (1941).

Sette, O. E.: Biology of the Atlantic Mackerel (*Scomber scombrus*). Part 1: Early life history, including the growth, drift and mortality of the egg and larval populations, *U.S. Fish and Wildlife Serv. Fishery Bull.*, **50**:149–237 (1943).

Slobodkin, L. B.: Conditions for Population Equilibrium, *Ecology*, **36**:530–533 (1955).

Smith, H. S.: The Role of Biotic Factors in the Determination of Population Densities, *J. Econ. Entomol.*, **28**:873–898 (1935).

Solomon, M. E.: The Natural Control of Animal Populations, *J. Animal Ecol.*, **18**:1–35 (1949).

Stevenson, J. C.: Distribution and Survival of Herring Larvae (*Clupea pallasi* Valenciennes) in British Columbia Waters, *J. Fish. Res. Bd. Can.*, **19**:735–810 (1962).

Takahashi, F.: Reproduction Curve with Two Equilibrium Points: A Consideration on the Fluctuation of Insect Population, *Res. Population Ecol.*, **6**:28–36 (1964).

Thompson, W. R.: Biological Control and the Theories of the Interactions of Populations, *Parasitology*, **31**:299–388 (1939).

————: The Fundamental Theory of Natural and Biological Control, *Ann. Rev. Entomol.*, **1**:379–402 (1956).

Tibbo, S. W., and T. R. Graham: Biological Changes in Herring Stocks Following an Epizootic, *J. Fish. Res. Bd. Can.*, **20**:435–449 (1963).

Tuck, L. M.: The Murres, *Can. Dept. Northern Affairs and Natl. Resources, Can. Wildlife Ser.*, **1**, 1960.

Turnbull, A. L., and D. A. Chant: The Practice and Theory of Biological Control of Insects in Canada, *Can. J. Zool.*, **39**:697–753 (1961).

Utida, S.: Studies on Experimental Population of the Azuki Bean Weevil 5. Trend of Population Density at the Equilibrium Position, *Mem. Coll. Agr. Kyoto Imp. Univ.*, **51**:27–34 (1941).

Varley, G. C.: The Biological Control of Agricultural Pests, *J. Roy. Soc. Arts*, **107**:475–490 (1959).

Watt, K. E. F.: Studies on Population Productivity II. Factors Governing Productivity in a Population of Smallmouth Bass, *Ecol. Monogr.*, **29**:367–392 (1959).

————: Mathematical Population Models for Five Agricultural Crop Pests, *Mem. Entomol. Soc. Can.*, **32**:83–91 (1963).

————: Density Dependence in Population Fluctuations, *Can. Entomol.*, **96**:1147–1148 (1964a).

————: Comments on Fluctuations of Animal Populations and Measures of Community Stability, *Can. Entomol.*, **96**:1434–1442 (1964b).

————: Community Stability and the Strategy of Biological Control, *Can. Entomol.*, **97**:887–895 (1965).

Zwölfer, H.: The Structure of the Parasite Complexes of Some Lepidoptera, *Z. Angew. Entomol.*, **51**:346–357 (1963).

4

MANAGEMENT THEORIES DERIVED FROM THE PRINCIPLES OF ECOLOGY

4.1 MAXIMIZATION OF PRODUCTIVITY

No matter what type of natural resource man manages, the object is to optimize the harvest of useful tissue (that is, to obtain maximum production compatible with stability of production). All resource management sciences (agriculture, forestry, fisheries, wildlife management, range management, and so on) have evolved elaborate bodies of theory and procedural know-how for obtaining maximum harvest. Nevertheless, the problem of achieving this maximization is more complicated than might appear. A first difficulty is that all populations have an age structure, and the size of the harvest depends on the age structure as well as the size of the population. We must also be very specific about what we mean by a maximum harvest. Do we wish to maximize the biomass harvest, the number of animals or plants harvested, or the number of animals or plants in some size or age category? The strategy for maximizing the numerical harvest of 1-year-olds might be very different than that required to maximize the biomass harvest of organisms 7 years of age or older. There is also some difficulty about the whole philosophical approach to the problem of maximizing harvest. Should harvesting policy be designed to maximize production of living tissue in the harvested population, or should it be geared to minimizing wastage due to natural mortality? These two philosophies lead to very different strategies. If we wish to maximize production, then we should thin out the reproducing segment of the population so that the offspring's growth will not be stunted because of intra-specific competition. Alternatively, if we wish to minimize wastage by natural mortality, we should harvest a number of immature individuals, which would otherwise be lost to natural mortality—we would compete with natural mortality forces for the chance of killing a member of the harvested population.

Before proceeding further, it is necessary to make a clear distinction between two concepts: *yield* and *productivity*. Yield refers to the individuals removed when a population is harvested. We may speak of numerical yield or biomass yield: the removal measured as a number or weight, respectively. Productivity does not measure the harvest, but rather the energy flux of the exploited population; it is usually defined operationally as the difference between the matter found in a population at one time and that found at a previous time. Sometimes it is measured as a weight, and sometimes as a number. For example, if the population at time t numbers 31, and that at $t + 1$ is 89, the productivity from t to $t + 1$ is 89 minus 31, or 58. Clearly this is not an accurate measure of energy flux since we do not know the actual amount of material produced by the population in the interval. The actual amount would only be known if the population was monitored continuously, so there would be a record of individuals born and died between t and $t + 1$. Productivity is different in different species, and depends on the biotic potential of the species. This, in turn, depends on the number of young per female, and the fecundity versus age curves for females of each species. The younger that females are when they first reproduce, the higher the biotic potential will be. We can illustrate the effect that different biotic potentials have on the yield obtainable from various species by using data from five different species studied in laboratory experiments on the harvesting problem. The species were unicellular algae (Ketchum et al., 1949); blowfly larvae (Nicholson, 1955); guppies (Sillman and Gutsell, 1957, 1958); *Daphnia* (Slobodkin and Richman, 1956); and flour beetles (Watt, 1955).

In each of these studies, the populations were harvested at different rates of exploitation (that is, different proportions of the entire population were removed each harvest in different experimental treatments). This makes it possible, through examination of the resulting data, to discover the particular rate of exploitation that results in maximum biomass productivity. The higher the biotic potential of the species being harvested, the higher will be the rate of exploitation that produces the maximum rate of biomass formation. The reason for this is that species with a very high biotic potential are capable of reproducing at a much higher rate than the environment can normally support. Such species normally are kept in check through intraspecific competition pressures of the type outlined in Sec. 3.1. They exhaust their food supply, and the resultant starvation lowers birth, survival, and growth rates. This means there is normally a great deal of wastage. By harvesting such species intensively, we remove this wastage. The greater the biotic potential, the greater this wastage would normally be, and hence the greater the rate of exploitation the population can sustain without being harmed. Of course, no matter how great the reproductive potential of a population, a rate of exploitation is finally reached at which any further increase will decimate the population. The yield produced at this rate, which has been called the "optimum yield" or maximum sustainable yield, is in fact a precise measure of the wastage the species is able to support. In the following five cases, we give the rate of exploitation for each which resulted in maximum biomass productivity. The rates are given as percentages of the population numbers removed per diem:

Blowflies	99 percent of adults
Daphnia	23 percent of total
Algae	13 percent of total
Tribolium	3 percent of total
Guppies	2 percent of total

These rates express the degree of competition normally occurring in the population. In the case of blowflies, the larvae are packed together in a wriggling mass on rotting carrion, and intraspecific competition is extraordinarily intense because the larvae cannot disperse since dead animals are usually scattered in pasture, range, or woodland.

The great wastage caused by competition suggests an approach to the problem of maximizing yield. Productivity determines the maximum sustained yield per unit time that can be removed. The ideal harvesting regime is one in which the sustained yield per unit time is exactly equal to the sustained productivity, less the remainder that must be left behind in the population to sustain the yield. Where a population is largely under the regulation of density-dependent factors, intraspecific competition is the principal factor regulating productivity. Until the optimum fishing or harvesting level is reached, increased yield, or rate of exploitation, actually increases productivity because thinning out the population results in decreased competition, and hence decreased biomass wastage. If, however, the rate of exploitation were increased beyond the maximum sustainable rate, the number of reproducers left in the population would not be great enough to reproduce individuals as fast as they were being removed by the harvesting operation, and the population would rapidly approach extinction. How high the maximum sustainable rate is for any population depends on the age structure of the population left behind after harvest and the frequency of the harvest, as well as the number of individuals left behind in the population.

Where we have wastage due to density-dependent factors causing starvation, we can formulate the optimum yield in mathematical terms as follows. Productivity from time t to $t + 1$ is treated as the dependent variable in an equation in which the independent variable is the difference between the stock in the population at $t + 1$ and that left behind after the previous harvest at time t.

$$\text{Max } (P_B) = \text{Max } [B_{t+1}(\mathbf{X})] - B_t \qquad (4.1)$$

where P_B = biomass productivity from time t to time $t + 1$
  B_t = biomass at time t
  $B_{t+1}(\mathbf{X})$ = biomass at time $t + 1$ as function of vector $\mathbf{X}$ of variables X_i, $i = 1, \ldots n$, which govern biomass production over time t to $t + 1$

Equation (4.1) is shorthand for the following: The maximum possible value for biomass productivity over the time interval t to $t + 1$ is achieved by obtaining the maximum possible biomass at time $t + 1$. This is obtained by adjusting the vector $\mathbf{X}$ so that the variate-values taken by the independent variables regulating B_{t+1} are such as to produce the maximum value for biomass at time $t + 1$. This is achieved by adjusting the rate of exploitation and the distribution of exploitation intensity over age classes in the harvested population so that the number and age distribution of individuals left behind in the population at time t, after harvesting, is such as to ensure maximum possible increase in biomass in the following time interval. The number and age distribution of individuals left in the population is regulated by such devices as changing the mesh size of nets and the number of nets, changing the legislation on hunting or bag limits in the case of wildlife populations, or changing the seeding density in the case of plant populations.

This particular theoretical formulation of the optimum yield problem is not, however, universally applicable. In many populations, the rate of conversion of incident solar radiation into biomass is not limited by density-dependent factors, but rather by extrinsic, nondensity-dependent factors such as weather. In such cases, attempting to increase productivity by thinning is futile because the population rarely, if ever, gets dense enough for density-dependent factors to be limiting. As we shall show in subsequent chapters, many populations are regulated largely by climate. Even then, competition is the ultimate population regulator, in that if all other regulating factors fail, competition pressures will indeed come into operation, but this rarely happens.

Where a population is largely regulated by climate, *any* harvesting decreases productivity, because productivity is already as great as climate will allow it to be. Since we can harvest such a population, often intensively, without decimating it, the role of harvesting in the dynamics of the natural population is to compete for a chance of killing individuals that would otherwise be wasted to natural mortality caused by a rigorous climate. Clearly, then, (4.1) does not represent a reasonable formulation of the problem with which we are confronted in such cases.

The way out of this difficulty is to completely change the emphasis of harvesting strategy. Instead of thinking in terms of maximizing productivity, our goal becomes the minimization of wastage:

$$\text{Max}(Y) = B_t - \text{Min}(R_t) \qquad (4.2)$$

where Y = yield at time t

B_t = biomass present at time t, and

$\text{Min}(R_t)$ = minimum number of reproducing individuals left at t in order to guarantee replacement of Y by $t + 1$

This equation states that our harvesting strategy at any time is designed to remove as large a yield as we can without impairing the ability of the population to replace itself. Yield may be quite large since the population is able to support a high wastage to natural mortality. This has interesting consequences for the age structure of the stock we choose to include in the yield. For example, it might turn out that the most sensible way to exploit a population is to concentrate harvesting pressure on those age groups in which natural mortality is greatest, and relax it for those age groups in which natural mortality is least. A multiplicity of different strategies suggest themselves for various cases, and the difficulties involved in assessing the relative wisdom of various strategies may create a demand for computer simulation studies of the type described in subsequent chapters. Suppose, for example, that the optimum yield procedure is to leave behind only individuals that will reproduce between harvests. How many such immature individuals should be left behind in the stock? We cannot leave behind the bare minimum number required to reproduce the stock because this bare minimum will be reduced by natural mortality between harvest time and the time reproduction takes place. We must leave behind enough immature individuals so that after natural mortality has taken its toll, there will be sufficient reproducers. But natural mortality will kill a different proportion of the stock at each age from one year to another, depending on weather. Some type of simulation study to explore the consequences of a range of conditions on stock requirements and harvesting policy would seem therefore to be called for.

4.2 PEST CONTROL

We shall discuss the implications of principles from population ecology and community organization for pest control, since many extensive studies demonstrate clearly that both these fields are relevant.

As Turnbull and Chant (1961) have pointed out, the concept "pest" is applied to a great variety of situations that may differ from each other ecologically. A single codling moth larva destroys an apple, and a loss of 4 to 5 percent of the apples in a crop often eliminates the growers' margin of profit. In this case, a relatively small density of the pests, which exhibit small-amplitude fluctuations through time about a low mean population density, is ruinous unless controlled. The great economic impact occurs because the pest attacks only that part of the plant which is of economic value. On the other hand, defoliators in a coniferous forest can eat a sizable fraction of the foliage before having a marked effect on the potential value of the trees for pulpwood. Here we consider only the pest characterized by extraordinarily wide-amplitude fluctuations, because these are pests for an ecological reason, rather than a reason having to do with man's selection of a small but vulnerable segment of a plant's biomass as the only part of economic interest.

Typical pests with wide-amplitude fluctuations are the spruce budworm, of the coniferous forests in North America, and the moth *Bupalus piniarius* L., a pest of German forests. Each of these species can run through an extraordinarily wide range of densities over a 30-year period. This raises the question as to what determines the amplitude of the fluctuations in various species. In Fig. 4.1 we plot graphs of average values (solid lines) for N_{g+1} versus N_g, where all the animals die after one generation (as in many insects) and N_g represents the number per unit area in the gth generation. N_g and N_{g+1} are measured at the same point in the life cycle. The difference between an insect population largely controlled by density-dependent factors and one largely controlled by density-independent factors is reflected in the scatter about the solid line in the two graphs. Where population regulation is largely under the influence of density-dependent factors, the scatter about the solid line will be small (the range is indicated by the shaded bands in the top left panel of Fig. 4.1). Where population regulation is largely under the influence of density-independent factors such as weather, the scatter will be much greater (range indicated by the shaded bands in the lower left panel of Fig. 4.1).

Consider the implications of the width of the band enclosed by the shaded areas. Where this band is narrow (where density dependence is important) a given N_g value produces a narrow range of N_{g+1} values, and the ratio N_{g+1}/N_g will not oscillate wildly. Where the band is wide (density-independent

Management theories derived from the principles of ecology 59

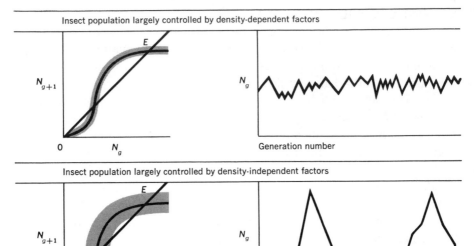

FIGURE 4.1

Effect of relative importance of density-dependent versus density-independent regulatory factors on amplitude of fluctuation of populations. (N_g represents population density in generation g.) The equilibrium points labeled E correspond to E in Fig. 4.5.

regulators are important), a given N_g value produces a wide range of possible N_{g+1} values, and N_{g+1}/N_g can oscillate wildly. The results are depicted in the two right-hand panels of Fig. 4.1.

What are the implications of these remarks for pest-control strategy? In the case of a largely density-dependent species, the response to mass mortality, as caused by an insecticide, will be quite predictable. There will be a great drop in N_g, but because of the resultant reduction in intraspecific competition, there will be an increase in the N_{g+1}/N_g ratio, and the population will show a "boomerang" effect. A year after using the insecticide, there will be lots of insects in the post-spray generation, and the insecticide manufacturers will still have a market. In the case of populations largely under the influence of density-independent factors, such as weather, the results of control are less predictable because of the wide range of N_{g+1} values that can be produced by a given N_g value. Suppose we spray such a population during a period when weather is unfavorable for the population. There will be a great reduction in N_g, but N_{g+1} will not reveal a strong boomerang effect because the N_{g+1} value for a given N_g will lie close to the lower edge of the shaded bands in the lower left panel of Fig. 4.1. The spray treatment will appear to be strikingly successful. Suppose, on the other hand, we spray a population largely regulated by weather at a time when the weather is favorable. N_g will be dropped to a much lower level because of the spray and, as expected by the solid line in the N_{g+1} versus N_g curve, there will be a boomerang effect and N_{g+1} will be greater than N_g after spray treatment. The difference between this case and that in which population regulation is largely by density-dependent factors is that here, because the weather is favorable, N_{g+1} for a given N_g will lie close to the top edge of the wide shaded band and will be very high indeed relative to the post-spray N_g. Hence, if the weather is favorable for a sequence of years, the net effect of spraying will be to *help* the population maintain itself at a high level. (Of course, while there is no gain in terms of the pest population, there is a gain in terms of the foliage that would have been eaten: the foliage is saved by spraying. In principle it is possible to use other methods of control to save the foliage *and* prevent the boomerang effect.) The use of insecticide only gains a respite for the foliage so the plants are still alive to endure another bout of defoliation in subsequent years, when the weather is favorable to the pests. Ultimately, of course, the weather changes, the insect-pest populations decline—and the proponents of in-secticidal control claim a great victory. This suggests a fascinating exercise to relate the timing of apparent final success in large-scale insecticidal control programs to weather conditions. If what I am saying is correct, then collapse of insect populations that have been sprayed for many generations without a decline in density should follow shortly after a change in weather conditions away from physiologically optimum conditions for the species.

What are the facts upon which the foregoing theoretical picture is based? First, the idea of relating narrow-band versus wide-band scatter in graphs (such as in Fig. 4.1), to narrow-amplitude fluctuation versus wide-amplitude fluctuation has come up repeatedly in studies by myself and others on the fluctuations of many species. (The same phenomenon will be mentioned again in connection with marine fish stocks in Chap. 5.) Then, it has been found that, for several insect species which fluctuate through a very wide range of densities, a high proportion of their variance is attributable to weather. This is true of the spruce budworm, which fluctuates through a 10,000-fold density range with 47 percent of the variance about the regression of log N_{t+1} on log N_t accounted for by weather (Morris, 1963, chap. 18.2); of the European spruce sawfly, which fluctuates through a 1,000-fold range in density with 40 percent of the variance about the same regression line accounted for by precipitation (Neilson and Morris, 1964); and is also true for Schwerdtfeger's (1941) data on *Bupalus piniarius* (Watt, 1963).

Far more convincing than arguments from ecological theory would be a detailed retrospective analysis of a pest-control program that has been reasonably well documented. I have chosen for this purpose the gypsy moth control program centered in New York State during the decade of 1950. The campaign has been summarized by Brown (1961), and the main features are brought out by Fig. 4.2. The question we are concerned with is, "Did the spray control program actually control the gypsy moth, or were year-to-year changes in abundance determined largely by one or more other factors?" It turns out that we can give a reasonably satisfactory answer to this question rather simply.

We obtain our raw data from two sources: Fig. 4.2 and the weather records of the Weather Bureau, United States Department of Commerce. From Fig. 4.2 we can read off A_t, acres defoliated in year t and acres sprayed by air in year t. From the A_t values we can calculate for each year A_{t+1}/A_t, which we will take as a measure of the trend index. Now, as explained in Chap. 3, there would always be a relationship between A_{t+1}/A_t and A_t even in a random number series, and hence the presence of such a relationship does not imply a causal density-dependent relation. So in order to make it easier to expose the effect of extrinsic factors operating on the population, we must first remove the effect of A_t on A_{t+1}/A_t. We do this by treating $\ln (A_{t+1}/A_t)$ as Y, the dependent variable in a linear regression analysis, and $\ln (A_t)$ as the independent variable. From now on, in the interests of brevity, we will designate measured values of the dependent variable as Y_o, for Y observed, and values calculated from the regression equation just described will be referred to as Y_c. Then $\ln Y_o - \ln Y_c$, or $\ln (Y_o/Y_c)$ is a measure of the trend index as corrected for the effect of the artifact (explained in Chap. 3). Having calculated this corrected measure, we can ask ourselves the question, "Which independent variable is most important in determining the trend index of the gypsy moth, one or more weather variables, or the aerial insecticide spray program?" On the basis of my experience with other insect species, a reasonable expression for the impact of weather on the gypsy moth in the area in question is the sum of the mean monthly temperatures for New York City for July, August, and September. Section 11.3 explains how a weather index in a city may be used as a measure of the weather in a large surrounding area provided the terrain is not hilly or mountainous. Here, we are specifically interested in the flat country of Long Island and the New Jersey coastal plain. Accordingly, in Fig. 4.3 our "corrected" trend index $[\ln (Y_o/Y_c)]$ is plotted against the resultant weather index; and in Fig. 4.4, against the number of acres sprayed. The lines drawn in Figs. 4.3 and 4.4 are the calculated regression lines.

Clearly, there is a rather convincing relationship depicted in Fig. 4.3,

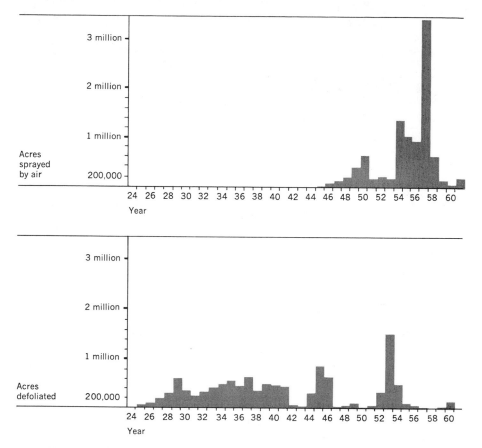

FIGURE 4.2

The struggle against the gypsy moth in the United States. Acreage showing substantial defoliation by gypsy moth larvae each year (below) is compared with acreage sprayed from the air (above), mostly with DDT at 1 lb per acre. Some suppression treatments used only ½ or ¾ lb of DDT per acre, and sevin has partly replaced DDT in recent years (Brown, 1961).

whereas in Fig. 4.4, there is not. The calculated lines in Figs. 4.3 and 4.4 account for 34.7 and 17.5 percent of the observed variance in our corrected trend index, respectively, and the regressions are significant at the 0.5 and 15 percent probability levels, respectively. What this means is that the aerial-spray program may in fact save foliage if it is timed correctly (before the gypsy moth larvae are large enough to consume a high leaf biomass per day per larva) but it does not have a demonstrable effect in reducing the population from one generation to another. The main factors in determining the genera-tion-to-generation ups and downs of the population are weather and density-

Management theories derived from the principles of ecology   **63**

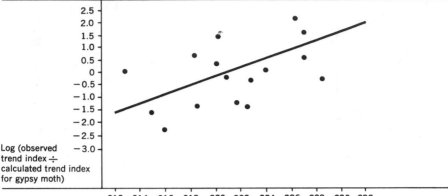

FIGURE 4.3

Relation between corrected gypsy moth trend indices and summer weather in New York City (weather data from Weather Bureau, U.S. Department of Commerce).

dependent factors. This entire argument becomes considerably more convincing when one studies detailed tables of year-to-year events in the gypsy moth situation. In Fig. 4.2, for example, while the gypsy moth population dropped in the face of extensive aerial spraying in 1957, it *rose* from 1950 to 1951 in spite of extensive spraying in 1950, which leads one to question the role of spraying as a causative agent in gypsy moth population dynamics. Second, the crucial summers, such as 1944 and 1952, for buildup of gypsy moth defoliation were unusually warm. This, of course, merely reinforces in a commonsense way what has been demonstrated statistically in discovering the highly significant regression of corrected trend index on weather.

Now where does this leave us with respect to pest-control strategy? This can be viewed as a game, in which we try to trick the pest species into annihilation, by eroding the homeostatic (boomerang) capability of the population, and it tries to trick us into thinking we have succeeded when in fact we have not. (Of course, the species does not "think" at all, but through eons of natural selection it has evolved homeostatic capabilities to deal with catastrophes, so that insecticidal treatment is merely one of a class of events the species is adapted to withstand.) The essence of this game lies in lag phenomena, that is, phenomena in which an effect resulting from a cause is only completely revealed a considerable period after the operation of the cause. When we spray a population, the *immediate* effect is mass mortality. But the surviving individuals face reduced competition pressure and hence

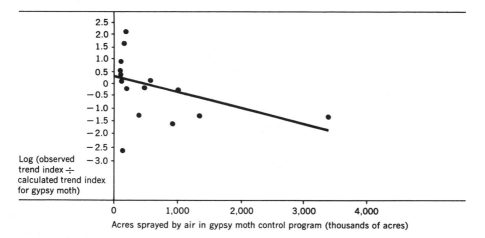

Log (observed trend index ÷ calculated trend index for gypsy moth)

Acres sprayed by air in gypsy moth control program (thousands of acres)

FIGURE 4.4

Relation between corrected gypsy moth trend indices and the extent of the aerial spray program.

have a higher probability of survival. By the beginning of the next generation (that is, after a lag, during which the effect of reduced competition gradually shows up), it will be apparent that the insecticidal treatment has not annihilated the population at all, as it appeared to have just after treatment. The population has, in effect, played a trick on us. Many kinds of control allow us to play a trick on the population by using the lag effect to our favor. Suppose we use biological control (parasites, predators, pathogens, mass release of sterilized individuals, or mass release of chemosterilants). The fundamental principle in all these treatments is the same. The affected, or doomed, individuals are left in the population *after* they have been committed to die, or *after* they have been committed to produce sterile offspring. In this way, individuals who are destined not to do the population any good in the future are still in the population, eating food, taking up space, and in general exerting competition pressure on individuals who will survive the treatment. In this situation, the population of pests has the worst of both worlds: it faces the same, or almost the same, competition pressure that there would be without control, so no anticontrol homeostatic mechanism is elicited, yet at the same time the population is being, or is about to be, controlled. This is the basic reason why such biological methods are so effective, when they work. However, to be effective, the biological technique of control must have parameter values that fall within certain ranges, as illustrated by the output from simulation analysis in Sec. 12.3.

Perhaps the most penetrating analysis of the fundamental biological

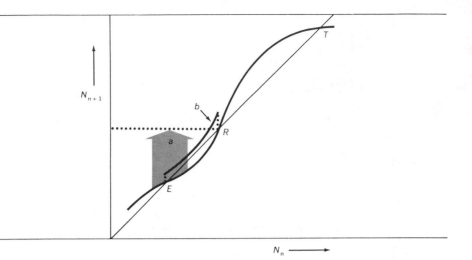

FIGURE 4.5

Reproduction curve having two equilibrium points and a population release point. In the figure, a is the upward variation of population from the curve passing over the height of population release point (R), and b is the upward variation of reproduction curve shifting the curve beyond the 45° line at the range from E to R (Takahashi, 1964).

nature of pests with wide-amplitude fluctuations yet published is that by Takahashi (1964). The basic essentials of his theory can be explained in terms of Fig. 4.5, which is reproduced from his paper. As explained in Sec. 6.1, a parasite or predator population has a fixed upper limit to the number of hosts or prey it can destroy per unit time; this is given by PK, where P is the parasite or predator density and K is the maximum attack rate per attacker (parasite or predator) per unit time. If the number of pests increases rapidly at such a rate that the attackers cannot keep up in numbers, we say the pests "escape" by simply supersaturating the attack capacity, PK, of all the attackers. Thus, in addition to having the equilibrium point E in Fig. 4.5, one of the wide-amplitude fluctuation type pests will have a release, or escape point, labeled R in Fig. 4.5. Above this point, only starvation is the ultimate control for the species. Most species do not escape from their attackers, so the upper limb of the curve, from R to T in Fig. 4.5, is lacking and their density-dependence is described by Fig. 4.1. Putting the matter somewhat differently, in the language used by Mozley (1960), the pest, or "weed," organism, because of a relatively high rate of reproduction, tends to be "transient." This is another way of saying "opportunistic."

Another implication is that since these wide-amplitude type pests have an ability to rapidly build up in numbers, they likely have what is referred to in Sec. 11.4 as a "*Drosophila* type" reproductive rate versus density curve, or have what is referred to in Sec. 6.2 as a high "condensation potential" or ability to withstand crowding (a term used by Frank, 1957). In other words, the species is unusually tolerant to increases in its own density. This being the case, it is only commonsense to attempt to control such species with biological-control agents (parasites or predators) with the same ability.

We will now consider the theory of pest-control strategies within the conceptual framework of a second body of ecological theory: community organization. Pimentel (1961) discusses the relationship between the complexity of community organization and insect population outbreaks. Evidence from many sources in the literature showed that outbreaks were less likely in mixed-species plant communities than in single-species stands. Accordingly, he planted two experimental plots: one with only *Brassica oleracea* L. (cabbage, collards, broccoli, Brussels sprouts, and kale) and one with this species plus five wild species of Cruciferae. The result was greater diversity of animal species in the mixed-species plantings, and no pest outbreaks, whereas there were pest outbreaks in the single-species stands. Extensive and detailed data of this type have been collected by Soviet ecologists as virgin steppes are converted to wheat fields in the program to increase Soviet wheat acreage. Table 4.1 illustrates the results: When steppe is converted to wheat mono-

TABLE 4.1

A comparison of total insect population in 1 square meter of virgin-grass steppe and of wheat field (Bey-Bienko, 1961)

Insects	Virgin steppe	Wheat field
Number of all species		
Homoptera	35	12
Heteroptera	38	19
Coleoptera	93	39
Hymenoptera	37	18
Others	137	54
Total species	340	142
Individuals of all species per square meter	199	351
Dominant and constant species:		
Number	41	19
Individuals per square meter	111.2	331.6
Percentage of the population	54.4	94.2

culture, the number of animal species of all types declines markedly, but the total number of individual animals increases. The simplification of the community diminishes the number of checks and balances and allows certain species to become abundant at the expense of all the other species. In Table 4.2, more detailed data are given from the same paper, illustrating how a particular species can build up in numbers in a monoculture, as many of the other species decline. The moral has been drawn by Pimentel (1961), Uvarov (1962), and many other workers: If it is economically feasible, monoculture and "clean culture" should be avoided. The more complicated and diverse the community of plants and animals, and the richer the pattern of trophic

TABLE 4.2

Densities of the more important insect species in virgin steppe and wheat sown for the first time (Bey-Bienko, 1961)

Insect species	Number/square meter		Ratio of increase (×) or decrease (÷)
	Virgin steppe	*Wheat field*	
Caeculus dubius	2.23	0.06	÷ 37.2
Sminthurus viridis	7.65	0.19	÷ 40.2
Ectobius duskei	2.83	0.03	÷ 94.3
Leptothorax nassonovi	16.48	0.03	÷ 550.0
Haplothrips tritia	1.07	300.40	× 280.8
Phyllotreta vittula	0.05	1.03	× 20.6
Hadena sordida	0.09	2.25	× 25.0

connections within the community, the more stable all species in that community will be, on the average, and the less trouble there will be with pest outbreaks. All ecologists who have ever studied the matter carefully have been struck by the relationship between pest outbreaks and habitat disturbed and simplified by man. This has not only been observed by entomologists, but also by Mozley (1960) in connection with pest snails, Frank (1957) in connection with rodents, and many others.

The theory of community organization also has implications for the strategy of biological control of pests. It has been noted (Sec. 3.3) that a high level of interspecific competition at one trophic level is associated with high species population stability at that trophic level. For example, competition between parasites and predators of a pest may prevent the whole complex of enemies from building up in numbers fast enough after a sudden buildup in the pest population. This is another possible explanation for the pest escape already mentioned. This makes the optimal type of biological-control agent to add to an already-existing complex one that does not compete directly with the parasites and predators attacking the pest (Watt, 1965). Direct competition can be avoided by using a parasite or predator operating at a stage in the pest's life cycle not already attacked. Alternatively, a direct competitor with a spectrum of physiological and behavioral attributes clearly superior to those of present members of the complex should be used (Sec. 12.3). In conclusion, some general features of pests should be noted:

1. A pest represents an abnormal, not a normal biological phenomenon. The vast majority of the 2 million or so species of living organisms are remarkably constant in numbers from year to year, and never eliminate their own food supply, so they are not limited by starvation.

2. A pest is a species operating without a long-evolved set of relationships with other species in its environment.

3. Because a pest lacks effective intra- or interspecific density-dependent regulating mechanisms, it is controlled by density-independent mechanisms, and we should always be alert for a climatic explanation for fluctuations in pest abundance. This has been found to hold true for lampreys (Sec. 5.3), grasshoppers (Edwards, 1964), many species of pest Lepidoptera, and is likely true for rodents and rabbits (Sec. 6.2).

4.3 LAND-MANAGEMENT POLICIES

As settlers from Europe reached each new continent, the immigrants were faced with a choice, though it is only recently that anyone has realized there was a choice. One option, and the one invariably chosen until recently, was to kill all native game, clear the land, and introduce intensive European-style agriculture. The other option is to introduce a scientific program of cropping the native fauna and flora. It has been taken for granted until recently that the former alternative was so attractive that in effect there was no choice.

The decision referred to is so important and far-reaching in its consequences that it deserves the most critical attention possible. If the alternative chosen is wrong, for any reason, then in view of present world food shortages, it constitutes a catastrophe, and should be reversed immediately. The decision usually made would be wrong if it turned out that the community of plants and animals at any point on the earth's surface represented an assemblage of species that maximized the utilization of incident solar radiation for that soil type at that point on the earth. If this were the case, then any change imposed by man away from the natural ecosystem would represent a change away from maximum energy flux. While it seems sensible that natural selection would operate on an ecosystem so as to select that group of plants and animals that maximizes the energy flux, data documenting this point are hard to come by. However, there are some data, principally from the American West, and from Africa.

In Table 4.3, we array estimates for prehistoric populations of large hoofed mammals in the United States, and counts of livestock in the United States in

TABLE 4.3

Numbers of large mammals alive in the United States in primitive times, and under intensive farming, in millions of head

Estimates for large hoofed mammals, primitive times*		Livestock on farms, 1959–1960†	
Animals	*Numbers*	*Animals*	*Numbers*
Buffalo	60–75	Cattle	96
Antelope	40	Pigs	59
Bighorn sheep	2	Sheep	33
White goats	1	Horses	2
		Mules	1
Total	103–118		
		Total	191

*After Seton, 1929.
†From *Statistical Yearbook 1961*, Statistical Office of the United Nations, New York, 1961.

1959 to 1960. The two total biomasses are quite similar. This is interesting, since the buffalo and antelope were probably most dense in plains country, where livestock stocking densities are low because of aridity and inadequate food. Presumably, if we were to raise buffalo and antelope with the same veterinary care, irrigation, food, and shelter now lavished on our livestock, we would have a good chance of raising more meat per annum per unit area than at present. Buffalo meat has recently appeared in supermarkets.

Much more impressive are data from Africa. Probably the most careful economic analysis ever undertaken of the relative profitability of cropping wild game versus cattle has been worked out for the Henderson Ranch in Southern Rhodesia (Dasmann, 1964; Matthews, 1962). Dasmann and Matthews showed that the maximum possible stocking density of cattle yielded a net profit only 78 percent as great as the net profit from a sustained yield of 13 species of wild game on the same acreage. Therefore it is worthwhile to consider exactly why native African game are more productive than introduced cattle. The following explanation is based on discussions by Huxley (1961), Ovington (1963), Pearsall (1962), and Talbot et al. (1965):

1. Native wild game have been selected by nature for eons to withstand extreme conditions and endemic hazards in their native habitats. These adaptations will be lacking in domestic cattle, sheep, and goats.

   a. In East Africa, the natural ecosystem on *poor* land is as productive as the human pastoral system on *good* land. The real explanation is in part nutritional: Wild game can do better on poor-quality vegetation.

   b. Wild game can remain in magnificent condition even under the alternating excessive drought and excessive rain characteristic of African plains, which would soon prove lethal to imported domestic cattle.

   c. The waters of East Africa at certain places and times become rich in fluorides which would kill people or imported cattle, and have a bone-softening effect on native cattle. Wild game are either immune or avoid toxic effects by seasonal migration.

   d. In at least a quarter of the entire African continent, tsetse fly infestations would kill cattle by trypanosome infection (parasitic *Protozoa*) to which wild game species are immune.

2. Native wild game make better use of incident solar energy because of their extreme diversification and specialization.

   a. Termites are a dominant element in the Central and East African conversion cycle; they take large quantities of vegetable material deep underground, from which it must mainly be extracted by deep-rooting trees and made available again in the form of foliage and seeds. Cutting trees breaks the cycle. There is no need to cut trees if native game, rather than cattle, are to be cropped.

   b. Natural selection selects as the climax of succession the community that maximizes the energy flux. Sec. 3.2 showed that forest is more productive than prairie or savanna. Only a cultivated crop is more productive (provided fertilizer is used), but cultivation soon ruins the soil in Africa. Monoculture also leads to instability. Hence the best animal resource is that which can coexist productively with trees.

   c. The enormous variety of African wild game uses the habitat much more efficiently than would ever be the case with domestic animals. For example, the buffalo, antelope, duikers, horse-antelope, dik-diks, gazelles, and impalas collectively use every available type of habitat: bush, woodland, swamp, forest, thickets, open plains, and rocky habitat. Giraffes can browse up to 18 feet, and tough-hided warthogs can graze under thorn bushes. Hippopotamuses eat coarse aquatic vegetation and elephants utilize a variety of habitats from hot, open lowland to high mountain forest. Every conceivable habitat is exploited: the equatorial rain forest by okapis and peacocks; open plains by zebras and gazelles; rivers and lakes by hippopotamuses; dense montane forest by bongos, giant forest hogs, and mountain gorillas; and rocky outcrops by klipspringers.

All levels of vegetation are utilized, and all types: grass, herbage, tree, bush, foliage, and fruit. The warthogs dig, and elephants use fruit, seeds, bark, and roots as well as all levels of foliage. The browsers are important in a marginal habitat because they use the material the trees reclaim from the termites. They also provide clearings where grass can grow, keep the bush in check, and prevent the country from turning into an impenetrable thicket.

Hippopotamuses and lechwes promote the conversion cycle in swamps and lake margins by fertilizing the water with their excretions, and help convert vegetational resources into fish protein.

Elephants aerate the soil, open up trackways through the forest, provide water for the whole community by digging, and make fruit available to impalas and baboons by knocking it off trees.

In addition, efficiency is kept high through elimination of the old, the sick, and the excess young. This function is served by lions, leopards, cheetahs, wild dogs, hyenas, servals, wildcats, jackals, foxes, civets, mongooses, eagles, vultures, and maribous. Omnivores are baboons and monkeys, insectivores are aardwolfs (an insectivorous hyaenoid), aardvarks, and scaly anteaters. A vast trophic engine with a multistoried canopy is at work and nothing is wasted.

A large proportion (at least over a third) of the African plateau is able to produce more animal protein by means of wild herbivores than by means of stock. In East Africa the standing crops per square mile are (Talbot, 1963):

livestock (mostly cattle) on	
moderately well-managed grassland	to 32,000 pounds
Masailand savanna	to 16,000 pounds
wild game (with a few sheep and goats) on	
similar savanna	to 90,000 pounds

3. The great variety of wildlife in Africa makes for great community stability. (If one species is wiped out by disease, many are left to exploit its

food.) The complexity of the trophic mesh prevents great instability in any species population.

4. The habitat is ecologically very "brittle." The soil is rich in iron and aluminum, but poor in the biologically important elements potassium, calcium, and sodium. The soil has low ability to utilize key elements because, compared with European agricultural soils, it has a narrow range of size and type of particles; this hinders the physicochemical processes that make the key elements readily available for plant growth. This soil type, combined with alternative excessive rainfall and drought, favors erosion. Slight overgrazing has serious consequences.

African soils and habitat therefore cannot stand up to the same type of treatment that led to successful agriculture and stock-raising in Europe. The assumption that other habitats could stand such practice led to the temporary or permanent ruin of large habitats in drier parts of South Africa, the shortgrass prairies of North America, the once-fertile lands of North Africa, and the western highlands of Scotland.

Similar comments apply to the other great African habitat, the tropical rain forest; the astonishing bulk of vegetation does not imply very rich soils. When the vegetation is removed, and the spongelike soil is cultivated, it rapidly deteriorates; the process is virtually irreversible.

The facts presented in this chapter lead inexorably to the following conclusion. No ecosystem should be altered from its natural state by man in the interests of increasing productivity unless it can be conclusively demonstrated by experiments that the alteration really does lead to higher productivity. (This assumes that higher productivity is the goal. We may have to sacrifice high productivity for other goals, such as quality of human existence and stability of production.)

It must be remembered, however, that there is a tremendous body of information on how domestic animals and plants should be handled and that there are tremendously improved varieties in existence. Any balancing of the merits of the two systems of culture would therefore have to take into account both the problems and the potentialities of development of native biota.

REFERENCES

Bey-Bienko, G. Y.: On Some Regularities in the Changes of the Invertebrate Fauna during the Utilization of Virgin Steppe [in Russian with English summary], *Rev. Entomol. U.R.S.S.*, **40**:763–775 (1961).

Brown, W. L.: Mass Insect Control Programs: Four Case Histories, *Psyche*, **68**:75–111 (1961).

Dasmann, R. F.: "African Game Ranching," The Macmillan Company, New York, 1964.

Edwards, R. L.: Some Ecological Factors Affecting the Grasshopper Populations of Western Canada, *Can. Entomol.*, **96**:307–320 (1964).

Frank, F.: The Causality of Microtine Cycles in Germany, *J. Wildlife Mgt.*, **21**:113–121 (1957).

Huxley, Sir J. S.: "The Conservation of Wildlife and Natural Habitats in Central and East Africa," UNESCO, Paris, 1961.

Ketchum, B. H., J. Lillick, and A. C. Redfield: The Growth and Optimum Yields of Unicellular Algae in Mass Culture, *J. Cell. Comp. Physiol.*, **33**:267–279 (1949).

Matthews, L. H.: A New Development in the Conservation of African Animals, *Advan. Sci.*, **18**:581–585 (1962).

Morris, R. F. (ed.): The Dynamics of Epidemic Spruce Budworm Populations, *Mem. Entomol. Soc. Can.*, **31**:1–332 (1963).

Mozley, A.: "Consequences of Disturbances," H. K. Lewis and Co., Ltd., London, 1960.

Neilson, M. M., and R. F. Morris: The Regulation of European Spruce Sawfly Numbers in the Maritime Provinces of Canada from 1937 to 1963, *Can. Entomol.*, **96**:773–784 (1964).

Nicholson, A. J.: Compensatory Reactions of Populations to Stresses and Their Evolutionary Significance, *Australian J. Zool.*, **2**:1–8 (1955).

Ovington, J. D. (ed.): "The Better Use of the World's Fauna for Food," Symposia of the Institute of Biology, no. 11, London, 1963.

Pearsall, W. H.: The Conservation of African Plains Game as a Form of Land Use, in E. D. LeCren and M. W. Holdgate (eds.), "The Exploitation of Natural Animal Populations," pp. 343–357, Blackwell Scientific Publications, Ltd., Oxford, 1962.

Pimentel, D.: Species Diversity and Insect Population Outbreaks, *Ann. Entomol. Soc. Am.*, **54**:76–86 (1961).

Schwerdtfeger, F.: Uber die Ursachen des Massenwechsels der Insekten, *Z. Angew. Entomol.*, **28**:254–303 (1941).

Seton, E. T.: "Lives of Game Animals," vol. III, part II, "Hoofed Animals," Doubleday, Doran & Company, Inc., Garden City, N.Y., 1929.

Sillman, R. P., and J. S. Gutsell: Response of Laboratory Fish Populations to Fishing Rates, *Trans. 22 N. Am. Wildlife Conf.*, **1957**:464–471.

——— and ———: Experimental Exploitation of Fish Populations, *U.S. Fish and Wildlife Serv. Fishery Bull.*, **58** (133):214–252 (1958).

Slobodkin, L. B., and S. Richman: The Effect of Removal of Fixed Percentages of the Newborn on Size and Variability in Populations of *Daphnia pulicaria* (Forbes), *Limnol. Oceanog.*, I (3):209–237 (1956).

Takahashi, F.: Reproduction Curve with Two Equilibrium Points: A Consideration on the Fluctuation of Insect Population, *Res. Population Ecol.*, **6**:28–36 (1964).

Talbot, L. M.: Comparison of the Efficiency of Wild Animals and Domestic Livestock in Utilization of East African Rangelands, *Publ. I.U.C.N.*, *N.S.*, no. 1:328–335 (1963).

——— W. J. A. Payne, H. P. Ledger, L. D. Verdcourt, and M. H. Talbot: "The Meat Production Potential of Wild Animals in Africa," *Commonwealth Bur. Animal Breeding Genet.*, (Edinburgh), *Tech. Commun.* no. 16, 1965.

Turnbull, A. L., and D. A. Chant: The Practice and Theory of Biological Control of Insects in Canada, *Can. J. Zool.*, **39**:698–753 (1961).

United Nations Statistical Office: "Statistical Yearbook," United Nations, Department of Public Information, New York, 1961.

Uvarov, B. P.: Development of Arid Lands and Its Ecological Effects on Their Insect Fauna, *Arid Zone Res.*, **18**:235–248 (1962).

Watt, K. E. F.: Studies on Population Productivity I. Three Approaches to the Optimum Yield Problem in Populations of *Tribolium confusum*, *Ecol. Monogr.*, **25**:269–290 (1955).

———: Dynamic Programming, "Look Ahead Programming," and the Strategy of Insect Pest Control, *Can. Entomol.*, **95**:525–536 (1963).

———: Community Stability and the Strategy of Biological Control, *Can. Entomol.*, **97**:887–895 (1965).

THREE | THE *Principles* OF RESOURCE MANAGEMENT

5 IMPACT OF THE ENVIRONMENT AND MAN ON POPULATIONS

This chapter is designed to illustrate how the broad principles outlined in Part Two operate in a number of natural resources, each of which is chosen because of the availability of data especially suitable to illustrate a particular phenomenon. The reader should be cautioned, however, that while the insights provided here are sufficient in some instances for prediction of economically important populations, they do not provide an adequate basis for management. For manipulation of populations, we require more detailed insight into large-scale population phenomena (Chaps. 6 and 11).

5.1 DENSITY-DEPENDENT AND DENSITY-INDEPENDENT FACTORS: OCEANIC FISHERIES

We will illustrate certain principles of the dynamics of exploited populations by using two groups of oceanic fisheries: those of the North Sea, and tuna and sardines in the Pacific Ocean.

The reason for our great interest in North Sea fisheries is that they illustrate the state that all the world's oceanic fisheries will be in in a few years, when fishing pressure and population productivity will be approximately in balance. Table 5.1 gives examples of three different stages of development of major oceanic fisheries. Obviously that of the United Kingdom cannot be fished much harder than at present; the Japanese fishery is growing slowly, and presumably is approaching equilibrium between catch and biological

TABLE 5.1

Examples of various growth phases in oceanic fisheries, catches in thousands of metric tons*

Year	Stabilized fishery: United Kingdom	Fishery approaching stabilization: Japan	Developing fishery: Peru
1947	1,172	2,205	30
1948	1,206	2,431	47
1949	1,158	3,642	45
1950	988	3,086	70
1951	1,085	3,666	127
1952	1,105	4,819	136
1953	1,122	4,521	147
1954	1,070	4,544	176
1955	1,100	4,912	213
1956	1,050	4,762	297
1957	1,014	5,407	483
1958	999	5,504	930
1959	988	5,884	2,152
1960	923	6,192	3,531
1961	902	6,710	5,243
1962	944	6,863	6,830
1963	961	6,694	6,900
1964	975	6,350	9,117
1965	1,047	6,879	7,462

*From table A-4, *Yearbook of Fishing Statistics*, Food and Agriculture Organization of the United Nations, Rome.

productivity; that of Peru is still growing very rapidly, and may be expected to grow still further.

One of the most striking characteristics of the United Kingdom fishery is the surprising lack of variability in the landed catch from year to year. This is demonstrated by Table 5.2, in which for five of the principal species caught, the year-to-year variation in all but herring is remarkably small. A convenient index for expressing variability in a number of species with rather different mean landed catches per annum is the coefficient of variation, σ/μ, where σ is the standard deviation and μ the mean. This index "corrects" the standard deviation for the mean by expressing variation as a proportion of the mean.

We naturally seek an explanation for the remarkable constancy of these data from year to year. One explanation might be that as fish populations increased, the number of boats used decreased because many more fish could

TABLE 5.2
United Kingdom, landed catch in thousands of metric tons*

Year	Plaice	Cod	Haddock	Whiting	Herring
1947	49	319	160	51	226
1948	39	343	130	54	270
1949	36	388	133	40	206
1950	33	327	128	36	180
1951	28	378	119	49	169
1952	29	384	110	55	211
1953	33	362	118	49	228
1954	33	360	126	45	198
1955	37	379	147	52	165
1956	34	393	150	50	137
1957	34	344	159	49	121
1958	35	342	144	53	112
1959	39	340	122	53	138
1960	36	314	136	40	111
1961	35	292	133	46	88
1962	36	321	137	49	91
σ/μ	0.13	0.09	0.11	0.11	0.33

*From table C-4, *Yearbook of Fishing Statistics*, FAO, Rome.

be caught per boat and the market could only absorb a fixed upper total catch. This explanation does not suffice because in fact the total tonnage of fishing craft in the United Kingdom remained nearly constant from one year to another. In fact, the Yearbooks of Fishery Statistics of the Food and Agriculture Organization of the United Nations indicate that during the period 1952 to 1962, the total tonnage of first class fishing vessels employed in the United Kingdom varied only 17 percent from the lowest year to the highest; in effect the data in Table 5.2 come close to being catch per unit effort data, rather than just catch data, and hence indicate great stability in the fish stocks in the sea, as well as great constancy in the landed catch. In order to explain this stability, we must examine the habitat and the biology of North Sea fish in some detail.

The North Sea has been one of the most productive and stable areas for fish production in the world. Wood (1956) indicates three reasons for this. Productivity diminishes with depth because the incident solar energy which powers photosynthesis is less likely to penetrate water, the greater the depth. The most productive oceanic areas occur then over shelves surrounding the continents. The North Sea does not exceed 100 fathoms (600 feet) in depth

Impact of the environment and man on populations   79

from the Straits of Dover to the latitude off the northern tip of Shetland, with the exception of a few deep pits in the northern North Sea, and a narrow gulley along the southwest coast of Norway. Because the area is influenced directly by the Gulf Stream, the water temperature is quite stable and moderate, ranging typically from about 15.5°C in the north to 17.5°C in the south in summer, and 5°C in the north to 4°C in the south, in the winter. Third, because of the preceding two factors, there is a very rich supply of plankton (tiny plants and animals floating or drifting near the ocean surface) and benthos (bottom fauna) to serve as fish food.

The North Sea fish species of commercial importance are of two types: demersal (bottom) and pelagic (surface). The two orders of demersal fish are those with soft fin rays (cod, haddock, whiting, and hake), and the flatfish (halibut, flounder, sole, brill, dab, plaice, and turbot). The flatfish are called Heterosomata in reference to their asymmetrical bodies: in the mature form, both eyes are on one side of the head, and the fish lie on the ocean bottom on their eyeless sides. The principal economic pelagic species are the family *Clupeidae* (herring, pilchard, anchovy, sardine), and the order *Percomorphi* (perch, mackerel, barracuda, mullet, and tuna). In these fish the dorsal and anal fins contain both spiny and soft fin rays.

Nothing in the habitat or the biology of the North Sea fish species, however, is adequate to account for the truly amazing stability of these species. For example, unless there were some regulatory mechanism operating at a population level, very small temperature variations in the water might induce sizeable oscillations in population size. Since these oscillations do not occur, we are led to look for some type of feedback-governing mechanism that exerts very tight control over population size. Such a mechanism is illustrated for plaice in Figs. 5.1 and 5.2.

In Fig. 5.1 we see that over a wide range of values for biomass of adult plaice stock (B_A), the number of progeny surviving to recruitment (R) was surprisingly constant. The explanation for this constancy is given in Fig. 5.2, where an extraordinarily high proportion of the variance in "prerecruitment survival of young plaice" [Beverton (1962) uses this term for the ratio R/B_A] is accounted for by the biomass of the adult stock. (The term "recruit" is used in fisheries research to refer to a young fish that is just large enough to be vulnerable to the fishing gear.) The implication of the very small amount of scatter about the line in Fig. 5.2 is that factors extrinsic to the population (for example, density-independent factors such as water temperature) are of slight importance in population regulation. Also, the small scatter indicates real density-dependence, as explained at the end of Sec. 3.1. As Beverton points out, the sensitivity of the density-dependent population-regulation

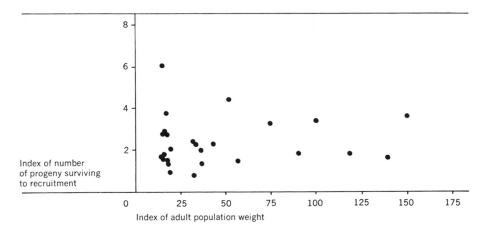

FIGURE 5.1

Relation between biomass of adult plaice population and number of progeny surviving to recruitment (Beverton, 1962).

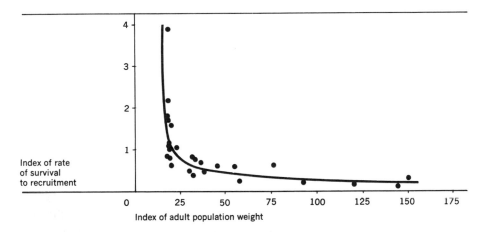

FIGURE 5.2

Index of survival to recruitment in plaice plotted against biomass of parent population (Beverton, 1962).

mechanism in plaice is remarkable, because in order to achieve it, 999,990 eggs out of every million spawned must die, but the extreme range of variation in egg deaths has only been 999,987 to 999,993. (These figures are deduced from a knowledge of the population density of recruits, and the population density of adult females.)

Before considering sardines and tuna in the Pacific, it seems worthwhile to indicate the current situation in the North Sea and North Atlantic fisheries.

Impact of the environment and man on populations 81

The demersal stocks of the North Sea and North Atlantic cannot be exploited much harder. There is no ground in the entire area that has not been thoroughly explored already, and any further pressure against the stocks will deplete them to the point where their recuperative power is impaired. Commercial species are in waters only 300 fathoms deep or less (1,800 feet), and all waters like this are being fished hard. The decreasing mean size of the fish caught suggests overfishing of most fish species.

Herring, on the other hand, could be fished somewhat harder. The decline in the British herring fishery (Table 5.2) is for economic, not biological reasons, and the decreasing United Kingdom catches have been compensated by the increasing European herring catches from the North Sea. The total European catch of herring and related species (sardines, anchovies, etc.) is remaining approximately constant, as shown by the following figures (from table A-6, 1962 *Yearbook of Fishing Statistics*, FAO, Rome).

Year	*Total European catches, herring and related species, millions of metric tons*
1957	2.60
1958	2.32
1959	2.53
1960	2.35
1961	2.41
1962	2.53

In contradistinction to the stable North Sea fisheries are the unstable sardine and anchovy fisheries off the coast of California, shown in Table 5.3. The explanation is to be sought in Fig. 5.3, which is the counterpart of Fig. 5.2 for plaice. The great contrast between these two sets of data is highly instructive. In plaice, survival rate is closely associated with the biomass of adult stock, whereas in sardines, some non-density-dependent factor is clearly very important. This factor is water temperature (Fig. 5.4). With the exception of the 1939 year class, there has been a striking relation between year-class strength and water temperature in the season in which the sardines are spawned. The bulk of the sardines are in their fourth year of life when caught (Marr, 1960, table 8). That is, the sardine year class spawned in 1949 would make its major contribution to the fishery in the 1952 to 1953 season. Figure 5.4 shows that the major break in temperature off the Scripps pier was the drop from 1948 to 1949. The major drop in the sardine fishery, as seen in Table 5.3, was from the 1951 to 1952 season to the 1952 to 1953 season.

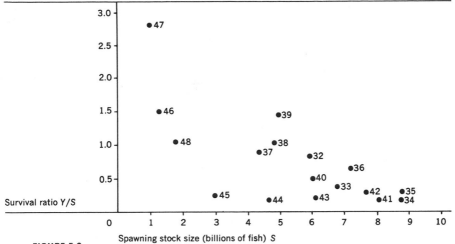

FIGURE 5.3

Pacific sardine survival ratios (year class divided by spawning-stock size) plotted as a function of spawning-stock size (Radovich, 1962).

TABLE 5.3

Total California landed catches of sardines and anchovies, in thousands of tons*

Season	Sardine	Anchovy
1943–1944	478	
1944–1945	554	
1945–1946	403	3
1946–1947	233	12
1947–1948	121	9
1948–1949	183	4
1949–1950	338	6
1950–1951	353	8
1951–1952	129	34
1952–1953	5	49
1953–1954	4	27
1954–1955	68	28
1955–1956	74	34
1956–1957	33	24
1957–1958	22	10
1958–1959	103	8
1959–1960	37	4
1960–1961	27	3

*From *California Cooperative Oceanic Fisheries Investigations Reports,* State of Calif., Dept. of Fish and Game.

Impact of the environment and man on populations   83

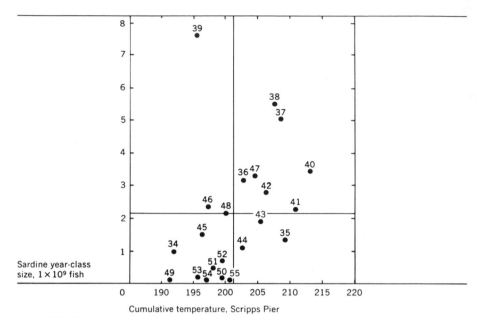

FIGURE 5.4

Relation between sardine year-class size and a measure of annual cumulative temperature (sums of monthly means, April to March) at Scripps Pier. Data are from Table XIV, U.S. Coast and Geodetic Survey (1956), and Mrs. M. Robinson, Scripps Institution of Oceanography (Marr, 1960).

Marr quotes data from E. H. Ahlstrom indicating that the incidence of anchovy larvae increased during the cold years of 1953, 1954, and 1955. Evidently the data in Table 5.3 reflect different temperature optima for anchovies and sardines, with that for anchovies being lower.

It is inevitable that the species composition of the fish community in an area will change in response to changing water temperatures, because different species of fish have different temperature optimums. Optimum temperature spectra for a variety of marine fish species are indicated in Fig. 5.5.

Sette (1960) has explained how ocean currents, as well as temperatures, regulate the survival of sardines off the west coast of North America. Larvae will only have a high probability of survival if they can migrate to the coastal nursery grounds, where conditions for development out of their planktonic phase are optimal. If, instead, they are swept into nonnursery areas by strong currents, survival will be very low. Years when the California current is strong and steady, moving southward down the California coast, then outward toward the central Pacific, sardine survival is low and year classes are weak. Years when the current is weak, the year classes are stronger (the population is more numerous).

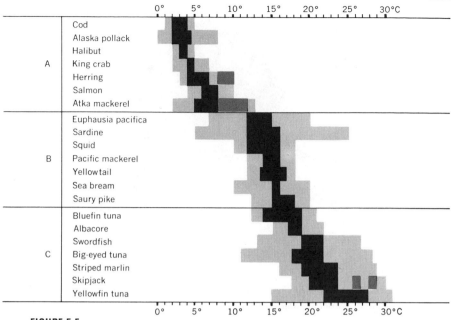

FIGURE 5.5

Optimum water temperature spectra of important fishes in Japan (Uda, 1957).

Despite the great effect of extrinsic environmental factors on sardine population, producing great scatter about the solid parabola in Fig. 5.6, there is some indication of a density-dependent effect. In 17 year classes plotted in Fig. 5.6, the 6 highest were all produced by intermediate-sized spawning stock. Suppose we ignore the remaining 11 year classes, and consider only the probable outcome of assigning the 6 highest year classes randomly to the spawning-stock categories: low, intermediate, high. The chance that all 6 would fall in the same category is only $(1/3)^6$, or 1/729. Therefore it is unlikely that stock size has no effect on recruitment.

Now what can we conclude from a comparison of North Sea plaice and California sardines? In general, we have the same picture as that discussed in Chap. 4 on pest control, and summarized in Fig. 4.1. Very stable species are controlled by a density-dependent mechanism. The bulk of the variance in numbers from time to time is accounted for by this mechanism or, putting it differently, extrinsic factors such as weather account for a small proportion of the variance in numbers from time to time. Very unstable species have only a small proportion of their variance in numbers from time to time accounted for by density-dependent mechanisms, extrinsic factors being relatively much more important. Furthermore, numerically unstable species population is in

Impact of the environment and man on populations 85

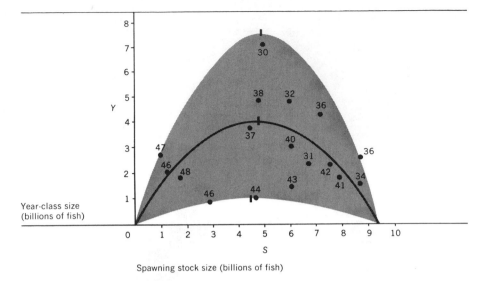

FIGURE 5.6

*Pacific sardine year-class size related to spawning-stock size, with the maximum
and minimum parabolas and the parabola of best fit (Radovich, 1962).*

an unstable environment, one only occasionally optimal for the species in
question. A stable species population will be in a stable environment. One of
the corollaries of these remarks is that we may expect to find the same species
stable in the center of its range of geographic distribution, and unstable at the
periphery of its range. This, of course, will not be true for those species living
in a range where weather and other extrinsic factors are unstable throughout
the *entire* range.

   The degree of scatter around the center parabola in Fig. 5.6 has an
important implication for fish-management strategy. Suppose we wish to
guarantee that the number of recruits into a fishery from a particular year
class will be at least minimum size Y, and we wish to guarantee this by leaving
behind in the water a "cushion," C. The cushion is a number in the spawning
stock in excess of the bare minimum required to produce Y. This number is
left behind after fishing to balance catastrophes due to weather. In both
graphs of Fig. 5.7, the spawning-stock size that must be left after fishing to
guarantee a recruit number of Y *under average conditions* is given by the
vertical line at 1. Under the worst possible weather conditions, in order to
obtain Y recruits, the stock size would have to be at 3 where there is a large
weather effect, but only at 4 where there is a small one. The more important
weather is in accounting for variance from year to year in the number of
recruits, the greater the cushion against catastrophe must be.

86   *The principles of resource management*

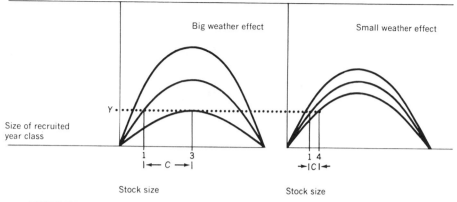

Big weather effect

Small weather effect

Y

Size of recruited
year class

$$\underset{1}{\vert} \underset{3}{\vert} \quad \underset{1\ 4}{\vert}$$

|← C →| →|C|←

Stock size   Stock size

FIGURE 5.7

Implication of the relative importance of weather in population regulation for the size of the "cushion" required to prevent sudden population collapse. Note that the scales for stock size are the same in both graphs.

The conclusion is that the optimal harvesting strategy for any exploited population will depend on the type of relationship that exists between the species in question and the factors in its environment.

Where a fishery is largely under the influence of density-dependent factors, we can use some simple mathematical reasoning to develop an optimal fishing strategy, using only data on catch and fishing effort. This has been done by Schaefer (1957) for the yellowfin tuna fishery in the eastern tropical Pacific Ocean. His theoretical analysis of the tuna fishery follows very simply from the logistic law of population growth:

$$\frac{dN}{dt} = mN(K - N) \tag{5.1}$$

This defines a parabola, and since

$$\frac{d^2N}{dt^2} = (mK - 2mN)\frac{dN}{dt}$$

the maximum is at $N = K/2$. Suppose we catch a number and weight of fish exactly equal to the annual rate of natural increase. This catch will be exactly equal to that required to keep the population in equilibrium, and it is therefore called the *equilibrium catch, C_e*. Since it is difficult to measure N, the population size for a large, widespread marine fishery $\bar{U}$, the mean catch per unit effort during a year is measured instead. Therefore (5.1) reduces to

$$C_e = a\bar{U}(M - \bar{U}) \tag{5.2}$$

in which a corresponds to m, and M corresponds to K.

Now fishing effort, F, is defined by

$$\bar{U} = \frac{C_e}{F} \qquad (5.3)$$

Substitute this for $\bar{U}$ in (5.2):

$$C_e = a\frac{C_e}{F}\left(M - \frac{C_e}{F}\right)$$

$$F = a\left(M - \frac{C_e}{F}\right)$$

$$\frac{F}{a} - M = -\frac{C_e}{F}$$

$$C_e = MF - \frac{F^2}{a}$$

The last equation defines a parabola. We can plot C_e against F for various years of the yellowfin tuna fishery and find the optimum yield (Fig. 5.8).

Only in the years 1950, 1952, and 1953 has the catch and fishing intensity been at about the maximum possible level. The maximum average annual equilibrium catch is about 193 million pounds, and corresponds to an average annual fishing intensity of 34,300 days, and an average catch per unit effort of 5,623 pounds per standard fishing day. For economic reasons, fishing intensity is often well below the optimum intensity.

5.2 DETERMINING WHEN A POPULATION IS OVEREXPLOITED: WHALES AND EXPERIMENTAL POPULATIONS

Whenever we increase the rate of exploitation of any harvested population (that is, increase the proportion of the stock removed at each harvest), certain changes occur in the stock, and in the subsequent catches. Up to a certain rate of exploitation, the ability of the stock to replace itself is not impaired. The rate of exploitation that can be sustained is very different for different species of animals, and depends on the biotic potential (reproductive rate and mean generation time). Beyond the critical rate of exploitation, the stock is decimated quickly. Thus it is important to know what symptoms to look for as warnings that the maximum possible rate of exploitation is being approached, or has been surpassed. It is also important to be able to distinguish changes brought about by heavy harvesting from those brought about by ruinously heavy harvesting.

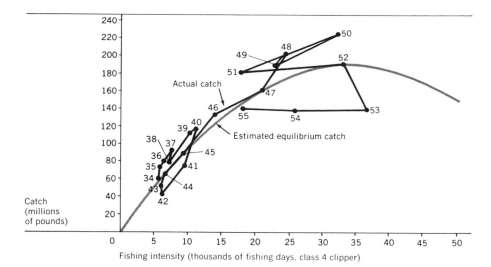

FIGURE 5.8

Relation between fishing intensity and total catch of yellowfin tuna in the total eastern Pacific. Points connected by solid line indicate actual values for each year from 1934 to 1955. Gray line is estimated functional relationship between fishing intensity and average equilibrium catch (Schaefer, 1957).

Two quite different sets of data are used to illustrate these problems. The first set comes from four laboratory studies conducted to determine how animal populations respond to various regimes of exploitation. The second set concerns whaling, because twice in history particular stocks of whales have been exploited hard enough to almost exterminate the population. This happened to sperm whales in the nineteenth century, and blue whales in the twentieth. Detailed statistics on international whale catches are reported annually in the *Yearbook of Fishing Statistics* published by the FAO, so it is easy to discover precisely how the stocks of whales are responding to exploitation.

The four laboratory studies were conducted on blowflies (Nicholson, 1954), guppies (Silliman and Gutsell, 1958), *Daphnia* (Slobodkin and Richman, 1956), and *Tribolium* (Watt, 1955). There is normally a great deal of wastage in laboratory populations because far more individuals are born than the intense intraspecific competition will allow to survive to maturity. Exploitation by the experimenter merely reduces the intensity of this competition among the remainder, and increases the biomass productivity (productivity from time t to $t + 1$ is defined as the difference between the biomass present at later and earlier times). The level of exploitation that produces maximum biomass productivity for the four species considered varies with the biotic potential, as illustrated by the figures on page 56.

Our principal interest is in determining if some underlying features are common to exploited laboratory populations, despite the great variation in biotic potential.

Age and size composition of a population serve as sensitive indices of the effect that various levels of exploitation have on the homeostatic potential of a population. Hence, age and size composition should be monitored closely, because one way of defining "ruinously hard exploitation" in biological terms is to call it "that level of exploitation which extends the homeostatic capability of a population beyond the point at which any homeostatic response can occur."

The age structure of the population removed was different in different experiments: Silliman and Gutsell included all age groups of guppies except fry in the exploited segment of the population; Slobodkin removed only newborn *Daphnia*; Watt removed adults, pupae, and half the large larvae in *Tribolium* cultures; and Nicholson removed only adult blowflies. All four studies show that increasing the rate of exploitation, up to the maximum rate sustainable by the population, increases the productivity of the age class at which exploitation is directed relative to the productivity of other age classes. Thus exploiting a particular age class, up to the maximum possible rate, makes that age class more abundant, relative to other age classes, than it would have been with less exploitation. If the optimum (that is, maximum sustainable) level of exploitation is exceeded, the age class falls off to extinction. How fast extinction occurs depends on the ratio of the exploitation rate to the biotic potential. We will first give the evidence in support of this conclusion, then explain why the phenomenon occurs.

A typical and clear-cut case is Watt's replicate number 44 (table 3, 1955). Adults, pupae, and half the large larvae were removed at each (30-day) census. The following figures reveal the essential features of the situation:

Age of replicate in days:	120	150	180	210
Rate of exploitation:	0.63	0.76	0.86	1.00
Adult numerical productivity (A):	126	171	82	15
Total numerical productivity (T):	157	174	64	24
A/T:	0.80	0.98	1.28	0.63

As the rate of exploitation increased to the maximum sustainable rate, the ratio A/T increased to a maximum, then plummeted as the maximum rate of exploitation was exceeded. It is this sudden drop in the ratio A/T that warns of impending catastrophe for the harvested population.

It is somewhat more difficult to interpret Silliman and Gutsell's data because they changed the treatment through time for each of their experimental populations. Each of their two "fished" populations had an exploitation rate of 25 percent for weeks 40 to 76, 10 percent for weeks 79 to 118, 50 percent for weeks 121 to 148, and 75 percent for weeks 151 to 172. They conclude that optimum exploitation rate is between 25 and 50 percent (their fig. 14), but their fig. 11 shows that at 25 percent both fished experimental populations had a dropping yield that might have continued to drop if they had not been switched to a 10 percent exploitation rate. Therefore I conclude that maximum yield from these guppy populations is between 10 and 25 percent. The following figures, summarizing data in Silliman and Gutsell's tables 2 and 3, illustrate the effect of different rates of exploitation on these populations:

Experimental period	Exploitation rate for experimental populations	Ratio of adults/total numbers in standing crop		
		Population A (experimental)	Population B (experimental)	Population C (control)
Weeks 40–76	25%	0.46	0.49	0.84
Weeks 79–118	10%	0.54	0.56	0.85
Weeks 121–148	50%	0.28	0.27	0.84
Weeks 151–172	75%	0.19	0.24	0.86

In the control population C, the number of adults represented 84 to 86 percent of the total population from week 40 to week 172. In experimental populations A and B, each of the four different exploitation rates produced a drop in the proportion of adults in the population. If my interpretation is correct, any fishing rate over 10 percent is harmful.

In Slobodkin's experiments, the more newborn *Daphnia* he removed, the greater the productivity, yield, and standing crop of newborn became, and the greater became the standing crop of newborn relative to the standing crop of adults.

The explanation for these observed changes in age structure of a population in response to exploitation is as follows. Consider first the case in which only adults are harvested. The greater the exploitation rate for adults, the fewer adults are left behind in the population, and less competition is directed against the subadult age classes. A higher proportion of the subadult stock will thus survive to adulthood, and productivity of adults will be greater relative to total productivity. This is true for exploitation rates up to the maximum sustainable rate. At supramaximal sustainable rates, however, we impair the ability of the population to replace the exploited adults, and adult numerical productivity drops rapidly.

By fishing young animals, as in the case of Slobodkin's experiment, we increase the probability of survival of newborn after the removal, and thus increase the productivity of newborn relative to total productivity. The final conclusion from these experimental studies, including Nicholson's, which showed similar patterns to Watt's, is that increasing the rate of exploitation increases the productivity of the age class which bears the brunt of exploitation pressure, relative to the productivity of other age classes, up to the maximum sustainable rate. Increasing the rate beyond this point produces a sudden drop in the relative productivity of the exploited age group. At this point we have been warned that the homeostatic capability of the population to respond to a stress has been eroded. In other words, we are removing an age class faster than reduction in competition within and against the age class can compensate for the removal.

Slobodkin and Richman have elucidated another means of detecting overfishing. When we increase the exploitation rate against a population, up to the maximum sustainable rate, we do not decrease survival of individuals in the population; the reverse is true. The following figures from Slobodkin and Richman's table 5 illustrate this point (from population H-15):

Exploitation rate	Median life expectancy, in days
25%	17.4
50%	19.6
75%	24.4
90%	32.2

When the maximum sustainable rate is passed, the median life expectancy must drop.

Before we attempt to use laboratory findings to interpret field data, one very important difference between the two situations must be pointed out. In the laboratory situations we have been discussing, the populations have been divided into a small number of age categories. In natural populations of fish, whales, seals, or other commercially harvested organisms, it may be possible to divide the population into a large number of year classes by reading scales, otoliths, or other hard parts. As noted in Chap. 3, animals typically have a reproductive level which is a changing function of age. The implication is that in natural populations with many ages of adults present, it is not sufficiently revealing to examine the change in ratio of adult numerical productivity to total numerical productivity, or the changes in median life expectancy, since these indices may mask the fact that a drop in the numbers of very old females is being more than compensated by an increase in the numbers of young females which are more fecund per capita.

To put the matter differently, suppose that the number of young that must be born each year to replace the exploited stock is given by R. Suppose that the number of females of age a present is N_a, the reproductive rate of a-aged females is F_a, and females reproduce between the ages of 5 and n.

$$R = \sum_{a=5}^{n} N_a F_a$$

But note that this sum may be made up in many different ways. A decrease in ΣN_a could actually increase $\Sigma N_a F_a$, as illustrated by the following example.

a	F_a	Initial situation (stock is underexploited)		Final situation (stock is exploited at maximum sustainable rate)	
		N_a	$N_a F_a$	N_a	$N_a F_a$
5	1.6	10,000	16,000	29,000	46,400
6	1.8	10,000	18,000	23,000	41,400
7	1.7	10,000	17,000	5,000	8,500
8 and over	1.3	30,000	39,000	1,000	1,300
	ΣN_a	60,000		58,000	
	$\Sigma N_a F_a$		90,000		97,600

This stock of animals has low biotic potential and high survival up to age 5. Increasing the rate of exploitation up to the maximum sustainable rate drops the median life expectancy and the productivity of animals 7 and older, but actually increases the productivity of the entire stock because there is a compensating increase in productivity of 5- and 6-year-olds, and the peak fecundity occurs at 6 years of age. Therefore while a shift in survival curve from the type in Fig. 5.9 to that in Fig. 5.10 indicates a population is being overexploited, a shift from that in Fig. 5.9 to that in Fig. 5.11 may not mean the population is in trouble. It is necessary to compute the change in

$$R = \sum_{a=a_1}^{n} N_a F_a$$

in the latter case.

With these remarks as a background, we will now consider the history of whaling. The order Cetacea, which includes whales, dolphins, and porpoises, also includes two rather different kinds of whales. The suborder Odontoceti, or toothed whales, includes the sperm whale, which may be up to 60 feet long (males) or 39 feet long (females). This is the whale on which the historically important nineteenth century New England whaling industry

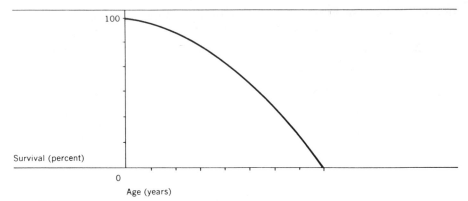

FIGURE 5.9
Survival curve in unexploited population.

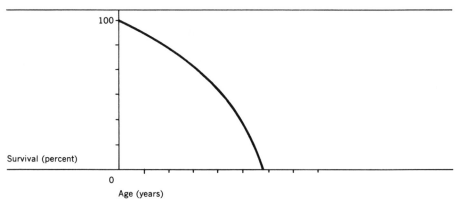

FIGURE 5.10
Survival curve in overexploited population.

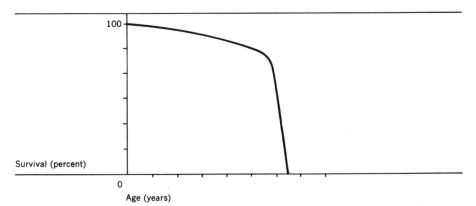

FIGURE 5.11
Survival curve in population in which exploitation at maximum sustainable rate has produced decreased survival of old adults but compensating increased survival of young adults.

was based. Five commercially important products were obtained from the sperm whale:

1. Sperm oil from the head, used as a fine lubricant and an illuminant.
2. Spermaceti, a spongy, oil-containing substance used in candles.
3. Whale oil, used for lubrication and cheap illumination.
4. Whalebone, used for stays, corsets, riding and carriage whips, and umbrellas
5. Ambergris, a product from the sperm whale's stomach used to manufacture perfume.

The sperm whale is the only member of the Odontoceti that has ever been commercially important. The suborder Mysticeti, the toothless, or baleen whales, however, contains many commercially important members: the blue whale, fin whale, sei whale, right whale, humpback whale, and bowhead whale. The largest of these, the blue whale, is about 23 feet long at birth, 52 feet long at weaning, and 77 feet at the second winter of life.

New England commercial whaling of the nineteenth century was based on the sperm whale; modern whaling is an international enterprise based largely on baleen whales; Japan, Norway, and the U.S.S.R. are the principal whaling nations. Since whaling has passed through two very different phases, we will discuss these separately.

Table 5.4 presents the basic data on the New England whaling industry of the nineteenth century, and shows the basic statistical pattern observed in all overexploited resources: The harvest per unit effort declines even though there is a sharp drop in total effort because of the declining profitability of the industry. The total number of sperm whales killed from 1835 to 1872 may be computed using information in Hohman (1928). The annual average number of vessels employed was 524. The average annual catch was 96,625 barrels of sperm oil and 172,448 barrels of whale oil. If we assume (after Hohman) 25 barrels of sperm oil is the average yield of one sperm whale, the average number of whales caught per season was 96,625/25, or about 3,870. Hohman assumes a 10 percent loss from sinking dead, without being caught, and other losses at sea. Thus 3,870 × 10/9, or 4,300 sperm whales, were killed annually for 38 years, or about 163,400 over the period. The peak catch was about 7,800 whales in 1847. This peak catch, compared with the 442 whales caught by United States vessels in the 1960 to 1961 season, gives a measure of the tremendous decline in importance of this industry for the United States.

Hohman believes the decline occurred for four basic reasons. First, increasing depletion of the whale stocks made necessary longer and more

TABLE 5.4

American whaling vessels, tonnage, and prices (Hohman, 1928)

Year	Total tonnage of whaling vessels, in thousands	Imports of whale oil, barrels, in thousands	Barrels/ton* afloat
1843	199	206	1.03
1846	233	207	0.89
1847	230	313	1.36
1848	210	280	1.33
1849	196	248	1.27
1850	171	200	1.17
1855	199	184	0.93
1860	177	140	0.79
1866	68.5	74.3	1.09
1871	69.3	75.1	1.09
1876	38.8	33.0	0.85
1881	39.4	31.6	0.80
1886	29.1		
1896	16.3		
1906	9.8		

*Note that catch per unit tonnage may not tell the whole story because ships may go much further on the average to make their catch as the stock declines. Also, catching power may not be a linear function of tonnage.

financially risky voyages, with constant danger of total loss of the whaling fleet because of the need to penetrate into areas where the fleet could be crushed by ice or become icebound. Second, petroleum and natural gas became more important as sources of lubricants and illumination, respectively, beginning in the 1860s, and Edison invented the incandescent light bulb in the late nineteenth century. Third, there was extensive destruction of whaling vessels in the civil war (1861 to 1865). Finally, 33 vessels with their cargoes and crews were trapped in the Arctic Ocean because of early fall ice formation in 1871.

Table 5.5 presents some statistics on the modern phase of whaling, in order to demonstrate that this is a truly international enterprise, conducted in several oceanic areas of the north and south. Modern whaling is conducted under the jurisdiction of the International Whaling Commission. The number of whales killed is fixed by international agreement. All countries together are assigned a quota for each season, which is measured in "blue whale units." By agreement, blue whale units are defined as follows: one blue whale = two fin whales = 2.5 humpback whales = six sei whales. Minimum lengths for

TABLE 5.5

The modern phase of whaling, 1960–1961*

Whaling areas	Whales caught
World total catch	63,484
Antarctic	41,289
North Pacific	10,705
Peru	3,602
Africa (Natal, Cape Province)	3,352
Australia	1,937

Countries engaged in whaling	Whales caught
Japan	19,891
Norway	12,829
U.S.S.R.	11,184
United Kingdom	4,551
Peru	3,602
South Africa	3,352
Netherlands	2,212
Australia	1,937
Falkland Islands	1,262
Brazil	1,083
Portugal	515
United States	442

*From *Yearbook of Fishing Statistics*, FAO, Rome.

killing have been set at 70 feet for blue whales, 55 feet for fin whales, and 35 feet for humpback whales. Recent world catches and quotas of whales are given in Table 5.6. I present this table to illustrate how misleading summary tables on the state of a multispecies resource can be. This table is expressed, not in terms of a biological unit, but in terms of blue whale units. Until 1963, Table 5.6 gives no evidence that the international whaling industry is facing a calamity, except for a slight decrease in the quota.

We shall now consider the various lines of evidence suggesting that the international stock of whales is being harvested at a rate far in excess of the maximum sustainable exploitation rate.

First, Table 5.7 makes clear that the species composition of the international catch is changing: the large blue whales are a diminishing proportion of the total catch. Since whalers always try to catch the largest whales they can find (Hamilton, 1948), Table 5.7 indicates that the most sought-after whale species is becoming difficult to find.

TABLE 5.6
Recent world catches and quotas of whales, in blue whale units**

Year	Total catch	Quota
1947–1948	16,364	16,000
1952–1953	14,866	16,000
1957–1958	14,850	14,500
1958–1959	15,300	15,000
1959–1960	15,511	15,000
1960–1961	16,433	
1961–1962	15,252	
1962–1963	11,306	15,000
1963–1964	8,429	10,000
1964–1965	6,987	8,000
1965–1966	4,091	4,500
1966–1967		3,500

*From *Yearbook of Fishing Statistics*, FAO, Rome, and International Commission on Whaling, London.

Second, the proportion of mature pregnant female whales has dropped alarmingly (Hamilton, 1948). As pointed out in Chap. 3, a species is in real danger of extinction when females and males cannot find each other, or are not stimulated to mate for some other reason associated with very low population densities. Hamilton gave the following figures to show that despite the relaxation in whaling during the second world war, females were having even more difficulty getting mated after the war than before.

Percentage of mature pregnant female whales

	Blue whales	Fin whales
1934–1935	28.7	33.1
1945–1946	20.7	23.2

Third, Hamilton pointed out that the mean length of whales is decreasing. Blue whales of 89 feet and over used to be common; they have become very rare. Blue whales have been exploited so hard that they have a much lowered probability of attaining adulthood.

Fourth, Laws (1962) has shown that the survival versus age curves in fin whales indicate that the stock of fin whales is being depleted at a very alarming rate. Two of his curves are reproduced in Fig. 5.12. This striking and alarming change in survival from 1955–1956 to 1956–1957 occurred in a specific area of the Antarctic. The curve has steepened markedly in 1 year, and Laws states that this can only be explained by selective mortality

TABLE 5.7

*Changes in species composition of international whale catches**

Year	World total catch	Blue whale catch†
1937–1938	54,902	15,035
1947–1948	43,431	7,157
1950–1951	55,795	7,278
1951–1952	49,832	5,436
1952–1953	45,009	4,218
1953–1954	53,642	3,009
1954–1955	55,074	2,495
1955–1956	58,126	1,987
1956–1957	59,056	1,775
1957–1958	64,586	1,995
1958–1959	64,489	1,442
1959–1960	63,717	1,465
1960–1961	65,811	1,987
1961–1962	66,026	1,255
1962–1963	63,579	1,429
1963–1964	63,001	372
1964–1965		20
1965–1966		1

*From table H-3, 1962 *Yearbook of Fishing Statistics*, FAO, Rome; earlier yearbooks were also used.
†28,000 blue whales were caught in 1930–1931. Despite the 4½ years during the second war in which each adult female could have raised one calf every 2 years, the stock is depleted.

of older whales. He points out that the truly striking difference between these two curves is not an artifact caused by an accident since both samples were collected in the same relatively small area by biologists working on the same factory ship, with the same catchers and men. Japanese biologists working in another area got almost identical results in these seasons. [The age of fin whales can be determined by the baleen plate (for young whales), laminations in ear plugs, and an assumption of 1.43 corpora per year in adult females.] (Laws, 1962.)

If we combine the observations on experimentally exploited populations and whale stocks, we arrive at the following conclusions about the symptoms of hard exploitation versus those warning of ruinously hard exploitation.

If a population is being exploited hard, but not ruinously hard, it will show decreasing mean age and size in the stock, indicating a dropping probability of survival to extreme old age, but the ability of the adult stock to

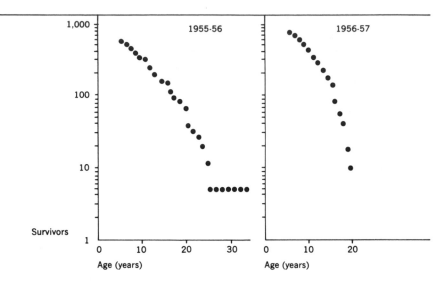

FIGURE 5.12

Survival curves for female fin whales in Antarctic Whaling Area I, illustrating the tremendous pressure exerted against large, old whales (Laws, 1962).

replace itself will not have been impaired. There will also be a change in the survival versus age curve, but this will not be such as to indicate impaired collective fecundity.

If a population is being fished ruinously hard, it will indicate this by the following symptoms (not all will be present in all cases):

1. A decreasing proportion of females pregnant.
2. A decreasing catch per unit effort.
3. A decreasing catch relative to the catch of related species.
4. A failure to increase in numbers rapidly after a respite from harvesting.
5. A change in productivity versus age curve which, when interpreted using a fecundity versus age curve, shows that the ability of the population to replace the harvested individuals has been destroyed.
6. A change in survival versus age curve which, when interpreted using a fecundity versus age curve, shows that the ability of the population to replace the harvested individuals has been destroyed.

Another general principle is that the lower a species' biological potential, the higher the survival in nature will be, and this implies less buffering on the part of the species to sustain heavy harvesting. Conversely, those species that produce a large number of offspring to compensate for a great deal of wastage (high mortality due to competition and/or extrinsic factors such as inclement weather) are the species that can sustain the heaviest rates of exploitation. In such cases, man merely competes with nature for the surplus biomass that would be wasted in any case. Animals with little ability to support a high rate of exploitation are whales and fish species that mature late in life and/or have few eggs: sturgeon, muskellunge, and lake trout. Animals with a great ability to support heavy harvesting reproduce at an early age and have a large number of young: herring, pilchards, sardines, anchovies, and smelt, for example.

One might assume that the calamitous situation in the whaling industry would make it easy for the International Whaling Commission to obtain agreement from the member nations to enforce strict conservation measures immediately. This is not the case. In whaling as in all other resource industries, economic considerations may result in destruction of the resources. As an example of the kind of situation that occurs when a resource is being annihilated, I conclude with a news item concerning the sixteenth annual meeting of the International Whaling Commission at Sandefjord, Norway, June, 1964. (Report by FAO, *International Marine Sciences*, November, 1964, p. 16, published by UNESCO, Paris.) While reading the following, one should bear in mind that as mankind continues to destroy resources, he systematically pounds nails into his own coffin because, as his numbers increase exponentially, the needs for resources become greater, not less. Destroying what we have rather than managing it intelligently is hardly the way to meet this need.

The Commission undertook in 1961 to bring, by July 1964, the regulation of pelagic whaling in the Antarctic into line with scientific findings concerning the state of the whale stocks and the sustainable yields that might be taken from them. To this end it appointed a Special Committee of Three (later Four) independent scientists to review with the Commission's Scientific Committee the existing data pertaining to the stocks and the effects of whaling on them. The reports of this work were considered by the Commission at its 16th meeting, together with proposals by the scientists for immediate drastic reductions in the permitted catches of fin- and sei-whales, and a complete ban on the catching of blue whales in the Antarctic, to allow all these greatly depleted stocks to begin to build up again to sizes such that they could in future sustain higher yields. The Commission acted on the proposal to prohibit the catching of blue whales. The proposal for effective conservation action with respect to the other species was, however, not acceptable to the four countries—Japan, Netherlands, Norway, and Union of Soviet Socialist Republics—at present conducting pelagic whaling operations in the Antarctic. (Since the meeting, the Netherlands has, by selling its own expedition, and with this, its national quota to Japan, withdrawn from this activity.) Thus, with four objecting out of the fourteen countries present at the meeting, the necessary three-quarters majority for the conservation proposal could not be obtained. The Commission therefore took no action in this regard—to the dismay of international and national organizations, governments, and individual scientists concerned—and the prospects for the future of the whaling industry based on the Antarctic stocks are dismal. Earlier the Director-General, and the governing bodies of FAO, among others, had repeatedly expressed grave concern at the developing situation in whaling, not only because of the loss of a potential yield of food and other products worth about $200,000,000 annually, but also because of the implications of a failure of the IWC for the efforts of other regional and specialized fisheries commissions and councils to work towards a rational management of the international exploration of fisheries resources of the high seas. Member countries are now therefore reconsidering their position, and the matter is coming up for discussion, and perhaps action in other fora.

Arrangements had been made for FAO to continue to co-operate with the Commission in making continuing assessments of the state of the whale stocks and predictions of yields. While, because of the Commission's failure to act effectively on its undertaking of 1961, these arrangements cannot now proceed, other arrangements are being made to ensure that such assessments will continue to be computed, and the results made public in future years, in the hope that it will eventually prove possible for the necessary conservation measures to be adopted, especially as there is now concern also for the rapid depletion of whale stocks in the North Pacific and elsewhere.

Meanwhile the pelagic whaling countries have worked out draft rules for an International Observer Scheme to supervise the application of the Commission's regulations, and are also negotiating, outside the Commission, an agreement on an overall Antarctic catch quota, although none are considering a quota low enough to prevent further depletion of the stocks.

5.3   A PEST: THE SEA LAMPREY IN THE GREAT LAKES

The sea lamprey problem is analyzed here for two reasons. First, it is one of the most studied and best-documented instances of the havoc that results when a pest invades a community, and it reveals certain fundamental features of the population dynamics of pests. Second, the details of the problem provide excellent illustrations of many of the homeostatic mechanisms discussed in Chaps. 3 and 4.

The lamprey *Petromyzon marinus* is not a fish but a more primitive form of vertebrate in the class Cyclostomata. The name of the class refers to the sucking, rasplike mouth of the adult, used to rasp holes in the victims to suck out blood and body fluids. The marine lamprey lives on the Atlantic coast of North America and Europe, but migrates in to fresh water to spawn. The marine form is 30 to 36 inches long as an adult. In addition, a 15-inch freshwater form of the lamprey has been known to inhabit four of the Finger Lakes of upper New York State for centuries (Wigley, 1959). This freshwater form has also been known in Lake Ontario since the beginning of history. Before November 8, 1921, when an adult marine lamprey was discovered in Lake Erie (Dymond, 1922). no *Petromyzon* had been found further inland than Lake Ontario. Presumably their path was blocked by Niagara Falls, but with the completion of the Welland Canal in 1829, and particularly with reconstruction of the canal in 1932, this obstacle was removed (Zimmerman and Bright, 1942). Canals are useful to lampreys for two reasons: not only can they swim through, but they are often towed through, as they have a frequently observed habit of attaching themselves to the hulls of ships with their sucking mouths.

The following discussion of lamprey biology is drawn from monographs by Applegate (1950), Lennon (1954), and Wigley (1959). Adult lampreys migrate upstream into rivers to spawn in April, May, and June. Each group spawns only once because the adults degenerate and die after spawning. The biotic potential is staggering: Applegate determined a mean of 61,500 eggs per female in marine lampreys in Lake Huron tributaries, and Wigley found that the average-sized female in Lake Cayuga had 43,000 eggs. In 1961, about 70,000 adult lampreys were trapped in Lake Superior tributaries by United States and Canadian government agencies (Great Lakes Fisheries Commission, 1964). Under ideal conditions, this number could produce $70,000/2 \times 61,500$ adults. If these killed only one fish each, the loss would be 2 billion fish, whereas the largest poundage of all fish species ever removed from Lake Superior by Canadian vessels was 8 million pounds, in 1915 (Frankland, 1955). Clearly, lampreys are capable of annihilating fish in the Great Lakes. The larvae live as blind, filter-feeding ammocoetes for several years at the bottom of the streams in which they are spawned. Various workers have tried to determine the length of this larval period through analysis of the curves of frequency versus length obtained from samples of larvae. Applegate thought there were four age groups in his samples; Wigley found seven. For reasons I will give later, I think five or six is the correct number. In the last year of larval life, the larvae transform to the adult, parasitic phase, and migrate downstream between October and April. The adults migrate back upstream to spawn, 12 to 20 months after transformation.

The circular, sucking-rasping mouth of the adult lamprey contains about five concentric rings of teeth curved inward toward the center of the mouth. These are used to rasp a hole in the victim's skin. In addition, the adult marine lamprey has two buccal glands containing a liquid which acts as an anticoagulant against fish blood.

It is an extremely maneuverable swimmer. It sights a fish, opens its mouth (normally closed for better streamlining), and charges the fish from the side, top, or bottom. A large fish will survive several lamprey attacks, but even the largest, most vigorous fish eventually succumb. All fish go into a frenzy to remove the lamprey, but gradually subside. When lampreys attack other lampreys, the victim lamprey removes the attacker by tying itself into a knot at the point where the attacker strikes. Lampreys are initially repelled from human swimmers by the high body temperature. After a person has been in the water long enough to become chilled, lampreys will attack, and will cause wounding unless removed. They are a particular menace to long-distance swimmers in the Great Lakes.

No fish in the Great Lakes are known to be free from lamprey attacks. They have wounded trout, whitefish, herring, smelt, suckers, carp, channel catfish, bullheads, pike, burbot, walleyes, yellow perch, smallmouth bass, rock bass, etc.

The rasping tongue, the anticoagulant, and the suction pressure used by the lamprey to create feeding attachments on fish cause injuries to many tissues and organs with consequent impairment of function in the host. Fish that do not die from the lamprey attack are more subject to attack by fungi, parasites, or predators. A lamprey-attacked fish is unfit for commercial marketing. Many attacked fish appear to be in a state of shock: pale, immobile, with reduced respiration (slow, weak gill movements).

The spread of marine lamprey in the Great Lakes is indicated by the following records (from Applegate, 1950):

1921: One adult found in Lake Erie.
1927: Two adults found in Lake Erie.
1934: First lamprey found in Lake St. Clair.
1936: Four specimens found at widely scattered localities in Lake Michigan.
1937: First spawning run discovered in Lake Huron (first report of any kind in this lake).
1945: First two specimens discovered in Lake Superior.

In 1946, a survey was made of spawning runs; they were discovered in 68 Michigan streams. By 1949, 24,643 marine lampreys had been taken from a spawning run in one northern Michigan river. There were 20,000 to 110,000 eggs per female, depending on size. Obviously by 1949 the lamprey had enough biotic potential to virtually annihilate many Great Lakes fish species.

The lamprey-control program was initiated when the United States Fish and Wildlife Service began trapping lampreys in mechanical traps in 1948. Since then, the control program has gone through two major phases, the first based on electrocution of the downstream-migrating ammocoetes, and the second based on selective larval lampricides. Both phases depend on the assumption that *all* lamprey spawning is done in streams.

TABLE 5.8

Differential effect of lamprey on lake trout and whitefish in the Great Lakes

Year	Production of lake trout in Lake Michigan in thousands of pounds (four states combined)*	Production of lake trout in Lake Huron in thousands of pounds (Canadian data only)†	Production of whitefish in Lake Huron in thousands of pounds (Canadian data only)†
1935	4,873	4,255	1,936
1936	4,763	4,314	1,499
1937	4,988	3,901	1,664
1938	4,906	3,800	1,587
1939	5,660	3,203	1,390
1940	6,266	2,726	1,098
1941	6,787	2,823	926
1942	6,484	2,197	782
1943	6,860	1,609	603
1944	6,498	1,140	537
1945	5,437	862	367
1946	3,974	731	428
1947	2,425	377	456
1948	1,197	344	929
1949	342	398	1,320
1950		415	2,701
1951		552	3,592
1952		588	5,559
1953		344	6,479

*From Hile, Eschmeyer, and Lunger, 1951.
†From Frankland, 1955.

First, the United States and Canada cooperated in an enormously expensive program of building electrified weirs on all tributary streams of Lake Superior. This program was a failure because ice jams and logs swept downstream in spring floods damaged the systems long enough to let some lampreys escape. These lampreys faced enormously reduced competition pressure as adults, and in any case a small proportion of many million lampreys is still enough to wreak havoc with the fisheries. The electrification program ran from about 1951 to 1957.

By 1958, selective larvicides for lampreys had been discovered (Applegate and King, 1962). Whether these work or not (to the extent that lampreys are controlled) has yet to be seen. In the meanwhile, a cause of alarm was the finding, in 1958, of large numbers of larval lampreys in bays at the mouths of rivers. Hansen and Hayne (1962) studied this phenomenon and found that marine lamprey larvae seemed to be congregated at the mouths of rivers, not spread all over lakeshores. These bay lamprey larvae were very much larger, on the average, than those in the rivers. This suggests that the bay larvae had not been spawned there, but rather moved out to the bay partway through their developmental period.

Hansen and Hayne also conducted scientific sampling of both bay and river larvae and found 136,800 to be the best estimate of the larvae in the river (Ogontz River, Michigan, 1959 data).

Enough of the background on the lamprey situation in the Great Lakes has now been presented to introduce a number of biological problems.

1. Lampreys have had a much greater impact on lake trout than on whitefish (see Table 5.8). Why?

2. Why is it that lampreys and lake trout have been able to coexist for centuries in the Finger Lakes, when lampreys have annihilated lake trout in the three upper Great Lakes in about 10 years after they arrived in each lake?

3. Since *Petromyzon* is more than twice as big in the Atlantic Ocean as in the Great Lakes, why do we see so little evidence of its operations there? What is the food of adult *Petromyzon* in the ocean?

4. What factors regulate the abundance of *Petromyzon* in the ocean? Does it have any natural enemies?

Question 1 is the simplest to answer. As Smith (1954) has shown, the net reproduction rate, R_o, is far less important than the mean generation time, T, in determining the intrinsic rate of natural increase, r. Since

$$R_o = e^{rT}$$

$$r = \frac{\ln R_o}{T}$$

r only changes in proportion to the log of R_o, whereas it changes inversely as T. Female whitefish reach maturity at age 3 or 4 and have 10,000 to 75,000 eggs; lake trout do not reproduce until age 7 and only have 5,000 to 15,000 eggs. The trout, then, not only have a quarter as many eggs as whitefish, they have a mean generation time almost twice as long, which depresses the intrinsic rate of natural increase about 1.4 times as much. The lake trout's homeostatic ability to bounce back from predation is thus much less than that of the whitefish. The late maturation hurts in another way, in addition to lowered biotic potential. The impact of lamprey on trout is a rapidly rising function of trout age (see Table 5.9). By age 7, when trout begin to reproduce,

TABLE 5.9

Impact of lamprey on lake trout in the Great Lakes as a function of lake trout age

	Age			
	3	*4*	*5*	*6*
Number of fish	908	341	103	8
Number scarred by lampreys	38	130	68	5
Number of lamprey scars†	45	192	95	10
Percentage scarred	4	38	66	63

*From Budd and Fry, 1960.
†Assume one scar represents one lamprey attack.

almost all the female trout have been killed. This explains why trout populations in the Great Lakes became extinct so quickly under the impact of lamprey predation, whereas after the initial onslaught of the predator, whitefish populations rose to the highest levels on record (Table 5.8).

The second question has some very interesting implications. Since 1875, it has been known that, beyond doubt, the lamprey in Cayuga Lake was *Petromyzon marinus* (Wigley, 1959). The lampreys in the Finger Lakes are the same size as those in the Great Lakes. One of two alternative possibilities must then be true. Either the lampreys in the Finger Lakes do not have as great an impact on the lake trout there as they do on the trout in the Great Lakes, because of a difference in behavior, or else the lamprey population in the Finger Lakes is regulated by some factor other than starvation. The first possibility can be eliminated quickly. In Table 5.10, we show the percentage of lake trout with lamprey scars for Cayuga Lake and South Bay, Lake Huron, separately for each length class.

On the basis of Table 5.10 we would have expected all the lake trout in Cayuga Lake to have been annihilated long ago, yet they have not been. Why? In Sec. 4.2, we theorized that many pests are characterized by unusually violent population fluctuations, relative to most species of animals. We argued, and presented evidence, that these wide-amplitude fluctuations occur whenever density-independent factors are much more important than density-dependent factors in regulating the abundance of a species; finally, we postulated that the reason for the importance of density-independent factors in pest population dynamics must be sought in one or both factors of high intrinsic rate of increase and unusually fast population buildup in response to favorable weather conditions. Now if all this is true, since lampreys are

TABLE 5.10

Comparison of impact of lamprey predation on lake trout populations in Cayuga Lake and South Bay, Lake Huron

Length class (inches)	Percentage of trout with lamprey scars	
	Cayuga Lake, 1950*	South Bay, 1957†
17.0–17.9	23	18
18.0–18.9	72	25
19.0–19.9	71	50
20.0–20.9	87	63
21.0–21.9	95	63

*Table 43, Wigley, 1959.
†Table IV, Budd and Fry, 1960.

"pests," we should suspect that their populations might fluctuate wildly, and that the populations are very much under the influence of weather.

Wigley (1959) has found detailed support for the ideas expressed above. First (Wigley, table 45), the number of marine lampreys in Cayuga Lake dropped sharply from 1949 to 1951, from a population of 10,000 to 15,000 to one of 4,435. This drop was reflected in a drop in the average incidence of lamprey scars on lake trout from 60 percent of the lake trout scarred in 1949 to 38 percent scarred in 1951. Clearly, the extrinsic factor operating on the lamprey population saves the trout in Cayuga Lake.

It turns out (Wigley, fig. 9) that water temperature in the streams during the lamprey spawning migrations is the key factor in regulating lamprey populations in Cayuga Lake. Once the pre-spawning adults have migrated upstream, a drop in water temperature to the low 40s "drastically retards activity" (Wigley). Wigley did not discover any important natural enemies of lampreys, though he considered a long list: sea gulls, raccoons, great blue herons, and water snakes, all of which eat adult lampreys; minnows, which eat the eggs; and leeches, trematodes, and nematodes, which infect the adults.

All these facts suggest that it would be worthwhile to seek an explanation for changes in lamprey numbers in the temperature during the spawning months (April, May, and June) in the year in which a year class of lampreys is spawned. The problem is, however, to determine the length of time elapsed between the time a year class is spawned and the time it attempts to spawn. Applegate believes this would be 5 years (4 as larvae, 1 as an adult), but Wigley believes this to be 8 years (7 plus 1). His data on Lake Cayuga lamprey populations (table 45) show 10,000 to 15,000 in 1949; 9,390 in 1950; and 4,435 in 1951. We can estimate the length of a lamprey's life by looking for the trio of years beginning 5, 6, 7, or 8 years prior to 1949 that shows a steady drop from year to year in the average temperatures in the Cayuga Lake area in April, May, and June. This trio of years should begin in 1941, 1942, 1943, or 1944. Mean monthly temperatures for April, May, and June, 1941 to 1947, are presented in Table 5.11 for Oswego, 40 miles from Lake Cayuga.

TABLE 5.11

Mean monthly temperatures for Oswego*

Year	April	May	June	April and May
1941	47.9	55.4	66.6	103.3
1942	47.3	57.4	64.8	104.7
1943	37.4	53.0	66.4	90.4
1944	39.6	59.2	63.6	98.8
1945	49.2	49.7	62.2	98.9
1946	43.1	53.6	63.2	96.7
1947	43.0	52.5	62.4	95.5

*From U.S. Department of Commerce Weather Bureau World Weather Records 1951–1950, Washington, D.C., 1959.

The only sharp drops in temperature from year to year are from 1942 to 1943, in April, and from 1944 to 1945, in May. When Wigley's data is re-checked, two points stand out: (1) In all three years, the bulk of the spawning run seems to have occurred in May; (2) Wigley's 95 percent confidence limits on population estimates are much closer together for 1950 and 1951 than for 1949, when his estimate was 7,374, with 95 percent confidence limits of 4,210 and 12,950 (from mark-recapture program). His 1949 estimate must be re-garded as imprecise. The best match between our temperature data and the lamprey-spawning-run data come from the following quartet of numbers.

1944	May mean temperature:	59.2
1945	May mean temperature:	49.7
1950	lamprey population:	9,390
1951	lamprey population:	4,435

The best guess we can make from these admittedly inadequate data is that adult lamprey populations in a spawning run are determined by the temperature of the spawning stream 7 years earlier (Wigley's population estimates are based on the number in the run the *following* spring).

We are on much sounder ground in dealing with the data of the Great Lakes Fisheries Commission, as collected under contract between the Commission and the United States Bureau of Commercial Fisheries. Here the data represent a much larger population, and are based on 26 Lake Superior streams. In Table 5.12, I have attempted to match data on adult spawning run population with data on temperatures in the area several years earlier. The lamprey adult spawning runs are most closely correlated with temperatures 6 years earlier, whether we make the comparison with April temperatures, or May, or June, or all three summed. Applegate's (1950) observations make it clear that lamprey spawning runs in northern Michigan streams extend well into June; therefore Fig. 5.13 plots catches of adult lampreys against the sum of April, May, and June mean minimum temperatures 6 years earlier.

Figure 5.13 suggests that prior to 1962, it is highly probable that the number of lampreys caught in a spawning run was almost entirely accounted for by the water temperatures in the streams in which they were spawned 6 years previously, and that by 1962 and 1963, some other factor was regulating lamprey populations (presumably the control program). This suggests that attempts to kill larval lampreys or adults probably did not make a real impact on the population until 1962 −6 years, or 1956.

TABLE 5.12

The relation between catches of adult marine lampreys in south shore Lake Superior streams, and the temperature at Marquette, Mich., in previous years

	Average minimum temperature, °F*					Catches of lampreys in 26 streams†
Year	April	May	June	Sum of preceding cols.	Year	
1950	26	39	49	114		
1951	33	44	49	126		
1952	36	42	53	131	1958	58,101
1953	31	41	52	124	1959	44,055
1954	31	37	52	120	1960	36,864
1955	39	43	55	137	1961	67,360
1956	30	37	52	119	1962	9,127
1957	33	42	50	125	1963	11,117
1958	34	40	47	121		

*From Climatological Data, National Summary, U.S. Department of Commerce, Weather Bureau.

†From 1964 *Report of Annual Meeting*, Great Lakes Fishery Commission.

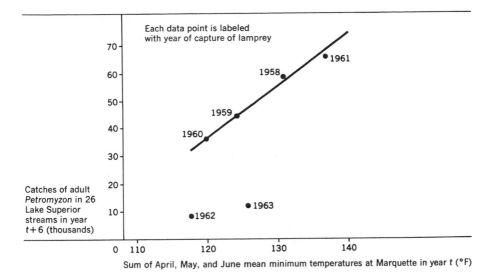

FIGURE 5.13

Relation between catches of adult sea lampreys in 26 south shore Lake Superior streams and the temperature at Marquette, Mich., 6 years previously.

Piavis (1961) studied the embryology of the marine lamprey and discovered that all embryos were dead 12 days after exposure to 50°F, and 5 days after exposure to 45°F. Table 5.12 shows that in the upper Great Lakes, the marine lamprey is subjected to water temperatures close to the limit of its endurance as an embryo. A review of the lamprey-control program in Lake Superior (Erkkila, Smith, and McLain, 1956), shows that although 43 control weirs were operating in 1954, some lampreys still escaped because of power failure to the electrocuting devices caused by flooding. By 1956, operation of the weirs had settled into an efficient routine, and use of larval lampricides had begun (Applegate and King, 1962).

What is the significance of this history for resource managers in other fields? First, when a pest has a high biotic potential, it is preadapted to withstand catastrophic mortality. A male and female *Petromyzon* will give rise to 60,000 young, so the mortality can reach

$$\left(\frac{60,000 - 2}{60,000}\right) 100 = 99.99667\%$$

and the population can still maintain its density. We need either a lethal chemical program that annihilates the population, which raises formidable problems because of the law of diminishing returns; or else we need some

Impact of the environment and man on populations 111

type of biological control such as genetic control, or control by pathogens, predators, or parasites that increase in intensity of effect as the pest density increases.

The lamprey study lends support to the general theory of pest regulation presented in the section on pest control. That is, pests are often unstable precisely because they are weather-regulated rather than regulated by density-dependent factors. Weather-regulation often happens where an animal is at the limits of its geographic range, as seems to be the case with the lamprey in the upper Great Lakes. In such instances, control that does not kill an extremely high proportion of the population, or is not density-dependent, merely elicits a homeostatic response. The pest is maintained at high densities until such time as an adverse change in the weather drives it down to sub-pest densities. Then proponents of the ineffective control claim victory. They can continue their claim until weather becomes favorable to the pest again—and its density rises.

Questions 3 and 4 are completely unanswerable at the moment. Mansueti (1962) reports that lampreys are very rarely found on fish in Chesapeake Bay in winter fishing. One fisherman could only recall two or three on striped bass in 30 years of fishing. The incidence on menhaden is much lower than one lamprey for every several hundred menhaden. If lampreys are found on Chesapeake Bay fish, they are quite small. Large *Petromyzon* 20 to 36 inches are found very rarely indeed. Why? To my knowledge, no one knows.

Various species of lampreys are extremely abundant in certain sections of the world's oceans. It is possible that they have a great impact on salmon stocks, for example, but the probability of an attacked salmon ever entering a spawning run is so slight that scarred salmon are rarely seen in spawning streams.

5.4 SUPPLEMENTING NATURAL REPRODUCTION: THE LOGIC OF HATCHERY OPERATIONS

Three factors determine whether it is worthwhile to supplement natural reproduction by release of hatchery-raised animals. First, are we dealing with density-dependent or density-independent regulation of recruitment? Second, how great is the "carrying capacity" of the environment? Third, how long will it take to recapture the released animals that had been raised in hatcheries?

Table 5.13 gives population estimates for a population of smallmouth bass under density-independent control. The great year-to-year variation in year-class strength is quite striking. Table 5.14 indicates the extent to which

TABLE 5.13

Estimates of total extant populations

Year of capture	Age, in years							Total
	4	5	6	7	8	9	10	
1947	12,200	7,460	4,150	730	680	350	60	25,630
1948	7,500	6,890	2,230	2,150	430	260	150	19,610
1949	2,700	2,810	1,390	1,400	660	180	60	9,200
1950	9,000	1,980	850	760	550	180	120	13,440
1951	44,900	3,190	150	430	380	160	120	49,330
1952	29,000	17,410	560	140	140	160	60	47,470
1953	36,800	12,260	4,000	420	60	70	30	53,640
1954	1,000	12,840	2,390	1,740	170	60	30	18,230
1955	1,000	190	4,760	1,220	690	70		7,930
Mean	16,000	7,230	2,280	1,000	420	170	70	27,160

this particular population is under control of temperature. Ninety-four percent of the variance ($=0.972^2 \times 100$) in recruitment of 4-year-olds from year to year is accounted for by temperatures from June to October in the year in which the recruits were spawned. Growth in weight during the second and third years of life is enormously dependent on temperature. The population is obviously living in an area close to the fringe of the geographical range of the species. Many years the habitat is so cold that there is almost no reproduction and almost no growth of juveniles. In a 9-year period, the strongest year class of recruits was 45 times the size of the weakest year class of recruits.

Another line of evidence brings out the great importance of temperature in regulating productivity in this resource (Watt, 1959). In 1951, 2,437 fingerling bass with a maxillary bone clipped for later identification were planted in South Bay, and 664 were planted in 1953. Only two bass with these clips have subsequently been caught. This is not surprising because only 1 percent or less of the extant population was planted. The dilemma in trying to supplement natural reproduction in a fishery largely under the influence of temperature is that in years when natural spawning is very successful, due to favorable weather, any planting of hatchery-reared fish makes a negligible contribution to the total stock. In years when the survival of native fish is poor, the survival of the introduced fish will be at least as bad. When a population is largely regulated by extrinsic factors, there is little motive for supplementing natural reproduction under either favorable or unfavorable conditions. (This assumes that mortality of stocked fish is the same as that

Impact of the environment and man on populations 113

TABLE 5.14

Correlation of recruitment, survival, and growth with various temperature indices (Watt, 1959)

Dependent variable	Statistics: regression coefficients in linear regression equation $Y = a + bX$ and correlation coefficient r	Values of statistics when independent variable is sum of temperature indices for stated combinations of months in year t			
		July to October	July to November	June to October	June to November
Recruitment at age 4 from year class spawned in year t	r	0.946	0.823	0.972	0.883
	a	7,434.	7,770.	7,159.	6,950.
	b	2,206.	1,545.	1,992.	1,539.
Instantaneous natural mortality rate from age 6 to 7 during year t	r	0.179	0.239	0.212	0.264
	a	0.4006	0.3824	0.3918	0.3738
	b	0.00553	0.0065	0.00613	0.00646
Growth in weight during year 1 (t)	r	0.029	0.116	−0.071	0.034
	a	0.702	0.695	−0.710	0.700
	b	0.000220	0.000776	−0.000556	0.000244
Growth in weight during year 2 (t)	r	0.749	0.689	0.796	0.772
	a	1.419	1.400	1.435	1.396
	b	0.03840	0.02611	0.03119	0.02493
Growth in weight during year 3 (t)	r	0.751	0.783	0.710	0.869
	a	5.926	5.938	5.902	5.907
	b	0.04024	0.03102	0.03165	0.02913
Growth in weight during year 4 (t)	r	0.171	0.064	0.142	0.060
	a	2.331	2.341	2.336	2.342
	b	0.01136	0.00332	0.00750	0.00272
Growth in weight during year 5 (t)	r	−0.471	−0.485	−0.462	−0.479
	a	2.874	2.855	2.848	2.838
	b	−0.03821	−0.02582	−0.03690	−0.02524

for natural spawning. If temperature influences the survival of eggs, not fry, then the introduction of fry might be successful.)

The extent to which we can usefully supplement the natural reproduction of populations controlled by intrinsic, or density-dependent factors, depends on the "carrying capacity" of the environment. Carrying capacity of a given habitat refers to its ability to support animal life. This ability is determined by incident solar radiation, available minerals, temperature, community organization, and community energetics. Because each environment has a fixed upper limit to its ability to support life, it makes little difference how many animals are stocked, provided enough are stocked. Any excess stocked over and above the carrying capacity will serve no useful purpose, because of the effect of competition on survival and rate of growth in individual animals. One of the best available sets of data for illustrating these points has been collected by Swingle and Smith (1939), and Swingle (1952) in connection with the stocking of ponds in Alabama. In May, 1936, a pond was stocked with 2 pounds, 5 ounces of bluegill fry per acre, and another pond was stocked with 180 pounds of year-old bluegills per acre. Analyses throughout the year indicated that the available food was equal in both ponds. When the two ponds were drained in November, the first had 105 pounds of fish per acre, and the second had 92 pounds per acre. Evidently, the two ponds each had a carrying capacity in the range of 90 to 110 pounds per acre. Where only 2 pounds, 5 ounces of fish were stocked, the biomass grew to the level supportable by the carrying capacity of the environment, whereas when 180 pounds were stocked, the biomass had to drop to the level supportable. In a second experiment reported in the same paper by Swingle and Smith (1939), two different stocking policies were applied in two successive years in a 1.8-acre unfertilized pond. The details are given in Table 5.15. Despite the very different numbers and weights of the stocked fish, the resultant biomasses on draining the pond in November were 293 pounds, 4 ounces the first year, and 296 pounds, 2 ounces the second. Some of the density-dependent mechanisms giving rise to this surprisingly well-regulated adjustment may be inferred by careful study of Table 5.15. Two types of compensatory mechanisms are clearly at work: those that regulate population density, and those that regulate growth rate of the individual animal. Furthermore, these two types of compensatory mechanisms are compensatory with respect to each other: the greater the number of fish produced, the less their mean weight will be. This is true for each species in the table where a comparison can be made.

A great deal of research has been conducted on stocking policies to determine the particular community organization that maximizes the efficiency of conversion of incident solar radiation into fish biomass. As might be

TABLE 5.15
Fish production in a 1.8 acre unfertilized pond with different rates of stocking (Swingle and Smith, 1939)

Species	Stocked fish, February				Recovered fish, November			
	1935		1936		1935		1936	
	Number	Weight	Number	Weight	Number	Weight	Number	Weight
Huro salmoides (largemouth black bass)			4	6 lb 1 oz			3	8 lb 3 oz
Micropterus coosae (red-eye bass)	29	7 lb 0 oz			31	15 lb 3 oz		
Helioperca machrochira (bluegill bream) and *Eupomotis microlophus* (orange-ear bream)	3,638	13 9	10	1 12	16,022	124 4	21,827	72 8
Pomoxis annularis (white crappie)			10	4 2			3,780	36 2
Ictalurus furcatus (channel catfish)			2	4 4			2	5 4
Ameiurus natalis (yellow bullhead)	298	11 6	10	4 4	1,312	75 13	2,060	99 9
Erimyzon oblongus (chub sucker)	520	8 10	200	4 0	4,704	78 0	2,733	74 15
Totals	4,485	40 9	236	24 7	22,069	293 4	30,405	296 9

expected, best utilization of a habitat occurs where there are several different species of fish to utilize the various food organisms available, and also keep each other sufficiently cropped to prevent stunted growth due to excessive competition. Swingle (1952) reports the following results from various stocking densities.

Number of fish stocked per acre	Average annual catch of fish, lb per acre
100 largemouth bass	42.2
150 bass + 1,500 bluegills	239.4
150 bass + 1,500 bluegills + 25 white crappies	282.2

In general, we would expect that the greater the number of kinds of species of fish in a given environment, the more efficient the overall food utilization would be, and the greater the productivity. Productivity has not been studied from this point of view, but standing crop, the biomass present at any time, has. Carlander (1955) showed that where we plot logs of pounds per acre against logs of number of species, we get a straight line, though with a lot of scatter. The data are from midwestern reservoirs (Fig. 5.14). The poor fit is because of the multivariable nature of the relationship, as indicated in the next paragraph.

The productivity, and also the standing crop of fish in lakes and reservoirs, depends on whether the water is oligotrophic (poor in food), or eutrophic (rich in food), how cold the water is, the type of fish, and its position in the trophic pyramid, relative to the incident solar energy. The closer the fish is to phytoplankton in the trophic pyramid, the greater the productivity and standing crop of that species will be. These points are illustrated in Fig. 5.15. Standing crops of trout are low because they live in cold, oligotrophic waters. The first eight species in Fig. 5.15 are, as adults, predators of other animals, including small fish. Gizzard shad, which have the highest standing crops of the species in Fig. 5.15, eat zooplankton and phytoplankton, and are therefore much closer to incident solar radiation in the food pyramid. Buffalo fish and carp are bottom feeders with a mixed diet of plants and animals (largely invertebrates), and bullheads are omnivores, eating invertebrates. Thus, the least productive fish, per unit area, either eat fish which in turn may have eaten small fish which eat zooplankton which eat phytoplankton, and are thus five steps from radiant energy, or else, like the trout, live in oligotrophic water. The most productive fish are only two or three steps removed from the sun's rays. Such considerations determine which kinds of animals we should stock in an area, how many, and which mixtures of species would most efficiently exploit available food materials in the environment.

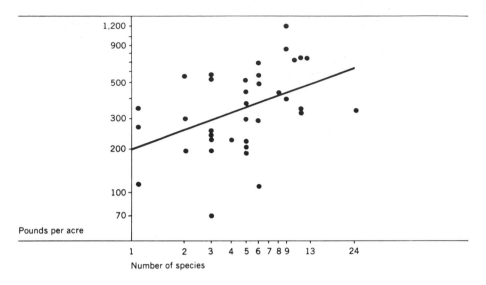

FIGURE 5.14
Relation between standing crops of fishes and numbers of species in midwestern reservoirs (Carlander, 1955).

In general, interest in planting young warmwater fish in lakes has waned. For example, Cooper (1948) reports that during the decade 1936 to 1945, the state of Michigan stocked bluegill fingerlings at an average of 35 per acre in 1,138 lakes; largemouth bass at 2.4 per acre in 687 lakes; and smallmouth bass at 1.9 per acre. To determine the contribution these stocking densities made to the fish stocks already in the lakes, 12 representative lakes were selected for seining. The lakes contained an average of 742 game and panfish species per acre. Clearly, stocking fingerlings in such cases makes a small contribution to total yield.

The situation with respect to stocking, as outlined above, has led to an entirely new concept as to how best to supplement natural reproduction: the "put and take" philosophy. This philosophy solves three problems simultaneously:

1. There is a tremendous demand for recreational fishing and hunting near urban areas.

2. This demand cannot be met by the number of animals supportable by the carrying capacity of the environment.

3. If we stock very young animals, most will die from natural mortality before anglers or hunters have a chance to kill them.

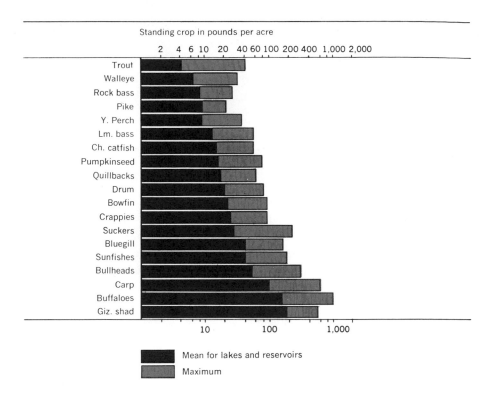

Standing crop in pounds per acre

Trout
Walleye
Rock bass
Pike
Y. Perch
Lm. bass
Ch. catfish
Pumpkinseed
Quillbacks
Drum
Bowfin
Crappies
Suckers
Bluegill
Sunfishes
Bullheads
Carp
Buffaloes
Giz. shad

■ Mean for lakes and reservoirs
▨ Maximum

FIGURE 5.15

Standing crops of named fishes in North American lakes and reservoirs. Usually these fish were in combination with other species. Figure furnishes rough approximation of the relative efficiency of the species listed (Carlander, 1955).

One answer is to raise animals until they are old enough to be beyond the age where losses to natural mortality are greatest, release them close to the beginning of the hunting season, and expect to have the bulk of the released animals caught in a few days. Jensen (1958) presents data on trout streams where more than half the released "catchable" (legal sized) rainbow trout were caught within 9 to 12 days of release. This is clearly a highly artificial situation: The fish are in the water such a short time that "carrying capacity" is not relevant.

Besadny and Wagner (1963) have made a statistical analysis of the value of stocking pheasant cocks in Wisconsin. As one might imagine, the importance of stocking depends on the number of wild pheasants shot per square mile. Where the population density of wild birds is low, stocked cocks constitute up to 51 percent of the birds shot. Where the population density of

Impact of the environment and man on populations   119

wild birds is high, stocked birds will only be 2 or 3 percent of the kill. Besadny and Wagner point out that stocking is only worthwhile where habitat is poor. Stocking in effect increases the carrying capacity of the natural environment: we raise the animals until some time before they are to be shot, and thereby allow a poor habitat to "support" a much greater population than it would otherwise be able to.

5.5   THE PROBLEM OF EXTINCTION: BUFFALO AND BIGHORN SHEEP

Extinction is a real problem in the management of any natural resource, no matter how plentiful. To demonstrate that even abundance of awesome dimensions is no guarantee against extinction, consider the case of the buffalo. Everywhere in the United States, one is impressed by the tremendous density of the most abundant large mammal, *Homo sapiens*. Yet, assuming an average weight of 120 pounds for this species, and a population of 200,000,000, the biomass is only 24 billion pounds. In contradistinction, population estimates of the buffalo in primitive times run as high as 75 million in the United States (Seton, 1929). If we assume an average weight of 1,000 pounds, the total biomass was 75 billion pounds. The diaries of early settlers indicate their amazement at the unbelievable mass of meat represented by these mammals, whose biomass dominated the landscape to about three times the extent of our own species at present (Roe, 1951; Seton, 1929). It is unlikely that any species of large animal has ever dominated its environment to a greater extent than the buffalo. Yet this awesome biomass dropped from a population of 40 million in 1830, to 200 in 1887, in the unbelievably short period of 57 years! In 1887, the foolish carnage still persisted, and 52 of the remaining animals were slaughtered (Roe, 1951). By the end of 1888, about all that remained of a staggeringly abundant natural resource was a pitiful herd of 26 animals in Wyoming. The buffalo history represents one of the sorriest chapters in the whole story of man's mismanagement of natural resources, and it seems worthwhile to examine the record with care, to see what it can teach us. Our data are drawn from Roe (1951).

There is controversy if the buffalo all over North America were a single species, *Bison bison*, or a number of variants so similar as to be essentially one species. At any rate, buffalo formerly ranged, in prehistoric times, all over North America. Fossils have been found in Alaska, Florida, California, Mexico, Texas, the Yukon, Ontario, Massachusetts (Cape Cod), Oregon, Georgia, Kansas, Kentucky, Arizona, Mississippi, Nebraska, South Carolina, Pennsylvania, and Washington.

Between 1730 and 1830, the march westward placed great demands on buffalo for food and clothing. From 1830 on, there was a systematic, unbelievably wasteful practice of mass slaughter of buffalo which typically utilized only the hides and tongues. It is extremely difficult to obtain good estimates on the number of buffalo killed in the nineteenth century. We can, however, be reasonably sure of figures (Roe, 1951) for the Red River settlement in Manitoba. This colony first organized a buffalo-hunting expedition in 1820, with 540 carts proceeding to the hunting range. About 119,000 buffalo were killed. By 1845, about 321,000 buffalo were killed, and by 1865, about 710,000. An astonishingly high proportion of this vast kill was simply wasted. On the average, only three out of every ten animals killed were used for meat, and only about 75 pounds of dried meat was obtained per animal used. There were only 3,147 people in the Red River settlement in 1831, and these killed about 176,000 buffalo that year, or about 55 buffalo per person per annum. Our present North American livestock resources, with only about one head of cattle per two people, would not last long under Red River management policies! Even with the vast amount of meat wasted, the consumption of buffalo per person per annum was amazing. If only 75 pounds of meat was used per animal, and only three out of ten were used for meat at all, this still represents a meat diet per person per year of

$$75 \times \frac{3}{10} \times 55 = \text{about } 1{,}200 \, \text{lb}$$

Since the average person would be hard pressed to eat about $3\frac{1}{3}$ pounds of meat a day, 365 days a year, some explanation is still needed for the great carnage. Perhaps the answer is the two or three dogs each person owned, which were mainly fed from the buffalo carcasses.

Many scholars have gone to great effort to determine exactly how many buffalo were killed each year in North America between 1870 and 1880, at the height of the carnage. One method of calculation is to add up the number of hides known to have been shipped by all railroads, the number of buffalo believed to have been killed and wasted by white men, and the number killed by Indians. For 1873, perhaps the heaviest year of the slaughter, Roe estimates as follows.

Hides shipped by rail	754,000
Buffalo killed and wasted	754,000
Buffalo killed by Indians	405,000
Total kill	1,913,000

All experts feel this estimate is low. Another method of calculating the annual kill in the period 1868 to 1874 is to examine railroad freight records and eyewitness reports on the volume of buffalo bones shipped East to be ground up for fertilizer. From these data, Roe gets the following estimates, which are probably more realistic.

Year	Number of buffalo killed
1868	1,500,000
1869	2,500,000
1870	5,000,000
1871	6,000,000
1872	7,000,000
1873	6,000,000
1874	3,000,000

After considering the population dynamics of the buffalo in the wild state during the nineteenth century, Roe estimates that in 1830 there were 40 million buffalo remaining, with a 50:50 sex ratio. He also assumes that 90 percent of the females calved each year, but only 18 percent of those survived. The addition to the herd each year in the wild state would be

$$40,000,000 \times 0.5 \times 0.9 \times 0.18 = 3,240,000$$

This number of buffalo could not be harvested every year because no allowance would have been made for natural mortality of adults. Data on modern buffalo (Fuller, 1962) (see Sec. 12.1) show a 10 percent per annum natural mortality in adult buffalo. If the rate were the same in the nineteenth century, the calving rate would not have replaced annual adult losses to natural mortality. Suppose the natural mortality was then 5 percent per annum. This is a reasonable assumption since the incidence of mortality caused by tuberculosis may have been well below the 5 percent per annum reported in a modern wood buffalo herd by Fuller (1962). Then we can write a simple equation allowing us to solve for the maximum sustained yield. If total mortality every year is to balance total recruitment of new calves and if risks are competing throughout the year with constant instantaneous hunting (H) and natural mortality (M) coefficients, then, taking

$$0.05 = 1 - e^{-M}$$

and

$$\frac{3.24}{40} = 1 - e^{-(M+H)}$$

gives

$$H = \log 0.950 - \log 0.919$$

and

$$\text{"Catch"} = \frac{NH}{M + H}[1 - e^{(M+H)}]$$

$$= 4 \times 10^7 \left(\frac{\log 0.950 - \log 0.919}{-\log 0.919} \right) \frac{3.24}{40}$$

$$= 3{,}240{,}000 \left[\frac{\log 0.950 - \log 0.919}{-\log 0.919} \right] = 1{,}272{,}700$$

This ignores the fact that natural mortality rates are age-specific. Also, it ignores the possibility that mortality rates are specific for sex within age. Fuller (1962) has noted that in modern buffalo, incidence of tuberculosis is more than four times as great in females than in males. Although he had a small sample, a chi-square test showed a significant difference. If buffalo were present in large numbers today, and a management strategy was being developed, such details as age- and sex-specific natural mortality rates, age-specific fecundity rates, and the sex ratios of the calves would all be built into a mathematical model to show how best to manage the resource.

To conclude, there were three principal reasons for the virtual annihilation of the buffalo.

1. There was a total absence of any attempt at management.
2. Seven-eighths of the animals killed after 1830 were females; thus the reproductive potential of the population was destroyed.
3. The buffalo was an extremely stupid and unaggressive animal. Males weighing a ton or more were herded into flimsy corrals that any three bulls could have smashed into matchwood. Hunters were allowed to ride unmolested through herds that could have stampeded and killed them.

It is interesting to compare this history with that of the cougar. Ever since the white man came to North America, he has tried to exterminate the cougar. Yet cougars are still found in many parts of North America, from New Brunswick to Palo Alto.

Bighorn sheep provide a somewhat different perspective on the problem of extinction. Never as abundant as the buffalo, they live in relatively inaccessible terrain, and have therefore had much less trouble with hunting pressure. The species has, however, difficulty remaining extant, for a combination of reasons (Buechner, 1960). Prior to the advent of the white man, bighorn sheep occupied rocky areas in 15 western states, British Columbia, and Alberta, and probably numbered 1.5 to 2 million. Now there are only 15,000 to 18,000 in isolated pockets in 11 states and Canada.

The great decline was not due to hunting alone. There were four causes of reduction.

1. *Hunting*: In the spring of 1875 about 2,000 sheep hides were removed from Yellowstone National Park.

2. *Scabies*: This lethal disease caused by scab mites appeared in each area soon after the arrival of domestic sheep. From 1880 to 1881, thousands of bighorn sheep died of scabies in Wyoming.

3. *Competition from domestic sheep*: This may have predisposed bighorn sheep to scabies epidemics.

4. *Civilization*: This restricted the winter range of the sheep, and increased the possibility of starvation in heavy snow.

Bighorn sheep populations in the United States are currently limited by severe competition for food. The sheep are restricted to inaccessible, high-altitude rocky areas, where there is deep snow in the wintertime (except for a few sites where they are protected from hunting, as in the Black Hills of Dakota). The populations get good browse in the summer, but the harsh winter conditions cause lambs to be born with a lack of vigor, and many die from pneumonia. The maximum annual permissible harvest is about 1,000, or 2,000 if young rams with unimpressive horns for trophies are shot.

Harvesting only rams actually increases the biotic potential of the population (Buechner, 1960). This can be demonstrated by reasoning as follows. Suppose population growth to be described by the simple equation

$$N_t = N_o e^{rt}$$

What happens to the value of r, the intrinsic rate of natural increase, if only rams are harvested? Suppose hunting changes the sex ratio from 100 females: 100 males, to 100 females: 25 males. Assume that there will be no decrease in pregnant ewes even though the ram population has been decreased by 75 percent. Since r is a measure of reproductive rate per capita per unit time less death rate per capita per unit time, killing three-quarters of the rams without killing females multiplies per capita r by $200/125 = 1.6$ times. (Note that N_o drops to 5/8 its prehunt value, if only adults are considered.)

Bighorn sheep therefore represent the opposite situation to buffalo, since hunting in effect increases the reproductive rate per capita; in buffalo, as hunting pressure increased, the reproductive rate per capita was being systematically forced down because females were being selectively killed to a much greater extent than males. In algebraic terms, buffalo were decimated because the N_o in

$$N_t = N_o e^{rt}$$

was diminished with no compensatory increase in r. Many harvested species, such as pheasants and the deer discussed in the next section, are able to withstand diminished N_o because there is a compensatory increase in per capita r.

The important conclusion is that optimum yield must be defined as the maximum sustained yield that can be removed from a resource *without impairing the ability of the resource to replace, by reproduction, the biomass harvested.* In most resources, the optimum yield varies constantly in response to changing weather conditions, site factors, interspecific competition, predation, disease, and factors intrinsic to the population (density-dependent factors). It is precisely this multiplicity of factors, and the resultant difficulty of defining optimum management strategy, that leads to dependence on complicated systems analysis and simulation procedures of the type explained in the last part of this book.

Some readers will be surprised that, as extinction approached, the ferocity with which buffalo were harvested increased rather than leveled off. This is reminiscent of the history of whaling and sealing, of certain fisheries, and the current world petroleum situation. Whenever man utilizes a resource, indications that the resource is in danger of being wiped out typically lead to the highest rate of exploitation allowed by technology rather than strict adherence to a quota system dictated by interest in conservation. There is a rational explanation for this ferocious exploitation, but the explanation is based on a fallacy, which is in need of urgent public debate. Any agency investing money must always consider the relative profitability of alternative investment policies. Since money will earn interest if left alone, the net profit from any venture, t years in the future, must be divided by the term $(1 + r)^t$, where r is the per annum compound interest rate, in order to express the relative profitability of various time periods in equivalent monetary units (present dollars). A unit of profit is thus more valuable now than at any time in the future, and the further into the future it is received, the less valuable it is. It is precisely this economic principle that leads to short-sighted thinking with regard to conservation. The solution is to impose the constraint on renewable resource management discussed in the previous paragraph.

5.6 PRESSURE ON ONE SEX: DEER MANAGEMENT

Section 5.5 discussed the destruction of an exploited population resulting from harvesting pressure directed against females. A quite different set of problems arises when exploitation is directed entirely against adult males, which can happen when hunters are interested in horns for trophies, or the males are larger and more brightly marked, or are the preferred target for some other reason.

One of the best examples of this set of problems is the deer population in California, where there is "bucks only" legislation. Furthermore, California deer have been the subject of a number of comprehensive and thoughtful monographs.

A succinct summary of the present deer situation in the United States was published by *Sports Afield* in October, 1963. In 1963, the United States deer population was estimated at 13 million. This is about four times the estimated 1933 population. Kills were more than a million a year in the decade 1948 to 1958, more than 1.5 million in 1958 to 1961, and 2 million in 1962. Evidently, deer populations can increase in the face of about a 20 percent kill, if this kill is appropriately distributed by age and sex. By contrast, the California kill is less than 9.5 percent per annum. Although California has a high human population and a high deer population (roughly 1.1 million), it only has 8 percent hunter success. Oregon, with about 750,000 deer, has 53 percent hunter success, and 139,700 deer were killed in 1962 as opposed to 59,988 in California. The California kill could be heavier, and there would be less crop damage if it were.

In view of this situation, two problems are worthy of immediate attention. The first is to determine as accurately as possible how many deer there are in California. The second is to determine what proportion of the deer population can be harvested each year. Happily, both problems can be solved using the same technique (Longhurst, Leopold, and Dasmann, 1952).

First, we can estimate the population density of deer as follows. Compute the maximum possible sustainable exploitation (kill) rate of bucks 2 years and older a deer herd can yield when only such bucks are hunted. From this, and the number of bucks actually killed in the hunting season, we can compute a minimum estimate of the total population that could provide this sustained kill.

Second, the above calculation yields, as a side-product, a minimum possible estimate of the maximum possible harvest (see Table 5.16). Table 5.16 illustrates the valuable heuristic role played by simulation: We can gain insight into the operation of a process by playing games to simulate the process, using sample numbers. We will have much more to say about such game-playing in Chap. 12.

The question arises, is the 9.5 percent in the last row and last column of Table 5.16 a product of the particular set of numbers used in this game, or would any set of numbers yield 9.5 percent? We can answer this by using simple algebra. Let x represent the number of adult females in any year.

TABLE 5.16

Percentage kill resulting from shooting all adult buck deer in a theoretical herd*

Time	Adult bucks	Yearling bucks	Does	Fawns	Total	Percent kill†
Year 1: Prehunt	70	30	100	100	300	
Hunt removal	70					
Posthunt	0	30	100	100	230	30
Year 2: Prehunt	30	50	150	150	380	
Hunt removal	30					
Posthunt	0	50	150	150	350	8.6
Year 3: Prehunt	50	75	225	225	575	
Hunt removal	50					
Posthunt	0	75	225	225	525	9.5
Year 4: Prehunt	75	112	338	338	863	
Hunt removal	75					
Posthunt	0	112	338	338	788	9.5‡

*From table 10, Longhurst, Leopold, and Dasmann (1952). It is assumed that yearling bucks serve the does, and that each doe (including yearlings) raises a fawn.
†Kill is expressed as a percentage of the posthunt population, since it is the latter that is usually measured in late autumn counts.
‡In all succeeding years, if all adult bucks are removed, the percentage kill will remain at 9.5 percent of the herd.

Then the population 2 years later, using the assumptions on which Table 5.16 is constructed, will have the following composition.

Adult bucks	$0.5x$
Yearling bucks	$0.75x$
Does	$2.25x$
Fawns	$2.25x$
Total	$5.75x$

Under this rather artificial scheme, which assumes *no natural mortality*, the permissible kill per annum will stabilize at $0.5/5.75 = 8.7$ percent of the prehunt population, or $0.5/5.25 = 9.5$ percent of the posthunt late autumn population. If we assume that 75,000 adult bucks were killed in 1947, the population must have been $75,000/0.087 = 860,000$ prior to the hunting season. However, not all bucks are killed in the hunting season, so the proportion of the population killed is lower than 8.7 percent and the population must be more than 860,000.

Another method of estimating the population is to determine the relation between kill and population in areas where the deer have been counted, and then use this ratio as a guide in estimating populations in similar areas. When this is done, kill rates turn out to vary from 1:13 to 1:30. (This compares with our theoretical kill rate of 8.7:91.3, or 1:10.5.) A kill of 75,000 actually represents a population of well over 860,000, or more like 1,123,000 (Longhurst, Leopold, and Dasmann, 1952, table 11).

Thus under "bucks only" legislation, shooting must lead to eventual starvation of the herd unless other factors are limiting, because the population cannot be limited by hunting when only adult bucks are shot. This raises the question as to the relative importance of various mortality factors in regulating deer.

Hunting takes 75,000/1,123,000, or about 6.7 percent of the population per annum. The only important deer predator in California is the mountain lion. Golden eagles, black bears, wildcats, and coyotes are probably of minor importance (Longhurst, Leopold, and Dasmann, 1952). The authors suggest 1,800 as a reasonable estimate for the California mountain lion population, and give 30 to 40 deer per year per mountain lion as the estimate considered most reasonable by the majority of experts. If the correct figure is 35, the annual percentage mortality of California deer attributable to mountain lion predation is only

$$\frac{35 \times 1,800 \times 100}{1,123,000}$$

or 5.5 percent of the population.

Longhurst, Leopold, and Dasmann show that starvation, disease, and parasites are far more important than hunting or predation in causing deer population mortality. Starvation, disease, and parasites, in turn, are dependent on range conditions, which in turn depend on how close the rangeland plant associations are to the climax of plant succession, which in turn depends on the fire history of an area. Climax forest, which has a dense high canopy, does not support dense deer populations because not enough light penetrates to the forest floor to allow extensive growth of herbs and woody shrubs with adequate growth of sprouts at heights accessible to browsing deer. Therefore fire (in the northern part of California) pushes forest succession back from the climax state, makes clearings allowing penetration of incident solar radiation to support adequate growth of deer browse, and hence increases the carrying capacity of the land for deer. Another factor affecting plant succession is overgrazing by livestock. When this occurs in bunchgrass areas such as Modoc County, there is subsequent invasion of woody species, valuable as winter browse for deer. Controlled burning (which is not feasible in the southern part of the state) can also result in tender foliage, which is good deer browse. In summary, the carrying capacity of the range determines deer population density in California, not hunting. Furthermore, because of the complex pattern of interactions among the various factors determining the carrying capacity of the range, and the lag effects, cumulative effects, threshold effects, and complicated economic aspects of range factors, range management constitutes a very challenging problem in simulation for resource management strategy evaluation (see Sec. 12.5).

Low hunting pressure adds to the impact of mortality factors such as starvation and disease because present deer populations can increase by 20 to 30 percent per year, but only 4 to 9 percent per year is being harvested because of the "bucks only" legislation.

Controlled burning converts chapparal (thickets of dwarf oaks) to shrubland. In mature chapparal, leafage on the average plant is out of the reach of deer. The foliage of these tall plants blocks off light, so succulent herbs cannot grow in their shade. When such tall shrubs are burned, new sprouts develop from the root crown, and provide succulent leafage within reach of the deer. The deer keep the shrub pruned back so there is unshaded open ground around it. In this open area, annual plants that provide winter and spring food for deer can grow. It makes a difference which kinds of plants are available as deer browse. Those with the most protein seem to be best, and plants growing in burned areas have more protein than those in unburned areas; the deer in burned areas are heavier (Einarsen, 1946).

Increased deer hunting would do more than minimize wastage to starvation and disease. Reproduction is also affected by malnutrition, through the following mechanisms (Longhurst, Leopold, and Dasmann):

1. The number of ova produced by breeding does is affected by nutrition.

2. If the doe is successfully bred, the ova are fertilized, are implanted in the uterus, and develop into embryos. The percentage of ova reaching embryo size is known to be higher in well-fed deer than in those poorly nourished.

3. The percentage of the embryos that mature and produce healthy fawns is dependent on the doe's nutrition. Starving does may resorb their embryos, abort, have stillborn fawns, or bear fawns too weak to survive.

4. Even if the fawn is born, the undernourished mother may not have enough milk to feed it.

5. On poor range, the does start to breed at a later age than they do on good range (Taber and Dasmann, 1958).

In summary, in California, man has cut down the population of deer predators through hunting, but has not replaced them as population regulators because of well-intentioned, but ill-conceived laws. Deer could be better managed by environmental control, and by increasing hunting pressure.

The complexity of deer management in particular, and management of chapparal country in general, makes them particularly suitable for computer simulation studies, shown in Sec. 12.5.

Some readers may wonder why there is any need for complex systems models (see Part Four) when some information can be obtained from a simple algebraic model. The answer is that hunting occurs over a short-enough period so that natural mortality can be assumed to be minor, and all we can claim for the algebra is a rough estimate of the maximum possible impact that hunting can have, given a rather artificial set of assumptions. The model as constructed does not allow for density-dependence, effects of winter weather, change in species composition or availability of food, or interactions with predatory or competitive species. Since there is no postulated negative feedback (density-dependent regulation), the model implies that in the absence of hunting there will be no population control, which is of course impossible. To obtain real insight into optimum strategy for managing deer, a systems model is required.

REFERENCES

Applegate, V. C.: Natural History of the Sea Lamprey, *Petromyzon marinus*, in Michigan, *U.S. Fish Wildlife Serv. Spec. Sci. Rept., Fisheries*, no. 55, 1–237 (1950).

——— and E. L. King, Jr.: Comparative Toxicity of 3-trifluormethyl-4-nitrophenol (TFM) to Larval Lampreys and Eleven Species of Fishes, *Trans. Am. Fish. Soc.*, **91**:342–345 (1962).

Besadny, C. D., and F. H. Wagner: An Evaluation of Pheasant Stocking through the Day-Old-Chick Program in Wisconsin, Wis. Conservation Dept., Madison, *Tech. Bull.*, no. 28, 1963.

Beverton, R. J. H.: Long-Term Dynamics of Certain North Sea Fish Populations, in E. D. LeCren and M. W. Holdgate (eds.), "The Exploitation of Natural Animal Populations," pp. 242–259, Blackwell Scientific Publications, Ltd., Oxford, 1962.

Budd, J. C., and F. E. J. Fry: Further Observations on the Survival of Yearling Lake Trout Planted in South Bay, Lake Huron, *Can. Fish Culturist*, **26**:7–13 (1960).

Buechner, H. K.: The Bighorn Sheep of the United States, Its Past, Present and Future, *Wildlife Monogr.*, no. 4, 1–174 (1960).

Carlander, K. D.: The Standing Crop of Fish in Lakes, *J. Fish. Res. Bd. of Can.*, **12**:543–570 (1955).

Cooper, G. P.: Fish Stocking Policies in Michigan, *Trans. N. Am. Wildlife Conf.*, **13**:187–193 (1948).

Dymond, J. R.: A Provision List of the Fishes of Lake Erie, *Publ. Ont. Fish. Res. Lab.*, **4**:55–74 (1922).

Einarsen, A. S.: Crude Protein Determination of Deer Food as an Applied Management Technique, *Trans. N. Am. Wildlife Conf.*, **11**:309–312 (1946).

Erkkila, L. F., B. R. Smith, and A. L. McLain: Sea Lamprey Control on the Great Lakes 1953 and 1954, *U.S. Fish Wildlife Serv. Spec. Sci. Rept. Fisheries*, no. 175, 1–27 (1956).

Frankland, E. (ed.): "The Canadian Commercial Fisheries of the Great Lakes," Market and Economics Service, Department of Fisheries of Canada, 1955.

Fuller, W. A.: The Biology and Management of the Bison of Wood Buffalo National Park, *Can. Dept. Northern Affairs Natl. Resources, Wildlife Mgt. Bull.*, ser. 1, no. 16, 1962.

Great Lakes Fishery Commission: "Report of Annual Meeting, Great Lakes Fishery Commission," The University of Michigan Press, Ann Arbor, Mich., 1964.

Hamilton, J. E.: Effect of Present Day Whaling on the Stock of Whales, *Nature*, **161**:913–914 (1948).

Hansen, M. J., and D. W. Hayne: Sea Lamprey Larvae in Ogontz Bay and Ogontz River, Michigan, *J. Wildlife Mgt.*, **26**:237–247 (1962).

Hile, R., P. H. Eschmeyer, and G. F. Lunger: Decline of the Lake Trout Fishery in Lake Michigan, *U.S. Fish Wildlife Serv. Fishery Bull.*, **52** (60):77–95 (1951).

Hohman, E. P.: "The American Whaleman," Longmans, Green & Co., Inc., New York, 1928.

"International Marine Sciences," UNESCO, Paris, 1964.

Jensen, P. T.: Catchable Trout Studies in Region II, 1957, California State Inland Fisheries, Administrative Report no. 58-1, 1–39 (mimeo.), 1958.

Ketchum, B. H., J. Lillick, and A. C. Redfield: The Growth and Optimum Yields of Unicellular Algae in Mass Culture, *J. Cell. Comp. Physiol.*, **33**:267–279 (1949).

Laws, R. M.: Some Effects of Whaling on the Southern Stocks of Baleen Whales, in E. D. LeCren and M. W. Holdgate (eds.), "The Exploitation of Natural Animal Populations," pp. 137–158, John Wiley & Sons, Inc., New York, 1962.

Lennon, R. E.: Feeding Mechanism of the Sea Lamprey and Its Effect on Host Fishes, *U.S. Fish and Wildlife Serv. Fishery Bull.*, **56** (98):247–293 (1954).

Longhurst, W. M., A. S. Leopold, and R. F. Dasmann: A Survey of California Deer Herds, Their Ranges and Management Problems, *Calif. Dept. Fish Game Bull.*, no. 6, 1952.

Mansueti, R. J.: Distribution of Small, Newly Metamorphosed Sea Lampreys, *Petromyzon marinus*, and Their Parasitism on Menhaden, *Brevbortia tyrannus*, in Mid-Chesapeake Bay during Winter Months, *Chesapeake Sci.*, **3**:137–139 (1962).

Marr, J. C.: The Causes of Major Variations in the Catch of the Pacific Sardine, in "Proceedings of the World Scientific Meeting, The Biology of Sardines and Related Species," vol. 3, pp. 667–791, Food and Agriculture Organization of the United Nations, Rome, 1960.

Nicholson, A. J.: Compensatory Reactions of Populations to Stresses, and Their Evolutionary Significance, *Australian J. Zool.*, **2**:1–8 (1954).

Piavis, G. W.: Embryological Stages in the Sea Lamprey and Effects of Temperature on Development, *U.S. Fish Wildlife Serv. Fishery Bull.*, **61**:111–143 (1961).

Radovich, J.: Effects of Sardine Spawning Stock Size and Environment on Year-Class Production, *Calif. Dept. Fish Game Bull.*, **48**:123–140 (1962).

Roe, F. G.: "The North American Buffalo," University of Toronto Press, Toronto, Canada, 1951.

Schaefer, M. B.: A Study of the Dynamics of the Fishery for Yellowfin Tuna in the Eastern Tropical Pacific Ocean, [English and Spanish], *Inter-American Trop. Tuna Comm. Bull.*, **2** (6), 1957.

Seton, E. T.: "Lives of Game Animals," vol. III, part II, "Hoofed Animals," Doubleday, Doran & Company, Inc., Garden City, N.Y., 1929.

Sette, O. E.: The Long Term Historical Record of Meteorological, Oceanographic and Biological Data, *Calif. Coop. Oceanic Fisheries Invest. Rept.*, **VII**:181–194 (1960).

Silliman, R. P., and J. S. Gutsell: Experimental Exploitation of Fish Populations, *U.S. Fish Wildlife Serv. Fishery Bull.*, **58**:215–252 (1958).

Slobodkin, L. B., and S. Richman: The Effect of Removal of Fixed Percentages of the Newborn on Size and Variability in Populations of *Daphnia pulicaria* (Forbes), *Limnology Oceanog.*, **1** (3):209–237 (1956).

Smith, F. E.: Dynamics of Growth Processes, XIII, in "Quantitative Aspects of Population Growth," pp. 277–294, Princeton University Press, Princeton, N.J., 1954.

Swingle, H. S.: Farm Pond Investigations in Alabama, *J. Wildlife Mgt.*, **16**:243–249 (1952).

——— and E. V. Smith: Increasing Fish Production in Ponds, *Trans. N. Am. Wildlife Conf.*, **4**:332–338 (1939).

Taber, R. D., and R. F. Dasmann: The Black-Tailed Deer of the Chaparral, *Calif. Dept. Fish Game Bull.*, no. 8, 1958.

Uda, M.: A Consideration on the Long Years' Trend of the Fisheries Fluctuation in Relation to Sea Conditions, *Bull. Japan Soc. Sci. Fish.*, **23**:7–8 (1957).

Watt, K. E. F.: Studies on Population Productivity, I. Three Approaches to the Optimum Yield Problem in Populations of *Tribolium confusum, Ecol. Monogr.*, **25**:269–290 (1955).

——— Studies on Population Productivity, II. Factors Governing Productivity in a Population of Smallmouth Bass, *Ecol. Monogr.*, **29**:367–392 (1959).

——— The Conceptual Formulation and Mathematical Solution of Practical Problems in Population Input-Output Dynamics,, in E. D. LeCren and M. W. Holdgate (eds.), "The Exploitation of Natural Animal Populations," pp. 191–203, Blackwell Scientific Publications, Ltd., Oxford, 1962.

Wigley, R. L.: Life History of the Sea Lamprey of Cayuga Lake, New York, *U.S. Fish Wildlife Serv. Fishery Bull.*, **59**:561–617 (1959).

Wood, H.: Fisheries of the United Kingdom, in M. Graham (ed.), "Sea Fisheries, Their Investigation in the United Kingdom," pp. 10–79, Edward Arnold (Publishers), Ltd., London, 1956.

"Yearbook of Fishing Statistics," Food and Agriculture Organization of the United Nations, Rome, 1965.

Zimmerman, J. H., and F. F. Bright: "Our Inland Seas: The Great Lakes," Harper & Row, Publisher's, Incorporated, New York, 1942.

6 | IMPORTANT ECOLOGICAL PROCESSES

To maximize productivity of any renewable resources, it is necessary to manipulate phenomena of somewhat greater complexity than those considered in previous chapters. Section 6.1 deals with interspecies relationships of a type that can be important in regulating population productivity: predation. In Sec. 6.2, a new type of complexity is introduced: spatial heterogeneity with patterns that change through time in wavelike fashion. Section 6.3 discusses combinations of both these phenomena: relationships involving up to four species that operate as epizootic and epidemic waves in time and space.

6.1 PREDATORS AND PREDATION

Predators and predation are important in resource management for several reasons. When harvesting a resource that is subject to predation, we are competing with the predators and need to know how much damage they are doing, if any, and whether predator control by bounties, poisons, chemical sterilizing agents, or other means is called for. Alternatively, we may have the converse problem: too few predators. If rodents or insect pests are at extraordinarily high densities, a causative factor may be too few predators, or predators of the wrong type. We may then either relax mortality pressure on predators, or release new species of predators (for example, ladybird beetles to control scale insects). We thus have ample motivation for studying the process of predation.

The bounty system is a method of predator control with a long history. It is based on the assumption that we can decrease the population of a pest animal by offering rewards in exchange for proof of destruction of the animal. There are three motives for a bounty system: preventing losses to livestock, poultry, and certain fruit crops; protecting valuable game species against destruction; and insuring against the development of an enormous natural rabies reservoir (Jacobsen, 1945). However, as Jacobsen has pointed out, bounty systems typically have a number of built-in defects that render them ineffectual. In at least some cases which have been well documented, bounties have not resulted in extermination of the predator. Although Pennsylvania has had a bounty system since 1683, and $2 million was paid in bounties in 1915 to 1935 alone, only one predator, the wildcat, showed any evidence of being controlled. Jacobsen has noted that the bounty system can not work, in principle.

Typical bounty payments are $15 on mountain lions, $15 on wolves, and $6 on coyotes and bobcats (Utah H. B. #95, effective March 1, 1944). Long before a predator population has been reduced to an extremely low level, it will have ceased to be economically rewarding to work for bounties. In areas where predator control is important, such as the sheep-ranching areas of Utah, ranchers pay trappers a salary to trap coyotes, even though the bounty system is in force. There has also been a great deal of fraud practiced in connection with bounty systems. (Largest payments are in counties bordering a state, suggesting that scalps or hides are imported; scalps are manufactured from squirrel ears; trapped animals, particularly females, are released to "seed" predator populations and ensure constant source of bounty payments, etc.)

A natural question is, "Why don't predators annihilate their prey?" The big cats and other large predators are impressively efficient machines of destruction, so one would assume that they would denude the landscape of food, and then die from starvation. The answer is that intraspecific competition and an exceptionally rigorous and dangerous pattern of existence serve as highly efficient regulators of predator population growth.

A great deal of anecdotal information is available on the lives of large predators in Africa, illustrating the types of mechanisms that can regulate their populations (Wright, 1960). Statistics on nightly movements of a pride of two lionesses and eight cubs showed that the average move was 2,600 yards when the youngest cubs were 2 months old, and 2,800 yards when they were 20 months old. In other words, even when they are very young, lion cubs are subjected to the rigors of adult lion life. To illustrate one of the hazards implied by this statement, small cubs have been drowned trying to swim after adult lions across rivers in full flood. The reproductive rate in lions is sensitive to starvation, dropping from four or five cubs a litter to one or two. When the cub is 20 months old, the lioness's care ceases, and the cub must be self-sufficient. Many weaker cubs die at this age because of their inability to compete with stronger members of the litter for food.

Additional mechanisms of big-cat population regulation can be sought in behavioral data on encounters between prey and predators, and between different predator species. Wright noted that lionesses can have their necks broken or their skulls crushed in attacks on elephants. A great deal of such anecdotal information appears in the mimeographed reports of the Director of the Uganda National Parks. For example, interspecific competition between the big cats can be lethal. A large female leopard was killed and eaten by a lion; apparently the leopard killed a young water buck and carried it up into a tree. When the leopard came down, it was pounced on by the lion, killed and partially eaten (Trimmer, 1963). Lions may be killed in reprisal attacks by prey species. Trimmer (1962b) describes an adult lioness, accompanied by two cubs, trying to seize an elephant's calf, but being killed by a thorax wound from the elephant's tusk. In addition to the death of the adult lion, the one-adult pride of lions was left without its leader, thereby decreasing the probability of survival of the two cubs. Under certain circumstances, lions can be killed when attacking relatively small prey. Trimmer (1961) found a lioness that appeared to have been killed by a warthog. The lioness had a single wound in the throat, and her stomach was full of undigested warthog flesh. Trimmer surmised that having killed and eaten a warthog, the lioness was hunting again and, being gorged, was slow at the kill and hence sustained a wound from a warthog tusk. Quite apart from such direct threats to life, the great enemy of all predators is time: It may take the predator so long to search for, find, pursue, and catch prey that all but the fastest, strongest predators starve to death. The great importance of time in the predation process is acknowledged in the Holling model described briefly in this chapter, and in more detail in Sec. 11.6. A pride of four lions, three adult females and an adult male, were clocked when stalking and killing a buffalo (Trimmer, 1962a). It took them 45 minutes from the commencement of the stalk until the conclusion of the kill. This however, was only the pursuit time; to this must be added the great length of time it must have taken to discover the buffalo. Additional difficulties are created for the lions because typically the male does not participate in the stalk or kill, but only in the eating. An additional problem for predators is that prey may often outrun or outswim them, and escape. The predator will have expended a great deal of energy without return. In summary, there are a great variety of behavioral mechanisms keeping predator-prey population systems in a delicate state of balance. Predation is a difficult and extraordinarily time-consuming task for the predator, unless prey are extraordinarily abundant. (Herbivores are not affected by time because they are surrounded by their food.) Wright (1960) provides some statistics on the state of the balance in Nairobi Park. There is an average of 61 prey per square mile, and there are on the average 0.32 lions per square mile. The kill rate per lion per year is 10 to 50, depending on the age and size of the lion. Lions kill 3.2 to 16.0 prey per year from the 61 per square mile. Biomass utilization of lions is rather low, relative to that of very small predators, such as shrews. Lions only kill 0.11 to 0.13 pounds of food per day per pound of lion in the pride.

Another important matter concerning predation is the genetic and ecological effect on the prey population. Errington (1946) discovered, when studying the mink-muskrat system in the northcentral United States, that mink were not preying on muskrats randomly. The biological surplus that had not been able to establish regular home ranges bore the brunt of predation. Predators merely removed that part of the population that could not support itself. Muskrat population increases are inversely related to breeding-population density, so any effect of predators would be compensated for in any case. The most vigorous and aggressive muskrats are likely to be the ones to discover, occupy, and defend suitable burrows. If this is correct, then the muskrats preyed upon by mink were not only "surplus" in an ecological sense, but also in a genetic sense. Predation thus fulfilled a doubly useful role for the prey population. Predators maximize the energy flux per capita of prey in ecosystems, by selectively eliminating prey with lower-than-average ability to obtain food (slow, feeble, sick animals). This applies throughout the animal kingdom, with parasites, spiders, and birds increasing per capita efficiency of caterpillars and sawfly larvae; mink performing the corresponding service for muskrats; birds of prey for rodents; wolves for buffalo, and so on.

The great interest in predation (and the related phenomenon, parasitism) throughout the history of ecology has culminated, since 1958, in a great burst of activity on the quantitative, mathematical, and computer analysis of the phenomenon. It has been discovered that for a great variety of predators, and also Hymenopterous parasites, the predator or parasite attacks an increased number of hosts or prey, as the hosts or prey numbers increase. Solomon (1949) has called this response to increased prey or host density the *functional response*. It is illustrated in Fig. 6.1. He also recognized a *numerical response*, or increase in numbers of predators or parasites over time in response to an increase in numbers of hosts or prey. Figure 6.2 illustrates this response. Many sets of data show that the efficiency per predator or per

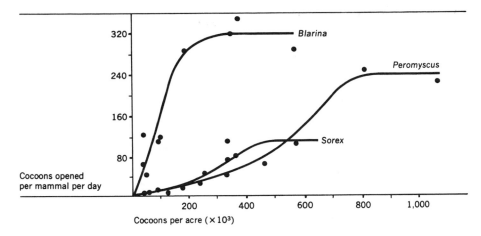

FIGURE 6.1

Functional responses of Blarina, Sorex, and Peromyscus (Holling, 1959).

parasite decreases with increasing predator or parasite density. This is illustrated in Fig. 6.3.

Thus the interacting system of predator-prey or host-parasite populations involves a number of compensatory mechanisms. The numerical response of the predators or parasites tends to prevent the prey or hosts from eating themselves out of food; the decreased per capita predator effectiveness at increased predator densities prevents the predators from annihilating their prey; the greater the number of prey, the more will be eaten, limited by the time it takes the predator to find and kill prey. A simple mathematical model has been proposed to account for the gross features of predator-prey and host-parasite systems (Watt, 1959).

Where N_A = number per unit area of attacked individuals
N_O = number of hosts or prey
P = number of predators or parasites
A = per capita effectiveness of predators or parasites
K = maximum per capita effectiveness of predators or parasites over measured period
a = searching efficiency of individual predator or parasite
b = intraspecific competition pressure among predators or parasites

$$\frac{dA}{dP} = -b\frac{A}{P}$$

or
$$A = aP^{-b} \tag{6.1}$$

Important ecological processes **137**

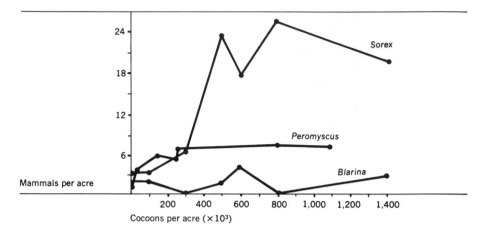

FIGURE 6.2

Numerical responses of Blarina, Sorex, and Peromyscus (Holling, 1959).

Where parasites or predators are not stimulated to search harder to find more hosts or prey (as in invertebrates), we have

$$\frac{dN_A}{dN_O} = PA(PK - N_A)$$

or

$$N_A = PK(1 - e^{-AN_OP})$$

or, from (6.1),

$$N_A = PK(1 - e^{-aN_OP^{1-b}}) \tag{6.2}$$

Where parasites or predators are stimulated to search harder to find more hosts or prey (as with vertebrates that exhibit evidence of learning), and there is a roughly sigmoid N_A versus N_O curve (as in Fig. 6.1),

$$\frac{dN_A}{dN_o} = PAN_o(PK - N_A)$$

and from (6.1),

$$N_A \doteq PK(1 - e^{-aN_o^2P^{1-b}}) \tag{6.3}$$

These equations can be fitted to data where we have N_A, N_O, and P by simply making plots on log-log graph paper. We arrive at the transformation to be plotted as follows. From (6.2),

$$1 - \frac{N_A}{PK} = e^{-aN_OP^{1-b}}$$

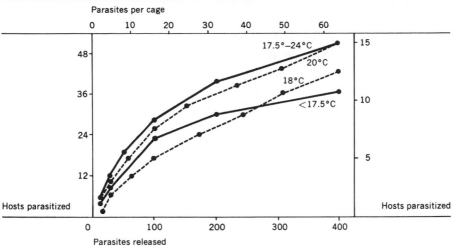

FIGURE 6.3

Comparison of the trends in the number of N. sertifer parasitized with increase in parasite density in laboratory cages (broken lines: 2 to 64 parasites released among 25 hosts) and on 25-square-foot plots of grass lawn (solid lines: 12 to 400 parasites released among 100 hosts) by females of D. fuscipennis at various temperatures (Burnett, 1956).

or
$$\ln \left(\frac{PK}{PK - N_A} \right) = a N_O P^{1-b}$$

or
$$\ln \left[\frac{\ln \left(\dfrac{PK}{PK - N_A} \right)}{N_O P} \right] = \ln a - b \ln P$$

Therefore, if we make an appropriate choice of K, we should obtain a straight line on log-log graph paper when we plot

$$\frac{\ln \left(\dfrac{PK}{PK - N_A} \right)}{N_O P}$$

against P. Similarly, to test (6.3), we plot

$$\frac{\ln \left(\dfrac{PK}{PK - N_A} \right)}{N_O^2 P}$$

against P on log-log graph paper. We may arrive at a good first approximation to K through our knowledge of any particular situation, because K

is the mean egg complement in the case of Hymenopterous or Dipterous parasites, or the maximum sustainable attack rate, in the case of predators. Alternatively, K may be obtained directly from graphs such as Fig. 6.1. The upper asymptote represents PK, so K is the asymptote divided by P. The model we have just presented, however, is not satisfactory in one major respect: it ignores the basic importance of time in host-parasite and predator-prey systems. The importance of this fact was noted in connection with the lion-buffalo system. It took 45 minutes for pursuit, but this time is trivial relative to searching time. The really basic factor determining the number of times a parasite or predator can attack is the total length of time available, and the time taken by various phases of the attack behavior cycle: search, pursuit, eating, and digestion. A new model has been put forward by Holling (1963; Sec. 11.6 in this book) dealing with this problem on its own terms. This model represents a major innovation over most mathematical models, in that time is treated as a dependent, not an independent variable. If we can calculate the length of time it takes a predator to search for, pursue, eat, and digest a sequence of prey, and if we know how much time was available for these functions, we know the number of events that occurred. Holling uses the following symbols:

T_D = time spent in digestive pause, after eating prey
T_S = time spent searching for prey
T_P = time spent pursuing prey
T_E = time spent eating prey

The relationship of these times to H_T, the hunger level that triggers searching, is illustrated in Fig. 6.4. The predator begins searching in the morning, and its hunger level rises until it eats the first meal. This meal does not satiate its hunger, so it immediately searches for the next meal. Several meals are searched for, pursued, and eaten without pause, until finally the hunger level drops below H_T, the hunger threshold. Now eating is followed by a digestive pause, during which the hunger level gradually rises to H_T, and searching begins again. For the remainder of the day, an equilibrium pattern is established. The pattern shown in Fig. 6.4 represents an average case; of course, by chance alone, the predator might encounter some prey very shortly after initiating a search, or a very long time after initiating it.

Various complications modify the predator-prey scheme given in Fig. 6.4. A predator with a "killer instinct" may attack far more prey than it can eat, and its attack threshold is so low that there is never a digestive pause. It will spend almost all its time in search and pursuit, and thus generates far more attacks per unit time than would otherwise be the case.

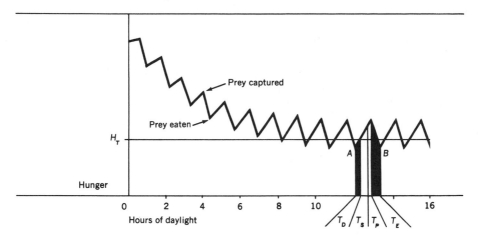

FIGURE 6.4

Schematic representation of the changes in a predator's hunger during 16 hours of daylight (Holling, 1963).
H_T = *hunger level that triggers attack*
T_D = *time spent in digestive pause*
T_S = *time spent searching for prey*
T_P = *time spent pursuing prey*
T_E = *time spent eating prey*

A major complication arises because the attack threshold may be dropped by the process of eating itself, as in a "feeding frenzy" of sharks, or the attack threshold of a particular predator for a particular prey may be dropped by learning.

One of the most interesting features of shark attacks on men is that they are often "clustered" in time and space. McCormick, Allen, and Young (1963) give a number of accounts to support this point.

Up to July 1, 1916, no one could remember a shark. having attacked a living swimmer off the New Jersey coast. Within the next 12 days, 5 swimmers were attacked, 4 fatally. The attacks occurred in a 70-mile stretch of shoreline, 3 of them 2 miles inland in a tidal river. Two days after the attacks, a Great White shark was captured six miles away from the scene of the last three attacks. It was 8½ feet long, and contained 15 pounds of human remains.

In 1931, three people were attacked by a shark or sharks near Havana, Cuba, two fatally, within nine days.

In 1899 at Port Said, Egypt, three boys were attacked within three hours (by a shark or sharks).

These incidents suggest that the great excitement produced in a shark by eating a meal lowers the attack threshold, and further attacks occur in rapid succession. It should be noted that many swimmers are approached but not attacked by sharks. There is speculation that the attack threshold for sharks against humans in the first instance is lowered by minute traces of blood in the water coming from a scratch or other small wound.

One of the most spectacular examples of a mammal having its attack threshold with respect to another mammal lowered by learning is related by Corbett (1948). A leopard developed such a strong preference for human food that it would go to great effort to obtain human food even if surrounded by other prey species. The mechanism by which it became a man-killer is fairly clear. In 1918 an influenza epidemic killed more than a million people in India, and in northern villages the inhabitants were dying faster than they could be disposed of. Bodies were simply carried to the top of a hill or cliff and tossed over. Instead of usually scarce food, the leopard suddenly was confronted by a superabundance of a new type of meat. Shortly after the end of the 1918 epidemic, the Man-Eating Leopard of Rudraprayag began killing. The intensity with which humans were selected is illustrated by the following anecdote. One night the leopard entered a room in which a 14-year-old boy was sleeping with 40 goats. There were no windows, and the one door was securely fastened. The leopard tore down the door with some difficulty (it was covered with deep claw marks) and killed and ate the boy. In the morning, not one of the 40 goats was even scratched.

The following table illustrates the way this particular leopard's taste developed.

Year	Number of humans killed
1918	1
1919	3
1920	6
1921	23
1922	24
1923	26
1924	20
1925	8
1926	14
	125

Other factors can lower attack thresholds. If mammals are crazed by rabies, or a persistently irritating wound, or are accompanied by young, they will attack when normally they would not. Most historical records of grizzly bear attacks on men indicate the bear was startled, cornered, or had already been shot at and wounded, for example. Young (1958) reports only five attacks by bobcats on humans in the United States since 1923, and in all cases, the bobcats had rabies.

6.2 LARGE-SCALE BIOLOGICAL WAVE PHENOMENA: FURBEARER PREDATOR-PREY CYCLES IN THE BOREAL FOREST, AND OUTBREAKS OF INSECT PESTS

We now consider the spatiotemporal character of population eruptions in different species within a community. Then we will consider the causative factor in population eruptions in various types of organisms, and point out the implications for strategy evaluation studies designed to show how best to deal with various eruptions and outbreaks. In Sec. 6.3 we will show that epidemic and epizootic waves can have certain basic features in common with furbearer and insect-pest eruptions. Sections 6.2 and 6.3 serve as the motivation for subsequent mathematical sections: 11.3, showing how to model weather; 11.7, on simulation of dispersal; and 11.8, which draws together the material and shows how to simulate epidemic and epizootic waves on a computer.

Miyashita (1963) has presented a comprehensive discussion of the nature and causes of outbreaks in insect pests, mainly basing his discussion on Japanese agricultural pests, but referring also to the international insect, bird, and mammal literature. On the basis of this exhaustive analysis of outbreak case histories, he decided that from a geographical point of view, outbreaks are of two broad classes: the *spreading-out type* and the *scattered type*. In the former, an infestation occurs first at one place, then spreads gradually from year to year; the total area infested increases from year to year, then rapidly decreases after the peak year. In the scattered type, infestations occur simultaneously at many places scattered over a vast area.. Miyashita stated that the spreading-out type seems to be connected with a widespread change in food conditions, whereas the latter type seems to be caused by a certain climatic condition prevailing over a vast area. Miyashita gives many examples of the scattered type in lemmings, German forest pests, and Japanese rice pests. Our concern here will be with the spreading-out type of population eruption; these create formidable mathematical difficulties for strategy evaluation studies, and raise interesting economic problems; also, some of the most economically important population eruptions are of the spreading-out, rather than the scattered type.

Miyashita's classification raises a number of interesting problems immediately: first, can it be demonstrated that the spreading-out type of outbreak is a real phenomenon, and is it a widespread phenomenon among animals that exhibit wide-amplitude population fluctuations?

In 1943, an outbreak of spruce budworm was discovered in an area of about 1,000 square miles in northern Ontario, just west of Lake Nipigon. This outbreak spread gradually until by 1948 about 15,000 square miles were heavily defoliated (Belyea, 1952). Population density of the spruce budworm larvae was expressed in terms of percentage defoliation of balsam fir trees. Figure 6.5 brings out the wavelike character of the spreading infestation. We shall observe this same pattern in epidemic and epizootic waves in Sec. 6.3. The peak densities are reached at the outer edge of the outbreak after the population densities at the epicenter have begun to subside. A spreading-out type of outbreak has a definite spatiotemporal pattern, not just a temporal pattern. That is, the progressions of events at different radial distances from the outbreak epicenter do not proceed in phase; rather they proceed in a wavelike fashion. This point is important for the way in which we structure our mathematical analysis of such phenomena.

This same spreading-out type of population eruption has been observed by Butler (1953) in cycles of Canadian mammals. Butler was interested in determining if such cycles of furbearers in Canada were produced by chance alone. If so, various regions of the country should *not* show peaks in furbearer populations in the same year. To investigate, Butler divided the Canadian fur-collection area into 63 sections, each as ecologically homogeneous as possible, and each with from one to six of the Hudson's Bay Company posts at which the furs were collected. For each of these 63 sections, he selected the peak population years, and in Table 6.1 we reproduce his summary table showing for each year and species of furbearer the number of sections with population peaks in that year, based on the production of pelts by trappers. Now if cycles are indeed due to chance alone, the expected number of sections with peaks in a particular year for any species is the mean number of sections with peaks per annum (that is, the column mean). A chi-square test can be performed to see if the distribution of numbers in each column deviates from this expected situation, with the mean in each row, more than would be expected by chance alone. For all six species in Table 6.1, and the two group totals, the chi-square test showed departure from a random distribution of section peaks that would have occurred by chance alone less than once in 10 thousand times. The interesting point that Butler discovered in preparing Table 6.1 for this test, however, was that the sections did not peak all at once, but started peaking in an epicenter group of sections from which peak densities radiated outward with the passage of years. This point is brought out more clearly in Figs. 6.6 and 6.7, in which maps of isophasal increasing density

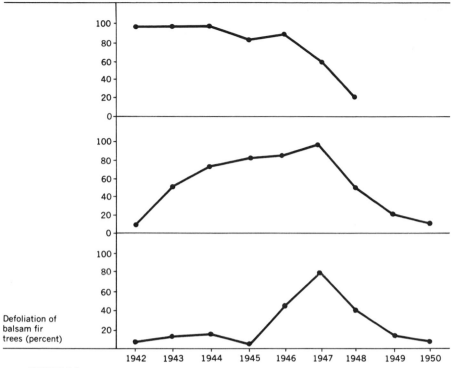

Defoliation of
balsam fir
trees (percent)

FIGURE 6.5

The wavelike character of a spreading type spruce budworm infestation (data from fig. 3 of Belyea, 1952).
Top panel: Population densities at the Chief Bay area, the outbreak epicenter.
Middle panel: Population densities at the Black Sturgeon Lake area, about 25 miles from the epicenter.
Bottom panel: Population densities at the southern border of the outbreak, about 50 miles from the epicenter.

lines are plotted for two different snowshoe rabbit population eruptions. Both eruptions began in the same general area of Canada: the northern part of the prairie provinces. This suggests another important question, which we shall consider later in this chapter: what determines the spatial location of an epicenter? If there is something unusual about the type of place that becomes an epicenter for eruption of a particular species population, what, precisely, is the unusual characteristic?

Before considering this matter, however, one more feature of Table 6.1 is noteworthy. The population peaks in the various species show a definite relationship to each other. This point is brought out more clearly by Fig. 6.8. The figure pools data from all of Canada. To perform a more critical inspection

Important ecological processes   145

TABLE 6.1

*Frequency of peak collections of various furs in 63 sections of Canada**

Year	Red fox	Fisher	Lynx	Group total	Mink	Muskrat	Group total	White fox
1916–1917	6	5	16	27	—	—	—	—
1917–1918	6	6	3	15	—	—	—	2
1918–1919	1	—	—	1	1	—	1	3
1919–1920	—	—	—	—	—	—	—	—
1920–1921	—	—	—	—	1	5	6	—
1921–1922	—	—	—	—	11	17	28	5
1922–1923	—	—	—	—	2	7	9	2
1923–1924	1	—	4	5	18	4	22	1
1924–1925	13	—	8	21	2	1	3	1
1925–1926	18	5	12	35	6	—	6	—
1926–1927	12	8	8	28	—	—	—	9
1927–1928	—	1	—	1	—	—	—	—
1928–1929	—	—	—	—	—	—	—	—
1929–1930	—	—	—	—	—	1	1	3
1930–1931	—	—	—	—	—	1	1	9
1931–1932	—	—	1	1	1	25	26	—
1932–1933	—	—	2	2	9	10	19	—
1933–1934	8	—	4	12	16	1	17	3
1934–1935	19	3	2	24	11	1	12	8
1935–1936	11	4	6	21	1	—	1	—
1936–1937	4	2	3	9	—	—	—	—
1937–1938	1	1	—	2	—	1	1	10
1938–1939	—	—	—	—	—	1	1	2
1939–1940	—	—	—	—	1	22	23	—
1940–1941	—	—	—	—	6	10	16	3
1941–1942	6	—	—	6	12	6	18	8
1942–1943	9	—	2	11	15	—	15	2
1943–1944	14	2	4	20	6	—	6	1
1944–1945	4	—	3	7	1	—	1	1
1945–1946	14	—	5	19	—	—	—	9
1946–1947	—	3	—	3	—	—	—	3
1947–1948	—	—	1	1	1	3	4	—
1948–1949	—	—	—	—	1	19	20	—
1949–1950	—	—	—	—	10	4	14	9
1950–1951	—	—	6	6	10	2	12	3
Total	147	35	90	272	141	142	283	97
Mean	4.2	1.0	2.6	7.8	4.0	4.1	8.1	2.8
Chi square	281	149	186	449	255	591	793	138

P in all cases < 0.0001

*From table I, Butler, 1953.

146 *The principles of resource management*

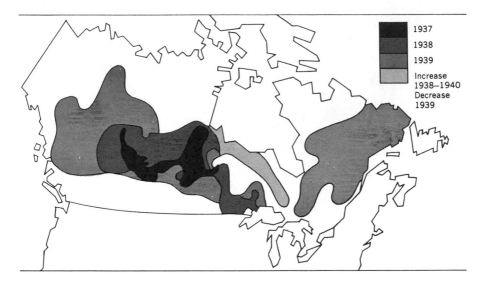

FIGURE 6.6

*Isophasal lines delineating areas of Canada in which observers reported increase
in the snowshoe rabbit population (Butler, 1953).*

of the data, Butler used the 63 sections of the country already mentioned.
In each section the data of each peak of the colored fox was noted; then the
peak years for other species were recorded as the number of years before or
after the respective fox peak. Analyzed this way, lynx peaks coincided with fox
peaks in 49 percent of the cases, and preceded fox peaks by 1 year in 36 percent. Fisher peaks coincided with fox peaks in 43 percent of the cases, and
followed them by 1 year in 57 percent. The speed of reaction of predator
populations to buildups in snowshoe rabbit populations occurs in the following
sequence: first lynx, then fox, then fisher. This sequence is explained by the
biology of the respective species. Lynx bear young the same year they breed,
and the young are mature and can breed the following year. Fisher have
delayed implantation, and do not bear young until a year after they are bred,
and these young do not produce a subsequent generation until they are
2 or 3 years old.

We now come to the problem of determining why the epicenters occur
where they do, and how the effect of the eruption at the epicenter is transmitted
to surrounding areas. Butler (1953), independently of Fujita (1954), indicated
his awareness of "*Drosophila* type" and "*Allee* type," or "*Tribolium* type"
curves for the effect of density on reproductive rate (see Figs. 11.7 and 11.9).
If in some sections of the country the endemic population levels of snowshoe
rabbits were 10 times as high, say, as in other sections, the eruption would

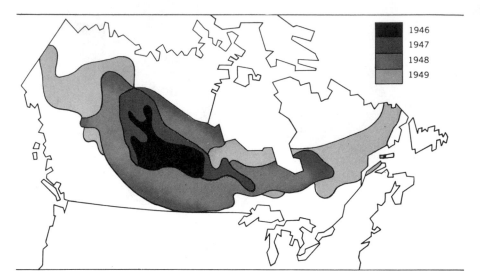

FIGURE 6.7

Isophasal lines delineating areas of Canada in which observers reported increase in the snowshoe rabbit population (Butler, 1953).

build up much faster in those (high-density) parts than in others, where there is an Allee type curve and the density at the epicenters is optimal for high reproductive rate. The effect of the eruption at an epicenter can easily be spread outward by migration (there is evidence that foxes can travel 300 to 400 miles in a few months). However, as Butler notes, the causation of these spatiotemporal patterns is complex, and meteorological factors are probably important; this complicates the mode of analysis since we need to know the way in which several different meteorological factors interact.

Keith (1963) has reviewed the literature on furbearer cycles in the boreal forest, and his remarks eliminate the possibility that the oscillations in population numbers are due to an oscillation intrinsic to the predator-prey interacting system. First, Moran (1953) noted that there is a synchronous fluctuation of animals (such as muskrat and mink) not ecologically dependent on snowshoe hares. Second, Elton and Nicholson (1942) reported that introduced hares on lynx-free Anticosti Island, Canada, were apparently cycling in phase with hares on the mainland. Third, Crissey and Darrow (1949) noted that snowshoe hares on Valcour Island, New York, declined well after the virtual elimination of all predators.

By exclusion, we are left with the hypothesis that these spreading-out types of eruptions may well be triggered by weather. In Chaps. 3, 4, and 5 we have seen that the most unstable animals, with respect to fluctuations in

148   *The principles of resource management*

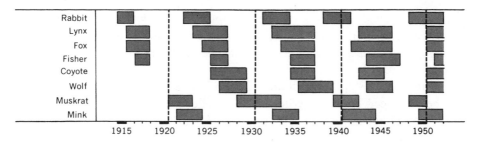

FIGURE 6.8

Relation between good years in the various species. Good years shown by gray rectangles (Butler, 1953).

population density, are allowed to be unstable because of the low importance of density-dependent factors relative to the importance of density-independent factors, such as weather. It is a small step from this notion to the concept that epicenters for outbreaks occur where they do because the epicenter locations have unusually unstable weather from year to year in those months and for those weather variables that are most critical for population buildup. This is not a new idea, but essentially only a slight restatement of a theory presented by many other workers.

For example, consider Uvarov's (1957) explanation of the outbreak dynamics of desert locusts. A characteristic of these epicenters is their climatic instability. The epicenters are in different places in different years. Because these locusts live in an extremely unstable environment, they have evolved the nomadism essential for survival. They reproduce when there is high temperature and humidity, and abundant vegetation that has sprung up in hitherto dry areas after rain. Such vegetation is mainly ephemeral, but after good rains it can be extremely abundant and lush. After the areas dry up, the locusts are forced to congregate in bands about wadi beds and other depressions. The adults have to leave such areas because food runs out; this causes migration. When migration is successful in discovering new food, there is further population increase, which leads to swarm formation. The odd feature in this situation is that immense swarms do not build up in those places in East Africa where swarms are able to survive for several years. It is instability of a habitat, per se, that leads to an outbreak. This instability is the main feature of epicenters.

The basic features common to all cases of spreading-out type outbreaks are as follows. First, an unstable habitat normally quite suboptimal for a particular species becomes optimal because of a change in conditions, and stays optimal long enough for the species to build up to enormous population densities. Second, because of these population densities, the population is subjected to tremendous density-dependent pressures caused by starvation or competition for living space. Third, to release this pressure, the population migrates outward.

This relationship between an unstable habitat and population eruptions has been noticed by many writers, whose collective experience covers a wide variety of organisms. It is a central thesis of Mozley (1960), on the basis of a lifetime of experience with pest snail hosts of bilharzia. Nevertheless, in order to support the thesis that unstable weather accounts for epicenters, we must provide evidence for the following steps in our argument.

1. We must show that the area of the epicenter is climatically unstable and that, at the time of population eruption, climatolgical maps show that the right type of weather to produce the eruption occurred at the epicenter site.

2. We must demonstrate statistically that the weather at the postulated epicenter did in fact account for a significant portion of the variance in population size from year to year at the postulated epicenter.

This is a tall order. There are few cases in which we have accurate records of the location of outbreak epicenters, since no one may even realize that there is an outbreak until after the population has spread well out from the epicenter. Epicenters often occur in remote areas, especially for furbearers and forest-insect pests, which provide some of the most striking examples of spreading-out type of outbreaks.

There is a case in which we have a great deal of descriptive and analytic information about an outbreak, due to the work of Belyea (1952), Elliott (1960), and Wellington and his associates (Wellington, Fettes, Turner, and Belyea, 1950; Wellington, 1954). The outbreak of the spruce budworm has already been referred to. Wellington et al. (1950) showed in their fig. 8 (reproduced here as Fig. 6.9) that the Lake Nipigon outbreak began in an area where, at the time of epicenter formation, there had been a 4-year period with an unusually low number of cyclones, relative to the long-term mean for that area. Since cyclones are roughly related to the incidence of humid air masses, relative scarcity of cyclones implies relative drought, which represents optimum conditions for spruce budworm (if accompanied by above-average temperature). Not only do Wellington and his associates provide enough data so we can determine *where* the epicenter should have occurred, but they also give us enough data to demonstrate the statistical validity of the relation between weather and population irruption. From their fig. 1 (Fig. 6.10), I have constructed Table 6.2. These data have been subjected to

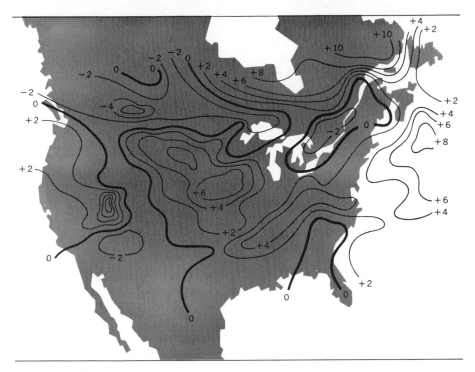

FIGURE 6.9

Deviations of the average annual numbers of cyclones for 1936–1939 from the long-term average, 1883–1947. Note the negative areas north and west of Lake Superior and east of Lake Huron, in contrast with the positive area north of Lake Huron (Wellington et al., 1950).

one-variable and two-variable regression analysis. An analysis of variance is given in Table 6.3, and observed defoliation values, along with values calculated from the two-variable regression, are plotted in Fig. 6.11. Table 6.3 illustrates an important and noteworthy result: neither defoliation the previous year nor June rainfall the previous year make a significant contribution to variance in defoliation, but the two together produce an F ratio significant at the 92 percent probability level. The reason for this odd result is brought out in Fig. 6.12, where defoliation is plotted against the previous year's June rainfall, and each point is labeled by year. There is a striking relation for the early year's points, or the later year's points, but not for the two sets together. This illustrates an important point in data analysis to which resource management analysts should be alert: The interaction effect of two variables may be much more important than the main effect of either of them taken separately. These data make it clear that climate can trigger a

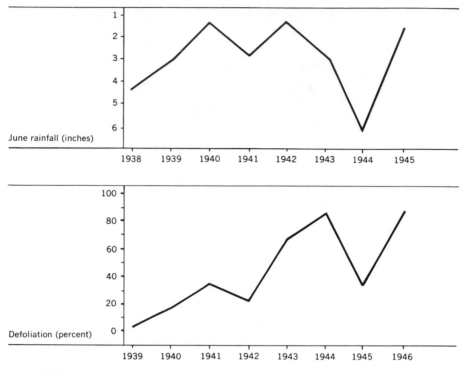

FIGURE 6.10

Association of the amounts of June rainfall during consecutive years with the percentages of defoliation by the spruce budworm. June rainfall for Cameron Falls, Ontario, is plotted with the amounts inverted. Defoliation at Black Sturgeon Lake, Ontario, is plotted with a lag of 1 year (Wellington et al., 1950).

spreading-out type of outbreak. For the spruce budworm, a large expanse of mature balsam fir trees is also a prerequisite; however, the trigger that determines the timing of the outbreak is climate.

Maercks (1954) demonstrated the existence of various relationships between weather and microtine cycles in Germany, analyzing data on the field mouse *Microtus arvalis* Palas. Weather affected cycles in two ways: by changing the amplitude of fluctuation and by changing the wavelength, or cycle period. The former seemed to be influenced by changes of rainfall and duration of sunshine. The latter seemed to be affected by deviations from normal weather conditions: mild winters prolonged outbreaks and long hard winters cut them off.

Figure 6.13 plots the peak years for Manitoba snowshoe hare eruptions and winter precipitation in Le Pas, Manitoba. The three peaks follow two or more winters of lighter than median snowfall, in all cases. However, more

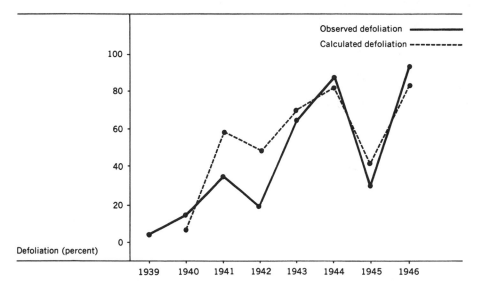

FIGURE 6.11

Observed defoliation of balsam fir by spruce budworm at Black Sturgeon Lake, Ontario, and defoliation calculated from two-variable regression in text (observed data from Wellington et al., 1950).

detailed district-by-district analysis of rabbit populations, precipitation, temperature, food availability, and burrows is needed for satisfactory analysis of this phenomenon, and such data are not available.

It now seems clear that the spreading-out type of outbreak is a reality and, at least in some cases, the trigger is weather. As I have pointed out

TABLE 6.2

The relationship between June rainfall at Cameron Falls, Ontario, and percentage defoliation of balsam fir trees by spruce budworm at Black Sturgeon Lake, Ontario

Year	Percentage defoliation	June rainfall previous year, inches	Percentage defoliation, previous year
1939	4	4.4	
1940	14	3.2	4
1941	34	1.6	14
1942	20	3.0	34
1943	64	1.5	20
1944	87	3.4	64
1945	33	6.2	87
1946	92	1.6	33

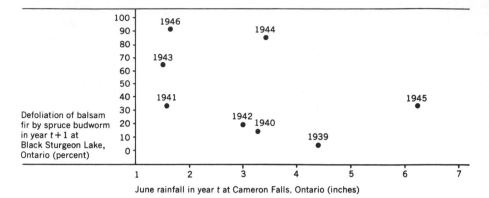

FIGURE 6.12

Effect of climatic instability on spruce budworm population instability (data from Wellington et al., 1950).

elsewhere (1964), this spreading, wavelike process means that an event at a particular place may be the result of some prior event at another place. This is probably the single feature of natural resource management strategy evaluation studies that produces the greatest analytic complexity; however, we can deal with it by using large computers (Sec. 11.7).

TABLE 6.3

Analysis of variance, regression of defoliation on defoliation and June rainfall in previous year, considered one at a time and together

Source of estimate	Sum of squares	Degrees of freedom	Mean square	F ratio
Linear regression of defoliation on defoliation, previous year	372.53	1	372.53	0.32
Residual	5,692.33	5	1,138.47	
Linear regression of defoliation on June rainfall, previous year	660.55	1	660.55	0.61
Residual	5,404.31	5	1,080.86	
Linear regression of defoliation on June rainfall and defoliation, previous year	4,556.53	2	2,278.27	6.04
Residual	1,508.33	4	377.08	
Total	6,064.86	6		

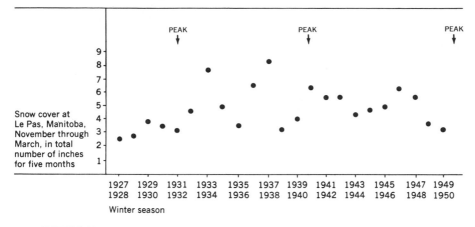

FIGURE 6.13

Relation between snowshoe hare eruptions in Manitoba and precipitation from November through March, at Le Pas, in northern Manitoba. (Precipitation data from World Weather Records, 1941–1950, U.S. Department of Commerce, Weather Bureau, Washington, 1959; hare data from Keith, 1963.)

A major remaining question is which factors combine to regulate the amplitude of fluctuations? Frank (1957) did a detailed study of *Microtus arvalis* in Germany. He found that three groups of factors combined to regulate the amplitude.

1. *Reproductive potential.* M. *arvalis* has wide-amplitude fluctuations because, to an amazing extent, the species is designed as a high-efficiency reproductive engine. A litter was dropped by a 33-day-old wild female. The litter size under optimum conditions in wild populations is seven. Litter weight is 53.2 percent of the mother's weight, both measured immediately after birth, in comparison with 33 percent in some other small mammals.

2. *Carrying capacity of the environment.* Frank concludes that for M. *arvalis*, the environment is optimal where there are "large, open, monotonous and uniform biotopes with extremely scant cover of trees and bushes." Amplitude of fluctuation is less where there is a higher proportion of woods, trees, and bushes. This is extremely interesting because a great number of authors have remarked that amplitude of fluctuations is greatest in animal species when the largest possible area was available to the species, and as large as possible a proportion of the area was occupied by the optimal food for the species. This has been mentioned by Dymond (1947) in connection with the uniformity of the environment in northern Canada and the oscillations of Canadian vertebrates; by Morris (1963) in connection with the spruce

Important ecological processes **155**

budworm; by Watt (1965) in connection with a great many species of forest *Lepidoptera*; and by Murray (1965) who noted that the highest densities of the 1957 great *Microtus montanus* outbreak in the northeastern corner of California were reached in cultivated land: alfalfa, clover, grain, and potatoes. Weather, of course, regulates the carrying capacity of the environment, and affects the reproductive potential directly, and via female mortality. Murray (1965) noted that an irruption may be truncated by heavy winter precipitation and extreme cold.

3. *Condensation potential.* Frank introduces this term to denote the ability of a species to tolerate extremely supranormal densities without deleterious side effects. In *M. arvalis*, as density increases, the females tend to live together in social group nests, so in effect there is a much smaller territorial requirement per female.

One more point about large-scale biological eruptions should be mentioned. Both types, spreading-out and scattered, occur in vertebrates and invertebrates. Murray (1964) noted that field mice irrupted in 1957 over much of the Great Basin, central and southern Oregon, northeastern California, and Carson Valley, Nevada. It is unlikely that all these irruptions were fed from a common source in such a short time, so it appears to be a scattered outbreak.

6.3 THE DYNAMICS OF EPIDEMIC AND EPIZOOTIC WAVES

We are concerned with the nature of epidemic and epizootic waves for four reasons. First, almost any resource management policy that affects the structure of a biotic community will, as a side effect, affect the probability of occurrence of epizootic or epidemic waves. Some of the effects are obvious; for example, draining swamps will decrease the density of, or eliminate, populations of mosquito larvae. More subtle effects are those that change the density, or spatial distribution, of small mammal populations. If the population density of small mammals drops very low, epizootic disease disappears. Elkin (1961) notes that epizootic disease disappears in the suslik *Citellus pygmaeus*, an important European and Russian secondary carrier of plague, when the suslik population density drops to one animal per hectare. Contrariwise, any policy that causes sharp buildup in small-mammal populations can lead to development of dangerous reservoirs of disease which can spread to humans.

Second, disease is becoming an important tool in the control of pests. After 67 years of fruitless efforts to control the Australian rabbit problem by other means, significant reduction of the vast hordes of rabbits was finally achieved with the virus myxomatosis in 1950 (Ingersoll, 1964). In DeBach (1964), there is a comprehensive statement of the effectiveness of disease in controlling insect pests. Bird and Burk (1961) give a singularly well-documented assessment of the effect of a virus in controlling an insect pest.

Third, one of the aims of a particular resource management policy may be to lower the probability of epizootic or endemic waves.

Finally, epidemiology can learn much from ecology, and ecology can learn much from epidemiology because basically both sciences deal with the same subject: the responses of individual organisms, populations, and communities to changes in the biotic and abiotic environment. By including epidemiological data within the area of their consideration, ecologists can considerably increase the total number of cases from which inferences can be drawn. An enormous body of epidemiological literature is primarily ecological in its point of view [for example, Burnet, (1953); Creighton, (1891); Elkin, (1961); Frost, (1941); Gill, (1928); May, (1958, 1961); Stallybrass, (1931); Tromp, (1963); Zinsser, (1960)].

Four things must be understood to completely comprehend the nature of waves of disease: the mechanism of transmission of the disease; the mechanism determining the temporal increase or decrease in incidence of the disease; the mechanism determining spatial movement of the wave of disease; and the mechanism by which extrinsic environmental factors can trigger sudden increases (outbreaks) in incidence of the disease.

May (1961) refers to diseases as being of three types: two-factor, three-factor, and four-factor complexes. In a two-factor complex, such as smallpox, only two species are involved: the pathogen and man. In a three-factor complex, there are three basic factors: a pathogen, a vector, and a host. An example is malaria: the pathogens are four main species in the genus *Plasmodium*; the vectors are adult female mosquitoes of the genus *Anopheles*; and the hosts, to which the infected blood of another host is carried by the mosquitoes, include man, monkeys, birds, and various rodents. Plague exemplifies a four-factor complex involving the plague bacterium, rodents, rodent fleas, and man. The mechanism of transmission of plague was worked out by the Indian Plague Research Commission in 1909. In summary, their findings were as follows (May, 1961, pp. 463–464):

1. Healthy rats contract plague from infected rats when the only apparent means of transmission is the rat flea.
2. Rats living in flea-proof cages can get plague after fleas from rats with plague are introduced into the cage.

3. Rats, guinea pigs, and monkeys do not get plague from plague-infected animals if no fleas are present.

4. If fleas are present, an epizootic wave tends to start immediately and spread in proportion to the number of fleas present, at a rate dependent on the time of year.

5. If successive instances of human infection are noted in a house, *invariably* there is evidence of a higher rat mortality than in houses with a single instance of plague infection.

6. *Human* fleas have been shown not to be vectors of plague.

7. Humans have been ruled out as carriers of *bubonic* plague for several reasons. Pus and excreta from infected humans is noninfectious. Nurses and doctors attending bubonic plague victims do not get infected. The great majority of plague patients had no contact with previous plague sufferers before falling ill. When bubonic plague appears in an area, in the great majority of instances, only one person is infected per house.

Clearly, in the bubonic version of plague, transmission is by rodent fleas, principally the rat flea *Xenopsylla cheopis*, which is brought into contact with humans by rats.

Bubonic plague, however, can switch from a zootic to a demic phase, pneumonic plague, if bubonic plague patients with secondary lung involvement transmit the disease to human contacts, in the form of fine droplets filled with the bacterium, which are inhaled. Once this mode of transmission is established, the disease can spread like wildfire, as it did between 1347 and 1350 in Europe, when 25-million people—a quarter of the European population at the time—died of Black Death.

Temporal trends in the incidence of disease

We will first describe the typical temporal pattern of an epidemic wave, and wave train, then explain what produces the waves. Creighton (1891, p. 662) gives the weekly bills of mortality for the Black Death, or plague, in London in 1665. These data are plotted in Fig. 6.14, and illustrate the typical temporal pattern of an epidemic wave. Note the rapidly accelerating rise to a maximum death rate, then the relatively sharp drop. The London population at January 1, 1665, was about 460,000; the average number of deaths per annum was 17,000, about 3.7 percent of the population. During the 1665 plague year, the total declared mortality was 97,306, or about 21.2 percent of the population.

The great London plague of 1665 was unusual, in that it was the third outbreak of plague in London in 62 years, and the severity of the plagues

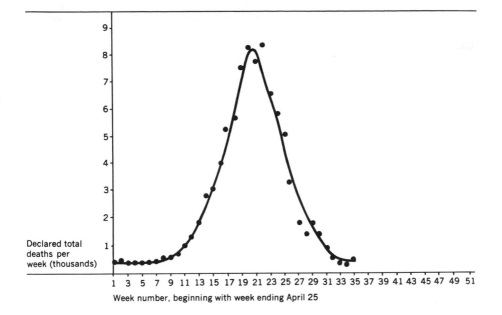

FIGURE 6.14
The Black Death in London, 1665 (data from Crieghton, 1891).

did not decline with successive epidemic waves. Creighton gives the following figures for the three London plagues.

Year	Estimated population	Declared plague deaths	Plague deaths as proportion of population
1603	250,000	33,347	0.13
1625	320,000	41,313	0.13
1665	460,000	68,596	0.15

Typically, with successive waves of an epidemic, the severity declines because of the development of immunity in the population, or selection for reduced virulence in the pathogen. For example, there were four main on-slaughts of plague in Europe in the fourteenth century. From various sources, the following rough data can be obtained.

Year	Proportion of population afflicted	Resultant deaths
1348	2/3	Almost all
1361	1/2	Almost all
1371	1/10	Many survived
1382	1/20	Almost all survived

Important ecological processes   159

Similarly, the following figures on plague in India (May, 1961, p. 483) indicate the lessening severity of a disease with the passage of time.

Year	Lethal cases of plague	Nonlethal cases of plague
1953	14,470	6,069
1954	5,639	1,031
1955	163	542
1956	68	263
1957	0	44
1958	0	26
1959	0	37

Many diseases have shown this pattern of either disappearing permanently, or ameliorating with the passage of time. Sweating sickness killed up to 80 percent of the English population in a series of epidemics terminating in 1551, after which the disease disappeared (Elkin, 1961). Scarlet fever and diphtheria show less severe clinical symptoms now than they did formerly (Elkin, 1961). Stallybrass (1931, p. 602) notes that there is clear evidence from India showing that in 20 or 30 years the rats acquire a racial immunity to *Pasteurella pestis*, the plague bacterium.

Burnet (1953) presents the modern view of the cause for the temporal pattern in waves of disease. The rate of spread of an epidemic at any moment is a function of the density of susceptible persons available and the number of sources of infection. As the entire population becomes immune or dies, the number of susceptibles drops off to virtually nothing. A typical epidemic has the following history.

At the beginning, the entire population is susceptible, and infection spreads rapidly. No immunity has developed, and each infected individual infects everyone else with whom he comes in contact. By the peak of the epidemic wave, half the population is immune, and the rate of new infections begins to drop. By the end of the wave, almost the entire population is dead or immune, and there are essentially no more susceptibles to be infected. This concept of the nature of the epidemic wave can be quantified, as in Chap. 11. Epizootic waves behave in the same fashion.

Spatial character of the epidemic or epizootic wave

Again, we shall illustrate the nature of a phenomenon characteristic of all diseases by means of plague, because it has been so thoroughly studied. Plague persists in geographic locations, or "foci" in wild rodents. These foci persist for long periods entirely independent of man and his activities.

The state of plague among mammals depends on the ratio of susceptibles to total population. If this ratio becomes very high because of explosive population growth or immigration, widespread epizootic waves are bound to occur. The resultant disproportion between a low number of rodents and a high number of fleas forces the fleas to search for substitute hosts. This can create a most dangerous situation for man.

One more complicating factor in epidemiology is that viruses, bacteria, and other pathogens are themselves biological, and therefore are evolving entities. Waning of plague could mean that the causative entity itself has become modified. Therefore plague is a most complex ecological and genetic phenomenon involving balance among four kinds of organisms: bacterium, flea, rodent, man.

Three times in history the balance has been such as to create plague pandemics. The first was "Justinian's plague" of the sixth century; the second (1346 to 1450) appeared in the Crimea, close to an enzootic center of sylvatic plague; and the third began in Hong Kong in 1897 (Stallybrass, 1931). We will illustrate the spatial character of the epidemic wave by outlining the history of the second great wave. (The most elaborate historical analysis I have seen of the Black Death of the fourteenth century is in Creighton, 1891.) A Latin manuscript by Gabriel de Mussis described the origin of the second pandemic at the siege of Caffa, a small Italian fortified post on the Crimean straits of Kertch. The plague broke out among the Tartar barbarians conducting the siege, which lasted three years. The Tartars died by the thousands every day, and finally dispersed. Two days after a ship containing Italians from Caffa arrived in Genoa, the plague broke out there. It spread throughout the Middle East, Egypt, and North Africa, invaded Sicily in 1346, Constantinople, Greece, and Italy in 1347, and by the end of 1347 was in Marseilles. By 1348 it was in Spain, northern Italy, eastern Germany, Paris, and England. It showed up in the western counties of England first, early in 1348. By November, 1348, it was in London, by 1349 it was in the midlands, and in 1357 it was still in the towns. In 1352, two-thirds of the academic community of Oxford died. Note the spread rate. We will reexamine this narrative when we consider the environmental factors that trigger increases in incidence of disease.

Now, let us consider the precise nature of a wave of disease in more detail. Such data are available from Bird and Burk (1961) who artificially disseminated a virus to control an outbreak of the European spruce sawfly. They measured the incidence of virus disease in larvae in 4 successive years after the virus was sprayed on 7 trees, at various distances. The data are plotted in Fig. 6.15 (for their area A). The general picture that emerges,

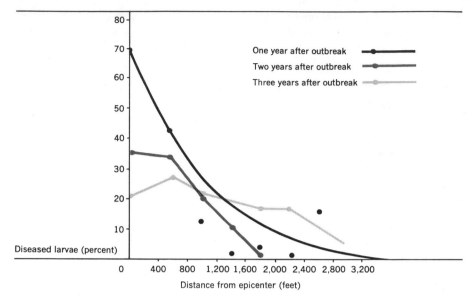

FIGURE 6.15

The spatiotemporal character of a wave of disease (data from Bird and Burk, 1961).

despite some small sample error, is that with the passage of time, the disease incidence declines at the epicenter, while simultaneously it may be rising some distance out. An idealized version of the situation is indicated in Fig. 6.16. The corresponding spreading characteristic of an epizootic wave for vertebrates is depicted in Fig. 6.17, illustrating the spread of a waterfowl cholera epidemic in San Francisco Bay. Thus epidemic or epizootic waves show a spatio-temporal process reminiscent of the spreading-out type of outbreak described in the preceding chapter.

The trigger for epidemic or epizootic waves

There are two different kinds of factors that can trigger a sudden increase in the incidence of disease: stress, and an increase in the ratio of susceptibles to immunes. A complication is that some underlying factors, such as weather, can operate through either of these causal pathways, depending on the range of values they are in. Mild weather can create an increase in populations, and the subsequent great increase in the ratio of susceptibles to immunes, through the birth of large numbers of young, can give rise to an epizootic wave. As the resultant supranormal population eats itself out of food, starvation and stress result, triggering an epidemic. Unduly harsh weather, on the other hand, can equally cause stress, and trigger the factors that produce an epidemic.

162   *The principles of resource management*

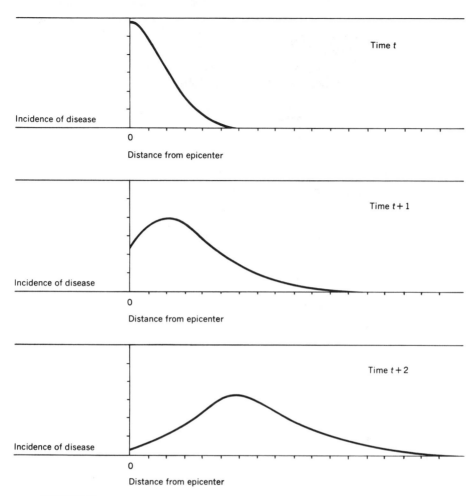

FIGURE 6.16

Idealized character of an epidemic or epizootic wave.

Migration is an important factor triggering epidemics. Population A, immune to disease A but not to disease B, mixes with population B which has disease B and is immune to it, and the results are explosive. All wars, which involve the concentration and mixing of great armies of men who would not normally be concentrated and mixed, have been epidemiologically explosive. Presumably this is what happened at the siege of Caffa. Vast hordes of un-washed, dirty-clothed Tartars were congregated in an area rich in fleas and known to be a focus of sylvatic plague in rodents, and the results were pre-dictable.

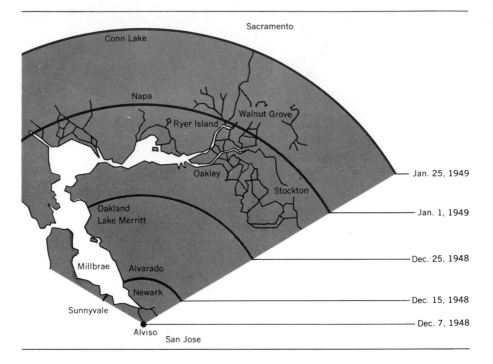

FIGURE 6.17

Northward spread of fowl cholera in wild birds during the 1948–1949 San Francisco Bay outbreak (Rosen and Bischoff, 1949).

Elkin (1961) notes numerous examples of epidemics associated with war. Justinian's plague occurred during a period of numerous wars, mutinies, and uprisings. The fourteenth century Black Death occurred when Europe was involved in a series of bloody wars, and great expanses of Europe and Asia were under the sway of Mongols and Tartars, with widespread population mixing of the type that triggers disease outbreak. In Table 6.4 (from Elkin's table 5), we see that disease was actually a much higher mortality factor in wars than fighting, up to this century.

There has also been a great deal of research on the relationship between climate and disease (Kutschenreuter, 1959; Sargent, 1960, 1964; Tromp, 1963). However, we probably have only a partial understanding of the mode of operation of climate in triggering epidemics, because of the complex way in which this triggering occurs. The following account is illustrative.

In 1957 an epidemic of influenza in the Netherlands began in mid-August, spread rapidly in September, and hit a peak from mid-September to mid-October, depending on the district. Except for May, August, and September,

TABLE 6.4

Comparison of war-related deaths to those from disease (Elkin, 1961)

Army, war	War-related deaths	Deaths due to disease	Ratio
British Army during war with France (1793–1815)	25,569	193,851	1:7.6
Russian Army during war with Turkey (1828–1829)	20,000	110,000	1:5.5
French Army during Crimean campaign (1854–1856)	20,193	75,375	1:3.7
British Army during Boer War (1899–1902)	7,534	14,382	1:1.9
Russian Army during war with Japan (1904–1905)	31,458	12,983	1:0.4
German Army during World War I (1914–1918)	1,061,740	140,302	1:0.1

average monthly temperatures for the Netherlands in 1957 were above normal, and were twice normal in January and February. June and the first week of July were exceptionally hot. From the middle of August till the end of September, the temperatures suddenly dropped to below average and the amount and duration of rainfall were almost twice the normal; the total rainfall in September exceeded the highest on record for 100 years. The hours of sunshine were thus far below normal (89 hours in September against 148 normally). Although 1957 was a very warm year (except in April, June, and December), all months had a considerable shortage of sun hours, giving a total shortage of about 100 sun hours for the year. There was a large percentage of healthy blood donors with high blood sedimentation rates (from May 10 until the end of October) and with low haemoglobin values (Tromp, 1963, pp. 528–529). High blood sedimentation rate is produced by low blood viscosity, which in turn is produced by a low level of mental stress (amongst many other factors). Low haemoglobin values are produced by stress due to cold (Tromp, 1963, pp. 315–317). It seems reasonably clear that this influenza epidemic was triggered by an unusual sequence of weather conditions, in which several months of unusually stress-free weather was followed by several weeks of weather that subjected the population to severe stresses.

If, in fact, many epidemics and pandemics are produced by this type of weather sequence, two important conclusions follow. First, few epidemiologists, or indeed few of any type of scientists, have the degree of data-analysis and conceptual sophistication required to build and test data against mathematical models to describe phenomena for which the independent variable is not a factor, but a pattern of temporal sequence of variate values. There may be quite specific causal relationships between weather and disease for a great number of different types of disease, but these may not be completely understood yet. Even in the present imperfect state of our understanding, weather is suspected of a causative role for many diseases, including malaria, plague, leprosy, pneumonia, smallpox, tuberculosis, cholera, and others (Gill, 1928).

Second, any major climatic upheaval could be expected to have calamitous side effects with respect to disease. It is interesting to note in this connection that Brooks (1954) described the period 1250 to 1400 as rainy and stormy; we have noted the devastating plagues in this period.

REFERENCES

Belyea, R. M.: Death and Deterioration of Balsam Fir Weakened by Spruce Budworm Defoliation in Ontario, *J. Forestry*, **50**:729–738 (1952).

Bird, F. T., and J. M. Burk: Artificially Disseminated Virus as a Factor Controlling the European Spruce Sawfly, *Diprion hercyniae* (Htg.) in the Absence of Introduced Parasites, *Can. Entomol*, **93** (3):228–238 (1961).

Brooks, C. E. P.: The Climatic Changes of the Past Thousand Years, *Experientia*, **10**:153–158 (1954).

Burnet, F. M.: "Natural History of Infectious Disease," Cambridge University Press, New York, 1953.

Burnett, T.: Effects of Natural Temperatures on Oviposition of Various Numbers of an Insect Parasite (Hymenoptera, Chalcididae, Tenthredinidae), *Ann. Entomol. Soc. Am.*, **49** (1):55–59 (1956).

Butler, L.: The Nature of Cycles in Populations of Canadian Mammals, *Can. J. Zool.*, **31**:242–262 (1953).

Corbett, J.: "The Man-Eating Leopard of Rudraprayag," Oxford University Press, Fair Lawn, N.J., 1948.

Creighton, C.: "A History of Epidemics in Britain from A.D. 664 to the Extinction of Plague," Cambridge University Press, Cambridge, 1891.

Crissey, W. F., and R. W. Darrow: A Study of Predator Control on Valcour Island, N.Y. State Cons. Dept., Div. Fish and Game, *Res. Ser.* no. 1, 1–28 (1949).

DeBach, P. (ed.): "Biological Control of Insect Pests and Weeds," Reinhold Publishing Corporation, New York, 1964.

Dymond, J. R.: Fluctuations in Animal Populations with Special Reference to Those of Canada, *Trans. Roy. Soc. Can. 5*, **41**:1–34 (1947).

Elkin, I. I. (ed.): "A Course in Epidemiology," Pergamon Press, New York, 1961.

Elliott, K. R.: A History of Recent Infestations of the Spruce Budworm in Northwestern Ontario, and an Estimate of Resultant Timber Losses, *Forestry Chron.*, **36**:61–82 (1960).

Elton, C., and M. Nicholson: The Ten-Year Cycle in Numbers of the Lynx in Canada, *J. Animal Ecol.*, **11**:215–244 (1942).

Errington, P. L.: Predation and Vertebrate Populations, *Quart. Rev. Biol.*, **21**:144–177, 221–245 (1946).

Frank, F.: The Causality of Microtine Cycles in Germany, *J. Wildlife Mgt.*, **21**:113–121 (1957).

Frost, W. H.: "Papers of Wade Hampton Frost, M.D., A Contribution to Epidemiological Method," The Commonwealth Fund, New York, 1941.

Fujita, H.: An Interpretation of the Changes in Type of the Population Density Effect upon the Oviposition Rate, *Ecology*, **35**:253–257 (1954).

Gill, C. A.: "The Genesis of Epidemics and the Natural History of Disease," William Wood & Company, Baltimore, 1928.

Holling, C. S.: Some Characteristics of Simple Types of Predation and Parasitism, *Can. Entomol.*, **91**:385–398 (1959).

————: An Experimental Component Analysis of Population Processes, *Mem. Entomol. Soc. Can.*, no. 32, 22–32 (1963).

Ingersoll, J. M.: The Australian Rabbit, *Am. Sci.*, **52** (2):265–273 (1964).

Jacobsen, W. C.: The Bounty System and Predator Control, *Calif. Fish Game*, **31**:53–63 (1945).

Keith, L. B.: "Wildlife's Ten-Year Cycle," The University of Wisconsin Press, Madison, Wis., 1963.

Kutschenreuter, P. H.: A Study of the Effect of Weather on Mortality, *Trans. N.Y. Acad. Sci.*, **22**: 126–138 (1959).

Maercks, H.: Über den Einfluss der Witterung auf den Massenwechsel der Feldmaus (*Microtus arvalis* Pallas) in der Wesermarsch, *Nachbl. Deut. Pflanzenschutzdienst* (Berlin), **6**:101–108 (1954).

May, J. M.: "The Ecology of Human Disease," MD Publications, Inc., New York, 1958.

————: "Studies in Disease Ecology," Hafner Publishing Company, Inc., New York, 1961.

McCormick, H. W., T. Allen, and W. E. Young: "Shadows in the Sea," Chilton Company–Book Division, Philadelphia, 1963.

Miyashita, K.: Outbreaks and Population Fluctuations of Insects, with Special Reference to Agricultural Insect Pests in Japan, *Bull. Natl. Inst. Agr. Sci.* (*Japan*), ser. C, no. 15, 1963.

Moran, P. A. P.: The Statistical Analysis of the Canadian Lynx Cycle. I. Structure and Prediction, *Australian J. Zool.*, **1**:163–173, 291–298 (1953).

Morris, R. F. (ed.): The Dynamics of Epidemic Spruce Budworm Populations, *Mem. Entomol. Soc. Can.*, **31**:1–332 (1963).

Mozley, A.: "Consequences of Disturbance; the Pest Situation Examined," H. K. Lewis and Co., Ltd., London, 1960.

Murray, K. F.: Population Changes During the 1957–1958 Vole (*Microtus*) Outbreak in California, *Ecology*, **46**:163–171 (1965).

Rosen, M. N., and A. I. Bischoff: The 1948–1949 Outbreak of Fowl Cholera in Birds in the San Francisco Bay Area and Surrounding Counties, *Calif. Fish Game*, **35**:185–192 (1949).

Sargent, F., II: Changes in Ideas on the Climatic Origin of Disease, *Bull. Am. Meteorol. Soc.*, **41**: 238–244 (1960).

————: The Environment and Human Health, *Arid Zone Res.*, **24**:19–32 (1964).

Solomon, M. E.: The Natural Control of Animal Population, *J. Animal Ecol.*, **18**:1–35 (1949).

Stallybrass, C. O.: "The Principles of Epidemiology and the Process of Infection," George Routledge & Sons, Ltd., London, 1931.

Trimmer, C. D.: Uganda National Parks Report for the Quarter ending 31 Dec. 1960, [mimeo.], 1961.

————: Uganda National Parks Report for the Quarter ending 30 Sept. 1962, [mimeo.], 1962a.

————: Uganda National Parks Report for the Quarter ending 31 Dec. 1962, [mimeo.], 1962b.

————: Uganda National Parks Report for the Quarter ending 30 June 1963, [mimeo.], 1963.

Tromp, S. W. (ed.): "Medical Biometeorology," Elsevier Publishing Company, Amsterdam, 1963.

Uvarov, B. P.: The Aridity Factor in the Ecology of Locusts and Grasshoppers of the Old World, *Arid Zone Res.*, **8**:164–198 (1957).

Watt, K. E. F.: A Mathematical Model for the Effect of Densities of Attacked and Attacking Species on the Number Attacked, *Can. Entomol.*, **91**:129–144 (1959).

————: Computers and the Evaluation of Resource Management Strategies, *Am. Sci.*, **52**:408–418 (1964).

————: Community Stability and the Strategy of Biological Control, *Can. Entomol.*, **97**:887–895 (1965).

Wellington, W. G.: Air Mass Climatology of Ontario North of Lake Huron and Lake Superior Before Outbreaks of the Spruce Budworm and the Forest Tent Caterpillar, *Can. J. Zool.*, **30**:114–127 (1952).

————: Weather and Climate in Forest Entomology, *Meteorological Monogr.*, **2**:11–18 (1954).

Wellington, W. G., J. J. Fettes, K. B. Turner, and R. M. Belyea: Physical and Biological Indicators of the Development of Outbreaks of the Spruce Budworm (*Choristoneura fumiferana* [*Clem.*] *Lepidoptera: Tortricidae*), *Can. J. Res.*, **28**:308–331 (1950).

Wright, B. S.: Predation on Big Game in East Africa, *J. Wildlife Mgt.*, **24**:1–15 (1960).

Young, S. P.: "The Bobcat of North America," The Stackpole Company, Harrisburg, Pa., and The Wildlife Management Institute, Washington, D.C., 1958.

Zinsser, H.: "Rats, Lice and History," Bantam Books, Inc., New York, 1960.

7 | CONFLICTS OF INTEREST

Many problems in resource management are complicated by a conflict of interest among groups who wish to use the same resource in different ways, or one group may regard the same resource as a boon, and another group may regard it as a pest. Practices designed to control one resource may have deleterious side effects on another. These difficulties are vividly demonstrated in the study of waterfowl and water in North America. Waterfowl sanctuaries could be used in other ways by farmers or by oil and gas interests. Waterfowl are of great interest to hunters (and firearms manufacturers), but are a pest to farmers of rice and wheat. Insect-pest control programs often kill waterfowl as a side effect. Waterfowl feeding areas are destroyed when swamps are drained, and can be created when dams increase acreage under water. Water resource management strategies affect power, navigation, irrigation, flood control, pollution, recreational use of wildlife and fish, and commercial fisheries. Any action enhancing one type of water utilization may be deleterious to another.

7.1 CONFLICTS OVER WATERFOWL

As a background to further discussion of waterfowl, we will consider their importance to hunters, and the importance of hunting to waterfowl mortality. Table 7.1 indicates the number of waterfowl killed by hunters, the number of hunters, and the average amount of time each spends hunting. Waterfowl

TABLE 7.1

Number of waterfowl killed by hunters in the United States, and number of active hunters, 1960–1961 (Crissey, 1961)

Waterfowl, number of hunters	Pacific flyway	Central flyway	Mississippi flyway	Atlantic flyway	Totals
Ducks	2,605,844	1,779,943	3,755,624	990,765	9,132,176
Geese	305,258	263,429	212,522	107,139	888,348
Coots	89,604	46,092	256,019	34,779	426,494
Totals	3,000,706	2,089,464	4,224,165	1,132,683	10,447,018
Number of active hunters	246,834	306,494	112,376	192,996	858,700
Average times hunted	4.175	3.772	3.988	3.728	

hunting is an important annual 4-day ritual in the lives of roughly 2 percent of the adult male population in North America. The seasonal contribution to the economy per waterfowl-hunter is about $80 in Canada and $46 in the United States (Benson, 1961). Most hunters would probably insist that this annual ritual has a value in their lives worth considerably more than their out-of-pocket expenses.

The *Waterfowl Status Reports* of the United States Fish and Wildlife Service indicate the hunting toll of the populations vulnerable to hunting pressure. Each year the Service estimates the kill, by state and by species, using a questionnaire survey mailed to hunters. A winter survey of the birds in each flyway is conducted by boats, cars, and aircraft. Crissey (1961) points out that we should interpret these data carefully, however. The winter survey does not constitute an estimate of the total flyway population since not all wintering areas of North American waterfowl are surveyed. John Gottschalk (personal communication) has pointed out numerous sources of bias in Table 7.2. However, since the mallard winters entirely within the United States, he believes the winter index for this species is more representative of the number of birds available to hunters in the fall than is true for most other species. Table 7.2 shows how the hunting kill in the Mississippi flyway has changed as a proportion of the estimates of total waterfowl populations. This table indicates a gradually rising trend in waterfowl populations with a gradually dropping kill. The final column indicates a quite striking drop in kill per unit of population. It does seem odd that there is a declining kill in the face of a

TABLE 7.2

Winter survey of waterfowl population estimates, and waterfowl kill estimates, Mississippi flyway*

Year	Index of waterfowl, in thousands (ducks, geese, and coots) (A)	Estimated total kill, in thousands (ducks, geese, and coots) (B)	$\dfrac{B}{A + B}$
1951	6,516	6,740	0.51
1952	4,924	11,913	0.71
1953	6,004	7,465	0.55
1954	6,309	5,724	0.48
1955	6,156	6,492	0.51
1956	8,365	6,554	0.44
1957	8,640	6,704	0.44
1958	7,804	6,268	0.45
1959	7,889	6,750	0.46
1960	7,885	3,944	0.33
1961	9,228	4,224	0.31
1962	7,975		
1963	8,678		

*From table C-10, U.S. Fish and Wildlife Service, "Waterfowl Status Report," 1963; kill data from "Annual Waterfowl Status Reports," 1953–1961.

rising population. In order to investigate this matter further, consider Table 7.3, which contains data on the number of hunters who purchased duck stamps in the Mississippi flyway, and the numbers in each of four species shot in that flyway each year. There was a decline in the number of duck hunters from 1956 to 1960, but there was a greater rate of decline in the kill statistics. We may assume that there is constant pressure by hunters on state legislators to keep bag limits as large as possible relative to the state of a population; therefore the rapidly declining catch per unit effort coupled with the more slowly declining hunter population seems to indicate that these four species of ducks are indeed declining, despite the winter survey data. Why? It is questionable whether an elaborate statistical analysis of published United States Fish and Wildlife Service Waterfowl Status Report statistics would be worthwhile because duck populations are clearly under the influence of a number of factors; we would lose a degree of freedom for each of them in any statistical analysis, and good data has only been tabulated for about a decade. (There are a number of difficulties with the published data: categories

TABLE 7.3

Mississippi flyway waterfowl-kill data*

Hunting season	Number of hunters purchasing duck stamps	Estimated kill by hunters			
		Mallards	Blue winged teals	Pintails	Redheads
1952–1953	980,665	2,825,090	754,270	241,640	
1953–1954	936,150	1,822,450	625,780	308,380	
1954–1955	918,685	2,578,470	537,639	319,669	130,025
1955–1956	1,012,762	2,715,919	266,052	146,019	126,744
1956–1957	1,016,338	2,775,452	407,538	242,933	75,871
1957–1958	1,004,255	2,587,149	276,559	199,943	140,538
1958–1959	924,628	3,043,796	240,020	180,507	54,464
1959–1960	694,516	1,279,578	243,284	120,455	29,805
1960–1961	701,535	1,548,751	253,101	125,893	4,085
1961–1962	575,798	854,000	42,100	80,200	1,800
1962–1963	448,544	406,800	44,100	47,800	2,000

*From U.S. Fish and Wildlife Service Special Scientific Reports: Wildlife.

of hunters have been defined differently in different years, some Pacific flyway data were lost in a fire, and figures on a given year may be revised considerably in consecutive annual reports.)

We know that there are other important population regulation factors for waterfowl besides hunting mortality. Waterfowl are dependent on the availability of certain foods, almost all of which are in turn dependent on the availability of water: pondweed, bulrush, smartweed, widgeon-grass, muskgrass, wild millet, wild celery, wild and cultivated rice, gastropods, crustacea, and aquatic insects. Waterfowl populations are thus extremely sensitive to the availability of ponds and marshes. At the worst of the great drought of 1936 in western Canada and the United States, North American waterfowl populations were at roughly a fifth their 1900 or 1944 level. In recent years, waterfowl breeding populations have been hit hard by drying-up breeding areas. Table 7.4 shows the number of water areas in July and the waterfowl breeding population sizes for the same years. A number of interesting points are brought out by the data. First, to a considerable extent the populations of the four species are in phase: blue-winged teals, pintails, and redheads all reached a peak for the decade in 1956, which was the second highest population year for mallards. Second, population trends do not

TABLE 7.4

*Number of water areas and trend in waterfowl breeding populations for southern Saskatchewan, Canada**

Year	Water areas, July	Index of breeding populations, in thousands			
		Mallards	*Blue-winged teals*	*Pintails*	*Redheads*
1952	855	1,536	265	1,374	37
1953	2,551	1,958	133	1,335	85
1954	3,037	1,912	264	1,275	73
1955	3,794	2,032	376	1,774	85
1956	1,753	2,381	376	1,905	149
1957	1,254	2,189	297	1,138	109
1958	763	3,000	203	748	62
1959	471	1,643	154	352	41
1960	916	1,590	133	575	51
1961	193	995	92	221	24
1962	246	674	39	216	58
1963	689	774	59	258	14

*From tables D-3 and E-14, U.S. Fish and Wildlife Service, "Waterfowl Status Report," 1961; and tables D-2 and E-14, "Waterfowl Status Report," 1963.

coincide with the trend in number of water areas, but follow with a 1-year lag, or longer (number of water areas reached a peak in 1955). Third, there is reason to believe that when water levels, and hence populations, are low, some factor or factors other than water level is playing a critical role in population regulation. Otherwise, why didn't the sharp rise in water areas in 1960 produce a rise in mallard or teal populations the following year? Perhaps hunting pressure is too great when combined with low water levels. Because of the observed lag, models to describe waterfowl populations must employ *difference equations*, not differential equations. That is, we have

$$P_{t+1} = f(W_t, P_t)$$

not

$$\frac{dP_t}{dt} = f(W_t, P_t)$$

where W_t = number of water areas at time t

P_t = waterfowl populations at time t

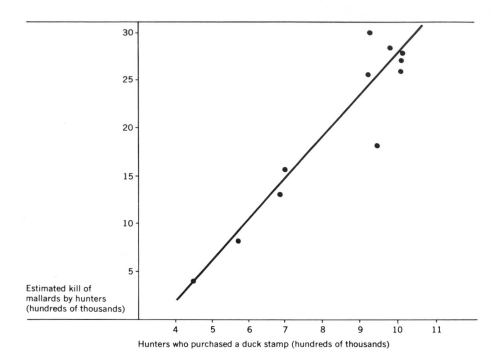

FIGURE 7.1

Estimated hunting kill of mallards in the Mississippi flyway as a function of number of hunters. (Plotted from data in Waterfowl Status Reports, U.S. Fish and Wildlife Service.)

Some more insight into the relative importance of hunting pressure and availability of water areas for waterfowl is provided by Figs. 7.1 and 7.2. Figure 7.1 is a plot of the mallard data from Table 7.3. If hunting were indeed the primary cause of mortality for mallards, we would expect a trend line rising from left to right and bending over to a horizontal asymptote on the right side, indicating dropping catch per unit effort caused by competition between hunters for a limited resource at high hunter densities. The absence of such an asymptote suggests that availability of birds (and numbers of hunters) is being limited by some third variable. The implicated third variable is revealed in Fig. 7.2. The relationship between size of mallard breeding population and number of water areas seems to have less scatter when numbers of breeding mallards are plotted against number of water areas the *previous* season. In summary, the really critical factor for waterfowl populations seems to be water areas available for breeding, though excessive hunting can have a major effect when shortage of such areas has lowered population densities to precarious levels.

174   *The principles of resource management*

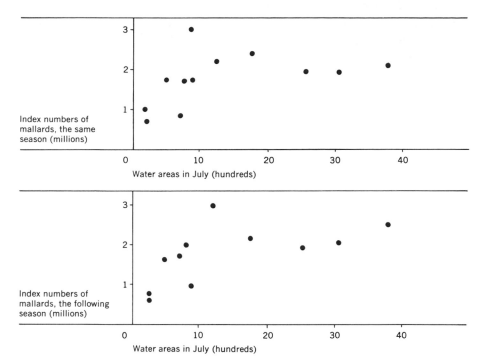

FIGURE 7.2

Southern Saskatchewan, Canada, mallard breeding populations as a function of number of water areas in July in Saskatchewan that season (top panel) and the previous season (bottom panel). (Plotted from data in Waterfowl Status Reports, U.S. Fish and Wildlife Service.)

There are several other types of mortality factors operating against waterfowl; several of these operate more severely against females than males and therefore can have a critical effect on the reproductive capacity of a population when it is low and in most need of a strong homeostatic capability.

1. During the time they incubate the eggs, females are exposed to ground and air predators, fire, and perhaps haying operations.

2. More adult females than males are shot in the north during the hunting season. This is because of delayed development of flight feathers and the incubation duties.

3. Females may be subject to greater hazard from botulism because of the time they spend incubating.

4. Females may be more affected by drought.

Conflicts of interest created by waterfowl are perhaps as intense in California as anywhere, because while California has been an important ancestral wintering ground for waterfowl, there is intensive cultivation of crops that are preferred foods of waterfowl. The following figures indicate the steady reclamation of marshland and conversion into rice acreage that has occurred in California.

Year	Rice acreage
1912	1,400
1920	162,000
1948	225,000
1954	485,000

This conversion deprived waterfowl of their ancestral wintering grounds, and at the same time created an ideal waterfowl habitat. Before rice fields are drained, they contain circulating water of just the right depth for waterfowl. Rice itself is the preferred food of grain-eating species of ducks. Hence we have two competing resources of great value: the rice, and the waterfowl.

Chattin (1949) has reported an annual kill as high as 3.3 million waterfowl in California. This represents $4 million, valued only as meat. However, hunters spend $46 each per season to kill about five birds each per season, so if this is a more accurate measure of hunters' valuation of the birds (including the worth of the recreation), the birds are worth 46/5 or approximately $9 each, and waterfowl are worth roughly $30 million per annum to the California economy.

On the other hand, Biehn (1951) reports impressive crop losses due to the birds. A thorough study of waterfowl damage to rice in California in 1942 showed a loss equivalent to 258,804 bags worth $905,000 (Horn, 1949). In the 1943 to 1944 winter season in Imperial Valley, a survey estimated losses to lettuce and irrigated pasture of more than $5 million due to waterfowl.

Losses have been kept down through the creation of waterfowl sanctuaries, where great expanses of waterfowl food are grown in ideal waterfowl habitat at state and Federal expense. (The Federal money comes from the Pittman-Robertson Act tax on sporting arms and ammunition and from Lea Act funds for purchase of lands to be used as waterfowl-management areas; the state money comes from part of the pari-mutuel fund, and from licenses and duck stamp sales, which support the California Department of Fish and Game.) It is noteworthy that the costs of providing waterfowl hunting are borne *in toto* by the hunters.

Another conflict of interest arises since the extensive sanctuary acreage is considered of great potential worth by oil and gas interests. In California, the following acreage has been put into duck sanctuary.

State (8 areas)	45,500 acres
Federal (10 areas)	172,000 acres
	217,500 acres

Oil and gas drilling rights have been leased on waterfowl refuges. From 1920 to 1952, only 11 leases were issued on Federal wildlife refuges (Day, 1959). From 1952 to December 2, 1955, 60 oil leases were granted on refuges under primary jurisdiction of the United States Fish and Wildlife Service. Another 214 leases were granted on lands under joint administration of the Service and other Federal agencies (Day, 1959). The United States House Committee on Merchant Marine and Fisheries issued a unanimous report in 1956, which said in part, "The fact that this activity took place while a suspension order was in effect for the ostensible purpose of reversing the regulations to provide greater protection to wildlife lands only aggravates the situation." The report was highly critical of the Secretary of the Interior, Douglas McKay, in whose department the issuing of the licenses occurred.

An important moral may be drawn from the waterfowl situation. Where there is a conflict of interests in a democratic society, it is necessary for each of the interested groups to be perpetually alert to ensure that its interests are being protected. It is necessary that great care and thought, and a real concern for and insight into the interests of each group, be incorporated into any large-scale systems planning for the management of natural resources (see Sec. 12.4).

7.2 SIDE EFFECTS OF WATER-MANAGEMENT PROCEDURES

Water resources are an important and revealing subject for illustration of conflicts of interest because of the many different interests involved, because water systems can be altered in many ways, and because of the wide variety of cause-effect pathways by which biological effects may be produced. There are four basic means by which man can have an effect on water. We will indicate for each of these basic means the types of effects that can be produced.

1. The creation of dams and regulation of impounded water have a variety of biological effects, some good and some bad. In Lake Torron, Sweden, after regulation, graylings found increased spawning area because in parts of the submerged area erosion had exposed suitable gravel bottom. Pike, on the other hand, spawn on bottom with vegetation. In regulated lakes, the high-water period is later than in unregulated lakes, and often does not reach the vegetated areas in time for pike spawning. One species thus benefits by regulation and one is hurt by it (Runnstrom, 1960). This leads to the more

general concept that the species composition of aquatic communities can be regulated by the way in which the curve for water level against time is changed by regulation. Wood and Pfitzer (1960) discuss the literature on such experiments, and conclude that water regulation can be so planned as to increase the proportion of desirable fish species in an impoundment. Conversely, of course, unplanned regulation can increase the proportion of undesirable species. Another aspect is that dams in tropical countries can create ideal environments for development of malaria and bilharzia, in which case the spread of these diseases must be effectively controlled by intensive fish culture (Maar, 1960).

2. The rate of water flow out of a dam, and the level of the impounded water from which the water flows, will affect all life downstream from the dam, by affecting stream temperature. The effect may be direct, with increased summer stream temperatures or decreased winter stream temperatures; or it may be indirect, when undercooling winter streams may allow ice to form on the river bottom and tear up or cover over spawning areas (Somme, 1960).

3. The creation or removal of obstructions in rivers can have a variety of sometimes surprising effects. Creation of dams often blocks spawning runs; removal of obstructions in freshwater rivers flowing into the sea may have catastrophic salinizing effects on freshwater resources (Meehean, 1960).

4. Pollution of water resources, and the destruction of all or part of the life therein, may occur via a variety of pathways. Increase in water temperature by industrial effluents can be just as lethal as chemical pollutants if the temperature is high enough.

To summarize, any action taken on a multiple-use resource can have a variety of biological effects, and these should be considered in prior planning. Precisely because there is such a variety of effects from any action taken, the sheer arithmetic involved in prior assessment of various management strategies is overwhelming. We conclude that computer simulation and optimization studies, and perhaps a "new mathematics," are the only feasible means available for this assessment. Since a variety of complexity has already been encountered in this book, we can proceed to the systems approach outlined in Part Four.

REFERENCES

Benson, D. A.: "Fishing and Hunting in Canada," Canadian Wildlife Service, National Parks Branch, Dept. Northern Affairs and National Resources, Ottawa, Canada, 1961.

Biehn, E. R.: Crop Damage by Wildlife in California, *Calif. Dept. Fish Game, Game Bull.*, no. 5, 1–71 (1951).

Chattin, J. E.: California Waterfowl Kill, State of California, Dept. of Fish and Game, 1–8 [mimeo.], 1949.

Crissey, W. F.: Waterfowl Status Report, *U.S. Fish Wildlife Serv. Spec. Sci. Rept. Wildlife*, nos. 26, 29, 33, 37, 45, 61; 1954, 1955, 1956, 1957, 1959, 1961.

Day, A. M.: "North American Waterfowl," 2d ed., The Stackpole Company, Harrisburg, Pa., 1959.

Glover, F. A., and J. D. Smith, Waterfowl Status Report, *U.S. Fish Wildlife Serv. Spec. Sci. Rept. Wildlife*, no. 75, 1963.

Horn, E. E.: Waterfowl Damage to Agricultural Crops, *Trans. N. Am. Wildlife Conf.*, **14**:577–579, 1949.

Maar, A.: Dams and Drowned-Out Stream Fisheries in Southern Rhodesia, in "Natural Aquatic Resources," pp. 139–151, International Union for Conservation of Nature and Natural Resources, Brussels, 1960.

Meehean, O. L.: Multiple Purpose Planning for Aquatic Resources, in "Natural Aquatic Resources," pp. 53–60, International Union for Conservation of Nature and Natural Resources, Brussels, 1960.

Runnstrom, S.: Hydro-Electric Power Stations and Fishing, in "Natural Aquatic Resources," pp. 61–68, International Union for Conservation of Nature and Natural Resources, Brussels, 1960.

Somme, S.: The Effects of Impoundment on Salmon and Sea Trout Rivers, in "Natural Aquatic Resources," pp. 77–80, International Union for Conservation of Nature and Natural Resources, Brussels, 1960.

Wood, R., and D. W. Pfitzer: Some Effects of Water Level Fluctuations on the Fisheries of Large Impoundments, in "Natural Aquatic Resources," pp. 118–138, International Union for Conservation of Nature and Natural Resources, Brussels, 1960.

FOUR | THE *Methods* OF RESOURCE MANAGEMENT

8 | FORTRAN PROGRAMMING

Part Four is predicated on the assumption that realistically complex problems in ecology and resource management can be solved only if computers are used at each step in the research program: analysis of sampling data, regression analysis and analysis of variance, tests of the suitability of various models and fitting of parameter values, and finally the development of management strategies through simulation and optimization. The complexity of the models required, and the volume of algebra required by statistical procedures involving large numbers of variables and iterative methods, preclude solution of the problems if any other approach is adopted.

8.1 THE POWER OF COMPUTER LANGUAGES

It is a main thesis of this book that computers and the languages used to address them are of tremendous importance to ecologists and resource managers for two reasons, only one of which is widely understood. The first reason derives from the well-publicized amazing properties of computer circuits and memory components. Informed laymen are now aware that the most powerful available computers can perform about 10-million additions a second, recall from fast-access core-memory any one of 256,000 numbers, and read and write on magnetic tape at 680,000 digits a second. The second reason stems from the conceptual nature of the various branches of mathematics. Each of these, such as the calculus of Newton and Leibniz, the geometry of Descartes, the algebra of Boole, and the set theory of Cantor, was designed to deal with a certain category of problems. The characteristics of certain branches of mathematics make them particularly suitable for dealing with situations where the variables can vary continuously along a range of values; these branches are referred to as infinitesimal mathematics. Other branches of mathematics are appropriate where variables can assume two or more discrete values; they are referred to collectively as finite mathematics. Complex phenomena in ecology and resource management involve both kinds of processes. To describe adequately the real logical structure of such problems, a type of mathematical language is required that utilizes the descriptive features of many different kinds of mathematics. Conveniently, the pseudoalgebraic codes used by scientists to present strings of instructions to computers have precisely this characteristic by design. This is true of Algol, Fortran IV, and PL/1, and all their related languages.

It is now necessary to explain the significance of pseudoalgebraic codes. For each type of developed computer circuitry, numerical or "machine language" instruction codes had to be developed also so the programmer could instruct the computer to perform a unit operation, such as "Reset the arithmetic register to zero and add to it the number stored in the physical location designated by numerical address number XXXX." Very soon it became obvious that this was far too time-consuming a procedure, since many real problems can only have their solutions described in terms of several thousand equations, which would require five to ten times as many machine-language instructions. The computer manufacturers expended enormous amounts of mathematical manpower to develop mathematical languages such as Fortran and Algol, which were problem-oriented, or user-oriented, rather than designed to trigger the circuitry of the computers. Sequences of instructions written in these languages are entered into the computer where compiler programs convert them to sequences of machine-language instructions that actually execute the operation. It was in the interests of the computer manufacturers to ensure that the user-oriented pseudoalgebraic codes were as powerful and flexible mathematical languages as could be devised. Thus languages such as Fortran contained the features of many different kinds of mathematical languages even when first developed, and additional features have been added since. The newest of the computer languages, PL/1, can perform an enormous variety of types of mathematical operations, and can describe many different types of processes. Precisely because the pseudoalgebraic codes were designed to be very powerful and flexible, they lend themselves to realistic description of complex biological processes to a greater extent than any other type of mathematical language yet devised.

8.2 APPLICABILITY OF FORTRAN TO BIOLOGY

It is not our intent to give an exhaustive introduction to the features of pseudoalgebraic codes; these can be obtained in many other publications. Rather, we shall point out those features of Fortran, the most widely used of the user-oriented computer languages, which correspond to particular phenomena in animals and plants.

It should be noted at the outset that the particular usage of the equal sign, =, in Fortran is peculiarly adapted to the nature of biological reality. In Fortran, the statement

$$A = B + C$$

does not mean A is equal to B plus C. Rather, it means, "Add the numbers in locations B and C to the arithmetic register, then store the sum in the location A." The equal sign does not express equality, it shows how to compute the entity A from the addends B and C. The significance of this unusual use of the equal sign for biology is clear when we note that B and C really refer to the time prior to the computation, and A is the product of the computation or, to put it differently, A only occurs *after* the computation is complete. This use of the equal sign corresponds to the fact that, in biological systems, the state of the system at any point in time is in part determined by the past history of the system, in an ecological as well as a genetic sense. Consider the simple difference equation

$$N_{t+1} = N_t e^r$$

This is one way of stating the well-known exponential growth law, and if NUMBER had been designated as a real number, it would be written in Fortran as

$$NUMBER = NUMBER*EXP(R)$$

Note that the symbol * indicates multiplication, thus reserving X for the variable X. This is not an equation, but a statement as to how NUMBER is produced from its previous value. This unusual use of the equal sign is basic to the use of pseudoalgebraic codes to describe the historical processes characteristic of biology. In each equation within the sequence of instructions designed to simulate a biological process, the right-hand side refers to a previous point in time, and the left-hand side refers to a subsequent point in time.

The computer can simulate a process in which certain events characteristically occur at certain intervals of chronological time by adding the number 1 to a variable, TIME, whenever a simulated time interval has elapsed. Suppose, for example, that we wished to simulate a biological process that had a diel cycle, in which activity only occurred from 8:00 A.M. to 10:00 P.M., the animal sleeping the rest of each day. This process could be simulated by having a variable, HOUR, set equal to zero at the simulated 8:00 A.M. After each hour's worth of activity has been simulated, the computer is programmed to add the digit 1 to the variable HOUR, then ask itself if HOUR is equal to 14 (that is, 10:00 P.M.). If the answer is yes, the computer enters a routine to simulate overnight events; otherwise, it cycles back to the beginning of the subroutine for simulating events in a daylight hour. Fortran deals with situations such as this, where one action is taken if a variable has one value, but another is taken if it has another value, by means of IF statements. In the situation just outlined, suppose the routine to simulate events during a daylight hour begins with statement number 100, but the routine to simulate events during a night begins with statement number 200. Then the routine to simulate events during a daylight hour would terminate with the following two statements.

$$HOUR = HOUR + 1.$$

$$IF(HOUR-14.)\ 100,\ 200,\ 200$$

When the computer operates on an IF statement of this type, it computes the quantity within parentheses, and determines if it is equal to a negative number, zero, or a positive number. The triplet of numbers following the second parenthesis tells the computer where to go to find its next instruction in each of these three cases. Thus, if HOUR is 13, HOUR-14 will be a negative number, and the computer will go to statement number 100 for its next instruction (that is, the first statement in the sequence of statements for simulating events in a daylight hour). If HOUR-14 is zero or positive, the computer will go to statement number 200 to find its next instruction (the first statement in the subroutine of instructions for use in simulating nighttime events). The IF statement, which is an example of a finite mathematical operation, has the most profound implications for biologists. The IF statement allows biologists to build an "either-or" capability into computer simulation programs to correspond to actual biological processes. Consider physiological and ecological thresholds, for example, in which the process only occurs if a variate-value is equal to or exceeds some threshold. We could describe this situation by the statement

$$IF(X-XTHRES)\ 1, 2, 2$$

The computer only proceeds to statement number 2 if X is equal to or greater than XTHRES; otherwise it goes to statement number 1.

Another important feature of the Fortran language is its ability to perform repetitive operations on subscripted variables by means of DO loops. In the simplest possible type of example, suppose that a area 100 miles by 100 miles had been divided into 625 four-mile × four-mile subsquares. Suppose each of these subsquares contained a number of deer P(I, J), in which P stands for population, I stands for the row number, and J stands for the column number of the square containing this population. Thus P(11, 23) stands for the population of deer in the 4-mile × 4-mile subsquare in row 11, column 23, of the 25 rows and 25 columns of squares in the big square. Now suppose we wished to describe the process by which exactly one deer was added to the population in each square. This can be handled succinctly by the three-statement sequence

$$DO\ 100\ I = 1, 25, 1$$

$$DO\ 100\ J = 1, 25, 1$$

$$100\ P(I, J) = P(I, J) + 1.$$

The first statement means, "Do repeatedly, down to and including statement number 100, the following sequence of statements, setting the value of I equal to the first of the three numbers (1), then subsequently repeat the operation with values increasing each repetition by the third integer (1), until the second integer (25) is reached." For each of these 25 repetitions, the computer performs 25 repetitions under instruction of the second statement. The third statement, the only operation included in this "loop," or cycle of operations, tells the computer to remove P(I, J) from memory, add 1.0 to it, then replace it in its original memory location. The presence of this DO loop feature in computer languages has several different types of significance for resource managers and ecologists. First, it is clearly a very easy way to program the highly repetitive type of algebra involved in multiple regression analysis, multiple analysis of variance, and simulation studies with a lot of repetition. Second, the use of subscripted variables opens up the possibility of simulating extremely complex types of processes where space or time must be subdivided, as in dispersal over an area, or when effects over time accumulate until the total effect of some cause is lethal.

Another important feature of Fortran IV is the functions MAX and MIN.

If we wish the computer to choose the largest value of the three numbers A,B,C, we use the notation

$$MAX(A,B,C)$$

Similarly,

$$MIN(A,B,C,)$$

selects the smallest of the three. Any number of numbers greater than 2 can be between the parentheses. This feature of Fortran IV is of critical importance to resource managers for two reasons. First, optimization of resource systems often involves selection of maximum or minimum values of some variable. Second, use of the MAX and MIN notation allows for extreme succinctness in the instructions; in some subroutines the number of Fortran statements required is halved by using this technique of selecting maximum and minimum values. Sections 11.7 and 11.8 provide illustrations.

In summary, the pseudoalgebraic languages for programming computers provide the biologist with a tremendously powerful descriptive tool because they combine the best features of algebra and calculus, matrix algebra (through the use of subscripted variables), and all the branches of modern mathematics that deal with partitioning, logical branching trees, and sets. More detail on techniques of applying these capabilities of such languages will be given in subsequent chapters.

9 SYSTEMS MEASUREMENT

Two different situations are encountered in estimating the variate value of any parameter in nature. In the first, discussed in Sec. 9.1, it is possible to see the population of entities being sampled, and so principal concern is with the efficiency of the sampling procedure. In the second, explained in Sec. 9.2, the population must be estimated by some inferential procedure, and the chief concern is with the logical foundations of the procedure, and use of techniques that do not violate the assumptions underlying the logical foundation.

9.1 DIRECT MEASUREMENT: THE SCIENCE OF SAMPLING

There are two objects in sampling: to obtain, first, the most accurate and, second, the most precise measurements possible for a given cost. In technical terms, *accuracy* measures the size of deviation from the true mean, whereas *precision* measures the size of deviations from the mean obtained by applying the sampling procedure repeatedly. Accuracy is low when bias of some type systematically forces the sampled variate values away from the population mean. Precision is low when the variance among means of subsamples is great. Ideally, we seek the sampling procedure with maximum efficiency. The most efficient possible statistic is that for which the error distribution tends to be normal, as the sample is increased, with the least possible variance.

At first glance, accuracy and precision might seem to be easily attainable desiderata. However, the modern development of sampling theory is testimony to the difficulty one has in obtaining accurate and precise measurements.

Two examples will indicate the difficulties, and exemplify the mechanisms that can produce bias and inefficiency.

It might be argued that merely increasing the size of the sample will eliminate bias. Suppose we estimate, using 3-inch-mesh gill nets, the relative abundance of various species and age classes within species of fish in a lake. Clearly, no matter how large a number of net settings we use to obtain estimates, we will overestimate the relative abundance of species and ages which have the appropriate diameter to be caught by the net. The estimates will be biased even with a very intensive sampling program.

Analysis of variance (ANOVA) shows significant variation at the 0.05 level in the level-to-level incidence of spruce budworm larvae within trees (Morris, 1955). Suppose we ignored or did not know this fact, and attempted to determine plot-to-plot differences in larval incidence by random sampling of trees without regard to level. We might do too much sampling at a level where the incidence was unusually high, and at other times, we might do too much sampling in a low-incidence level. Our sample means would therefore not be very repeatable, our precision would be low, and the sampling procedure would have low efficiency.

We will now consider in turn four specific techniques available for maximizing accuracy and precision.

1. Use of ANOVA to ensure that samples are representative of a population.

2. Determination of optimum sample size.

3. Maximization of efficiency for given sampling effort by means of stratified sampling schemes.

4. Use of cost functions to determine that mode of allocating effort within and between sampling units which produces maximum efficiency for a given cost.

Representativeness of the sample

In a study of population dynamics of the spruce budworm (Morris, 1955), a prerequisite for designing an unbiased, efficient sampling scheme was the determination of representativeness of various sampling units in the tree host. The object was to determine the mean larval density per 10 square feet of branch surface, within each balsam fir. Samples of larvae were obtained by clipping all secondary branches from one complete side of a main branch stem. Since branch size varies considerably within a tree, the width at mid-length and the length of the foliated part of each branch were measured. The product of these two measurements, in square feet, was arbitrarily referred to as *square feet of branch surface*. Obviously, this is not an accurate measure of branch surface or foliage surface, but it is a conveniently obtained correction factor that permits the direct comparison of insect populations on branches of various sizes.

Sampling units were selected according to the following plan, for determination of inter- and intratree variance. Codominant balsam trees were selected randomly in groups of four, each group representing a different forest condition or budworm larval population density. The living crown of each tree was divided vertically into four equal levels, and each of the levels was divided into north, south, east, and west quadrants. Two branches were sampled in each of the top eight sampling loci, because of the small size of the branches, and one in each of the bottom eight. All data from each locus were pooled rather than being treated as replicates. Thus the total number of degrees of freedom for ANOVA is 4 (for trees) × 4 (for levels) × 4 (for quadrants) = 64.

Before discussing the structure of ANOVA for such data, we must consider the four assumptions on which ANOVA is based.

1. Each variate value can be considered the sum of the following components:

    a. an overall mean (in this instance, an average population density for all trees, levels, and quadrants);

    b. a series of effects due, respectively, to "main effects" (in this instance, trees, levels, and quadrants), "two-factor interactions" (in this instance, trees × levels, trees × quadrants, and levels × quadrants), "three-factor interactions" (trees × levels × quadrants), and so on, if there are more than three main effects;

    c. effects due to fluctuations between random samples within loci (in this case, there is no replication so we have no source of variation of this type); and

    d. residual random errors. Nonrandomness of these residuals invalidates ANOVA.

2. Departure from the assumption of normality (variate values distributed in accord with the normal distribution) in theory violates the assumption underlying ANOVA, but simulation studies have shown that in practice ANOVA is very "robust" to violation of this assumption (that is, it is insensitive to it). (ANOVA is not "robust" to skew deviations, however.)

3. The variance of the random components in the analysis is assumed not to vary from class to class; violation of this assumption can be serious.

4. The "components of variance" must combine in an additive fashion as outlined under assumption 1.

Biological data typically violate all these assumptions to greater or lesser degree, because most biological entities (individual animals or parts of animals, colonies of cells, populations or communities) grow according to an exponential or logistic, not an arithmetic growth law. The extent to which ANOVA assumptions are violated will depend on how fast the entities are growing or can grow, and hence how wide a range of variate values is covered by the data. The wider the range of variate values covered, the more serious will be the violation of ANOVA assumptions. The simplest procedure to correct for this violation is to transform all raw data to another form. For most biological data, an appropriate procedure is to convert all data to logarithms. However, the log of 1 to any base is zero. To eliminate the difficulties associated with this fact, a more usual transformation is $X_{TRANS} = \log(X + 1)$.

The data in the present instance required this transformation (as will most population data, particularly from insects and other organisms whose populations cover a wide range of variate values from time to time and place to place). We will now consider the structure of models for ANOVA.

The precise structure of ANOVA models depends on whether we are dealing with crossed, nested, or mixed classifications. In the present example, where a north quadrant occurs in each of four height levels, quadrants and heights are crossed with respect to each other. That is, the north quadrant in the top level has a correspondence to the north quadrant in the bottom level. If we had no quadrants, but merely took four branch samples randomly within each level, the four replicates would be nested within each level. That is, sample 1 from level 1 should be no more similar to sample 1 from level 4, than to sample 4 in level 4. We would have a mixed model if we had a crossed classification with respect to levels and quadrants, but within level ×, quadrant loci selected 4 replicate samples at random. Here we would have nesting within a crossed classification, and hence a mixed classification. To illustrate the development of complex ANOVA models for testing representativeness of samples, we will use a hypothetical case in which in addition to trees, levels, and quadrants, replicate samples are taken within each level-quadrant locus of each tree.

Where $i = 1, \ldots, p$ trees
$j = 1, \ldots, q$ levels
$k = 1, \ldots, r$ quadrants
$\alpha = 1, \ldots, n$ replicate samples

we can assume we have $x_{ijk\alpha}$ observations of larval density. Each observation can be considered as the sum

$$x_{ijk\alpha} = \mu + \xi_i + \eta_j + \zeta_k + \beta_{ij} + \gamma_{ik} + \delta_{jk} + \lambda_{ijk} + \epsilon_{ijk\alpha}$$

where μ = overall mean

$\xi_i, \eta_j,$ and ζ_k = main effects due to trees, levels, and quadrants, respectively

$\beta_{ij}, \gamma_{ik},$ and δ_{jk} = two-factor interaction effects due to three possible two-factor combinations of trees, levels, and quadrants

λ_{ijk} = effect of three-factor interactions

$\epsilon_{ijk\alpha}$ = normally distributed random residual components

Now by definition, $E(\bar{x}) = \mu$, since $E(\bar{x})$ is the average value of $\bar{x}$ in the population.

$$E(\bar{x}_i) = \mu + \xi_i \tag{9.1}$$

$$E(\bar{x}_j) = \mu + \eta_j \tag{9.2}$$

$$E(\bar{x}_k) = \mu + \zeta_k \tag{9.3}$$

$$E(\bar{x}_{ij}) = \mu + \xi_i + \eta_j + \beta_{ij} \tag{9.4}$$

$$E(\bar{x}_{ik}) = \mu + \xi_i + \zeta_k + \gamma_{ik} \tag{9.5}$$

$$E(\bar{x}_{jk}) = \mu + \eta_j + \zeta_k + \delta_{jk} \tag{9.6}$$

$$E(\bar{x}_{ijk}) = \mu + \xi_i + \eta_j + \zeta_k + \beta_{ij} + \gamma_{ik} + \delta_{jk} + \lambda_{ijk} \tag{9.7}$$

Estimates of the various effects can be obtained from the following equations.

From (9.1),

$$\xi_i = E(\bar{x}_i) - E(\bar{x}) \tag{9.8}$$

From (9.2),

$$\eta_j = E(\bar{x}_j) - E(\bar{x}) \tag{9.9}$$

From (9.3),

$$\zeta_k = E(\bar{x}_k) - E(\bar{x}) \tag{9.10}$$

From (9.4),

$$\beta_{ij} = E(\bar{x}_{ij}) - E(\bar{x}) - \xi_i - \eta_j$$
$$= E(\bar{x}_{ij}) - E(\bar{x}_i) - E(\bar{x}_j) + E(\bar{x}) \tag{9.11}$$

From (9.5),

$$\gamma_{ik} = E(\bar{x}_{ik}) - E(\bar{x}_i) - E(\bar{x}_k) + E(\bar{x}) \tag{9.12}$$

From (9.6),

$$\delta_{jk} = E(\bar{x}_{jk}) - E(\bar{x}_j) - E(\bar{x}_k) + E(\bar{x}) \tag{9.13}$$

From (9.7),

$$\lambda_{ijk} = E(\bar{x}_{ijk}) - E(\bar{x}) - E(\bar{x}_i) + E(\bar{x}) - E(\bar{x}_j) + E(\bar{x}) - E(\bar{x}_k) + E(\bar{x})$$
$$- E(\bar{x}_{ij}) + E(\bar{x}_i) + E(\bar{x}_j) - E(\bar{x}) - E(\bar{x}_{ik}) + E(\bar{x}_i) + E(\bar{x}_k) - E(\bar{x})$$
$$- E(\bar{x}_{jk}) + E(\bar{x}_j) + E(\bar{x}_k) - E(\bar{x})$$
$$= E(\bar{x}_{ijk}) + E(\bar{x}_i) + E(\bar{x}_j) + E(\bar{x}_k) - E(\bar{x}_{ij}) - E(\bar{x}_{ik}) - E(\bar{x}_{jk}) - E(\bar{x}) \tag{9.14}$$

$$\varepsilon_{ijk\alpha} = \bar{x}_{ijk\alpha} - E(\bar{x}_{ijk}) \tag{9.15}$$

From Equations (9.8) to (9.15), inclusive, the deviation of any sampled larval population from the overall mean of samples can be expressed as

$$x_{ijk\alpha} - \bar{x} = (\bar{x}_i - \bar{x}) + (\bar{x}_j - \bar{x}) + (\bar{x}_k - \bar{x}) + (\bar{x}_{ij} - \bar{x}_i - \bar{x}_j + \bar{x})$$
$$+ (\bar{x}_{ik} - \bar{x}_i - \bar{x}_k + \bar{x}) + (\bar{x}_{jk} - \bar{x}_j - \bar{x}_k + \bar{x})$$
$$+ (\bar{x}_{ijk} + \bar{x}_i + \bar{x}_j + \bar{x}_k - \bar{x}_{ij} - \bar{x}_{ik} - \bar{x}_{jk} - \bar{x}) + (\bar{x}_{ijk\alpha} - \bar{x}_{ijk}) \tag{9.16}$$

From (9.16), we can write an equation partitioning the sum of squares for the total, that is,

$$\Sigma(x_{ijk\alpha} - \bar{x})^2$$

Note, however, that in deriving the formula for this sum of squares, all cross-product terms drop out, since the sum of the deviation of any set of observations from their mean must be zero. Also note that a term such as

$$\sum_{ijk\alpha} (x_{ij} - x_i - x_j + x)^2 = nr \sum_{ij} (x_{ij} - x_i - x_j + x)^2$$

because the summand does not depend on k and α. Therefore, from (9.16) we obtain

$$(x_{ijk\alpha} - \bar{x})^2 = qrn \sum_i (\bar{x}_i - \bar{x})^2 + prn \sum_j (\bar{x}_j - \bar{x})^2 + pqn \sum_k (\bar{x}_k - \bar{x})^2$$
$$+ rn \sum_{ij} (\bar{x}_{ij} - \bar{x}_i - \bar{x}_j + \bar{x})^2 + qn \sum_{ik} (\bar{x}_{ik} - \bar{x}_i - \bar{x}_k + \bar{x})^2$$
$$+ pn \sum_{jk} (\bar{x}_{jk} - \bar{x}_j - \bar{x}_k + \bar{x})^2$$
$$+ n \sum_{ijk} (\bar{x}_{ijk} + \bar{x}_i + \bar{x}_j + \bar{x}_k - \bar{x}_{ij} - \bar{x}_{ik} - \bar{x}_{jk} - x)^2 + \sum_{ijk\alpha} (\bar{x}_{ijk\alpha} - \bar{x}_{ijk})^2$$

It is computationally too difficult to obtain the component sums of squares in this form. Therefore we derive new formulas based, not on sums of squares of deviations, but on totals, squares of totals, and sums of totals squared as follows. The derivation uses the new symbols

$$N = pqrn$$

$$T_i = \sum_{jk\alpha} x_{ijk\alpha} = qrn\bar{x}_i$$

$$T = \sum_{ijk\alpha} x_{ijk\alpha} = N\bar{x}$$

$$\sum_i \bar{x}_i = p\bar{x} = \frac{T}{qrn} \qquad \text{whence} \qquad \bar{x}\sum_i \bar{x}_i = \frac{T^2}{Nqrn}$$

and

$$\sum_i (\bar{x})^2 = p\left(\frac{T}{N}\right)^2 = \frac{T^2}{Nqrn}$$

Then

$$qrn\sum_i (\bar{x}_i - \bar{x})^2 = qrn\left[\sum_i \left(\frac{T_i}{qrn}\right)^2 - 2\bar{x}\sum_i \bar{x}_i + \sum_i (\bar{x})^2\right]$$

$$= \frac{\sum_i T_i^2}{qrn} - \frac{2T^2}{N} + \frac{T^2}{N}$$

$$= \frac{\sum_i T_i^2}{qrn} - \frac{T^2}{N}$$

Proceeding in the same fashion, we may build up the whole set of computational formulas for our ANOVA, which then takes the following form (Table 9.1).

The reader may develop ANOVA computational schemes himself for testing representativeness of samples in any given situation, proceeding in the same fashion.

One last point about such tables needs to be mentioned, relative to the use and interpretation of F tests on the ratios of mean squares. We begin at the bottom of the table in testing for significant variance ratios, and work upward from highest-order to lowest-order interactions and finally to main effects. If one looks at any column of a set of F tables, he will notice that for a given level of significance, the greater the number of degrees of freedom for the denominator of the ratio, the lower the F value. This means it is easier to detect a significant difference if we increase the number of degrees of

TABLE 9.1
Analysis of variance: Trees, levels, quadrants, and replicate samples

Source of estimate	Sums of squares	Degrees of freedom	Mean squares
Main effects:			
trees	$\dfrac{1}{nqr}\sum_i T_i^2 - \dfrac{T^2}{N}$	$p-1$	All entries in this column are obtained by dividing the sum of squares entry by the appropriate number of degrees of freedom
levels	$\dfrac{1}{npr}\sum_j T_j^2 - \dfrac{T^2}{N}$	$q-1$	
quadrants	$\dfrac{1}{npq}\sum_k T_k^2 - \dfrac{T^2}{N}$	$r-1$	
2-factor interactions:			
trees-levels	$\dfrac{1}{nr}\sum_{ij}T_{ij}^2 - \dfrac{1}{nqr}\sum_i T_i^2 - \dfrac{1}{npr}\sum_j T_j^2 + \dfrac{T^2}{N}$	$(p-1)(q-1)$	
trees-quadrants	$\dfrac{1}{nq}\sum_{ik}T_{ik}^2 - \dfrac{1}{nqr}\sum_i T_i^2 - \dfrac{1}{npq}\sum_k T_k^2 + \dfrac{T^2}{N}$	$(p-1)(r-1)$	
levels-quadrants	$\dfrac{1}{np}\sum_{jk}T_{jk}^2 - \dfrac{1}{npr}\sum_j T_j^2 - \dfrac{1}{npq}\sum_k T_k^2 + \dfrac{T^2}{N}$	$(q-1)(r-1)$	
3-factor interaction:			
trees-levels-quadrants	$\dfrac{1}{n}\sum_{ijk}T_{ijk}^2 - \dfrac{1}{nr}\sum_{ij}T_{ij}^2 - \dfrac{1}{nq}\sum_{ik}T_{ik}^2 - \dfrac{1}{np}\sum_{jk}T_{jk}^2 + \dfrac{1}{nqr}\sum_i T_i^2 + \dfrac{1}{npr}\sum_j T_j^2 + \dfrac{1}{npq}\sum_k T_k^2 - \dfrac{T^2}{N}$	$(p-1)(q-1)(r-1)$	
Between samples	$\sum_{ijk\alpha}x_{ijk\alpha}^2 - \dfrac{1}{n}\sum_{ijk}T_{ijk}^2$	$N-pqr = pqr(n-1)$	
Total	$\sum_{ijk\alpha}x_{ijk\alpha}^2 - \dfrac{T^2}{N}$	$N-1$	

freedom in the denominator. Hence, if a high-order interaction is not significant, it can be pooled with the error term to make a stronger error term in subsequent tests as we work up through the table. (Some statisticians do not advocate such pooling; if we do pool, an almost-significant component is added to the error term, dragging down the F ratio on subsequent tests; if we don't, there are too few degrees of freedom in the denominator; judgment is necessary in deciding on an appropriate test strategy.)

In the example we have been considering, it turned out that levels and trees made significant contributions to the total variance from sample to sample. Therefore the sampling scheme was designed so that the same trees were sampled from season to season and year to year, and sample branches were selected so that the amount of foliage selected from each level was proportional to the amount of foliage in each level. This way, bias was avoided. By taking corresponding steps, the reader can ensure representativeness in any particular sampling scheme with which he is concerned.

Considerable judgment is required in interpreting implications of ANOVA for sampling design because cost factors and statistical factors must both be considered in arriving at an optimum procedure, as will be explained later in this section. In general, ANOVA can serve as a guide to sample design in that any "fixed" and statistically significant main effect should be considered as a possible stratification variable, since stratification would then remove that component of variance from sampling error. Gains resulting from stratification are reduced if the factor represented by the stratification variables interacts with a "random" factor. In the cited example where the random effect of trees and the fixed effect of levels were both significant, one should consider a stratified sampling scheme in which an independent sample of trees is drawn for each level stratum. Such a design is clearly inefficient in allocation of effort, but in terms of error variance, it is more efficient than sampling $\frac{1}{4}$ as many trees each at all four levels. Had there been interaction between trees and levels, this gain in statistical efficiency would have been smaller.

Tests of random main effects are virtually useless, and it would be dangerous to interpret nonsignificance as meaning that this effect is zero. If trees had tested nonsignificant one should not infer that all trees are alike and hence that one need sample only a single tree!

The conclusion reached here, that the same trees should be sampled on successive occasions, does not follow from the analysis of variance of measurements made on a single occasion. Neither does it follow that levels should be sampled proportionally. The latter is a question of allocation of sampling effort, and will be discussed in a subsequent section.

Determination of optimum sample size

There are two components to the problem of determining optimum sample size. One is the determination of the total amount of sampling effort, and the second is the determination of how this effort will be allocated, say, as between x units of effort on each of 20 trees per plot or $4x$ units on each of 5 trees per plot. Here we are concerned only with the former problem. Several different lines of reasoning can be used to decide how much sampling is required. Happily, the most universally applicable of these turns out to also have the simplest mathematical foundation. The standard error of any statistic is computed from the formula $s/\sqrt{n}$ (that is, the sample standard deviation divided by the square root of the sample size). For a given standard deviation, the standard error is a slowly dropping function of the sample size. This can be used to determine what sample size we wish. It might be decided that we can tolerate a standard error equal to 10 percent of the mean (a sensible error to aim for in most field work). If we know the sample mean, and the sample standard deviation, the required sample size is then calculated from the relation

$$\frac{\sigma/\sqrt{n}}{\bar{x}} = 0.1 \qquad \text{or} \qquad n = \frac{100\,\sigma^2}{\bar{x}^2}$$

More complicated means of determining the optimum sampling level can be developed once a systems model has been developed and we know the error attached to various parts of the model.

Maximizing efficiency through stratified sampling

Where costs of moving from sample to sample within a block, or from sample to sample between blocks, are equal or too small to be of consequence, the variance can be minimized by using stratified random sampling. The object of stratified sampling is to increase sampling efficiency by dividing the population into several internally homogeneous and nonoverlapping sub-populations that together comprise the whole population. Because of the homogeneity within strata, a small sample from each can be used to get a precise estimate of the mean for the stratum. But how do we allocate sampling effort to strata, in order to maximize efficiency?

This question was answered by Neyman (1934), who showed that the variance of an estimated population mean is smallest for a given total sample size when the sizes of the samples for the various strata are proportional to the standard deviations of those strata.

Cost functions to optimize allocation of sampling effort

Suppose that costs are important, and are different from one stratum to another, or between moving from block to block, as opposed to moving from tree to tree within a block of trees. In such cases, we can design the sampling scheme so as to obtain minimum overall variance for a fixed total sampling cost, or minimum overall cost for a fixed variance. The theory is developed in three steps as follows (Cochran, 1963).

Step 1   We derive a formula for the variance of the mean from a simple random sample.

Step 2   From that, we get a formula for the variance of the mean from a stratified random sample, in terms of the variances of the component strata.

Step 3   Using the formula derived in step 2, we derive a third formula showing how to allocate efforts so as to obtain the minimum variance for a fixed cost, or minimum cost for a predetermined variance.

We require the following symbols:

N = number of units in population (total number of sampling units in all strata)

n = number of units in sample

(For example, a unit, for which we want to make a determination of larval population density, may be 10 square feet of branch surface.)

N_h = total number of sampling units for stratum h

n_h = number of sampling units actually sampled in stratum h

c_h = cost per unit sample of sampling stratum h

c_o = irreducible fixed overhead cost

C = total cost

S_h = standard deviation of stratum h

W_h = stratum weight, = N_h/N

$\bar{X}_{st}$ = stratified mean per sampling unit

X = estimated population total for study areas

$\bar{X}_h$ = stratum mean

The object of stratified random sampling is to minimize the total variance, subject to the restriction that the fixed cost is

$$c_1 n_1 + c_2 n_2 + c_3 n_3 + c_4 n_4 + \cdots + c_L n_L = C - c_o$$

It can be shown that

$$\frac{n_h}{n} = \frac{W_h S_h / \sqrt{c_h}}{\Sigma(W_h S_h / \sqrt{c_h})} \qquad (9.17)$$

This equation says what the ratio n_h/n ought to be to minimize overall sampling cost. Thus, the allocation of sampling effort to any stratum should be

1. a rectilinear function of N_h/N; that is, the proportion of all units in stratum h,

2. a rectilinear function of the standard deviation of stratum h,

3. inversely proportional to the square root of the cost of sampling stratum h; that is, the greater the cost of sampling h, the smaller the sample should be.

We shall illustrate the gain in efficiency produced by stratification by using a problem in aerial counting of caribou (Siniff and Skoog, 1964). In game populations, high population density strata may have variances 100 times those of low population density strata. Hence allocation of sampling effort based on even a rough estimate of relative variance will produce improved efficiency over random sampling.

In this particular study, W_h represents the proportion of the total number of sampling units in each stratum, and S_h represents the estimated number of caribou in each stratum. The number of 4-square-mile sample units selected for aerial counting from each of six strata was made proportional to $W_h S_h$. The reasoning behind using number of caribou instead of standard deviation for S_h is that the latter would not be known prior to sampling, but there would presumably be a functional relationship between numbers and standard deviation (or variance). After sampling, it turned out that the logarithm of the variance was a rectilinear function of the logarithm of sample mean.

For each of six strata, the following were calculated:

$$N_h, \; n_h, \; \bar{X}_h, \; s_h^2, \; W_h, \; W_h s_h^2, \; \frac{W_h s_h^2}{n_h}, \; \frac{W_h^2 s_h^2}{n_h}, \; \bar{X}_h W_h, \; W_h \bar{X}_h^2$$

From all six strata, the variance and mean of the population estimate were calculated as follows.

$$s_{xst}^2 = \Sigma \frac{W_h^2 s_h^2}{n_h} - \frac{\Sigma W_h s_h^2}{N} = 75.03 \qquad (9.18)$$

$$X = \bar{X}_{st} N = 54{,}452$$

Suppose that instead of using stratified random sampling, we had used simple random sampling. In this case the variance would have been

$$s_{ran}^2 = \frac{N - n}{nN}\left[\Sigma W_h s_h^2 - \Sigma \frac{W_h s_h^2}{nN} + \Sigma \frac{W_h^2 s_h^2}{n_h} + \Sigma W_h \bar{x}_h^2 - (W_h \bar{x}_h)^2\right] = 186.81$$

Thus, stratified random sampling was 186.81/75.03, or 2.49 times as efficient as simple random sampling in this instance. Clearly, stratification could greatly reduce costs in a large-scale sampling program.

Two-stage sampling

Cost functions can also be used to maximize sampling efficiency in two-stage sampling. For example, suppose we wish to determine the incidence of larvae on trees, and we wish to determine the optimum number of sample units per tree. Here we may think of the trees as constituting a sample from a population of trees, and we wish to take a subsample from the subpopulation of sample units within trees. Let C_t be the cost of sampling one primary unit, or tree; C_s be the cost of sampling a secondary unit, or sample unit; s_t^2 be the variance component for trees; and s_s^2 the variance component for samples. Then (from Cameron, 1951), the optimum number of samples to choose per tree is given by

$$N_{opt} = \frac{C_t}{C_s} \frac{s_s^2}{s_t^2} \tag{9.19}$$

Morris (1955) illustrates one possible application of (9.19). Costs were measured in man-hours. The cost of moving from one tree to another, C_t, was 1.25, and the cost of drawing and examining one sample unit, C_s, was 7.07. Thus, where $s_t^2 = 0.0348$, and $s_s^2 = 0.0170$,

$$N_{opt} = \sqrt{\frac{1.25}{7.07} \times \frac{0.0170}{0.0348}} = 0.3$$

Since fractional units cannot be drawn, it was decided to take one sample unit per tree. If N_{opt} lies between two integers N and $N + 1$, N_{opt} is taken to be $N + 1$ if $N_{opt}^2 > N(N + 1)$, but N if $N_{opt}^2 \leqslant N(N + 1)$ (Cameron, 1951).

9.2 INDIRECT MEASUREMENT

Where it is technically feasible to sample a population, because it can be photographed from the air against a background of snow, or because the individuals are sufficiently stationary so we can remove groups of them from trees, or count them where they live, the methods of the previous chapter can be applied. However, most animals cannot be sampled this way: They move too fast to be counted, they live in an environment in which they are hidden from us (water, litter on a forest floor, burrows among the roots of pasture grasses), or they are so mobile that we can never be sure if we are counting the same individuals repeatedly or a succession of different individuals. In such cases, we must resort to some indirect method of estimating population density, such as mark-recapture methods, catch-per-unit-effort methods, or methods based on change in composition of the stock. Before explaining any of these methods, some warnings seem worthwhile, all of which are the products of the wisdom which invariably accompanies hindsight.

1. With one exception (the virtual population technique), none of the methods of indirect population estimation given in the literature is to be trusted without carefully checking to determine if the assumptions on which it is based are being violated. The assumptions almost always are violated, and most of the material in this chapter is therefore devoted to an explanation of tricks that can be used to correct for these difficulties. A principal reason for the violation is that different individuals in a population behave very differently toward any device designed to catch them: some are "trap happy" and some are "trap shy." Therefore a technique of population estimation may only be estimating the size of that segment of the population stupid enough, smart enough (in the case of live traps that constitute a warm, safe home with lots of good food), or with the appropriate behavior patterns to be caught.

2. Another important factor determining whether an animal will be caught is its age. Animals of different ages have different tendencies to move around, disperse, and be captured by any particular type of gear. All population estimates should estimate the strength of each age group separately: Any estimate that pools data on all age groups will typically be so meaningless as to be worthless.

3. Any research program designed to indirectly estimate population sizes should be designed so that the same answers can be obtained by several different methods, all based on quite different assumptions. The novice researcher has started on the long, hard road to truth when he has applied four of the indirect-population-estimation techniques given in the literature to determine the number of 6-year-olds in his population, and has arrived at estimates of 604; 8,239; 23,928; and 106,000. Precisely such widely discrepant estimates are to be expected in many situations. If the budget is too small to allow for a tag-recapture program; for scale, otolith, tooth, eye lens, or other method of age determination; and for catch-per-unit-effort program, all concurrently, it may be too small for the research to be worthwhile at all,

except in special situations (for example, in small, stationary populations).

4. Much of the mathematical literature on models for indirect-estimation procedures is devoted to the development of formulas for estimating confidence limits, error estimates, and so on. In my opinion, effort expended by the biologist to compute error estimates is misguided, unless he has similar estimates from several different methods based on different techniques, in which case the error estimates may not give much new information anyway. For this reason, I am not discussing error estimation in this chapter, in order to free more space for discussion of techniques for detecting and correcting for violations of assumptions related to the biology of the population. I am confident that this point of view will seem sensible to anyone with considerable *field* experience in this area. In other words, in this field, problems created by low repeatability are so small by comparison with problems caused by bias, that only the latter are worthy of serious consideration.

The plan of this chapter will be to introduce the basic theory underlying each of six different types of population estimation. We shall attempt to give the reader some intuitive insight into the nature of the estimation process by explaining the procedures as games, which the reader can play for himself by using cloth balls (chenille pompons available from drapery stores or trimmers' suppliers) and plastic baby-bathtubs. By learning about indirect population estimation through game-playing, the student gains insight into the statistical characteristics of the various estimation procedures, and also is introduced to the notion that there is value in gaming, or simulation, as a means to understanding complicated processes. After explaining the basic theory, we will show how the methods break down in practice, and then illustrate procedures by which the violations of the assumptions of the methods can be handled.

Method 1 The best known of the indirect methods of population estimation is also the oldest, dating from Petersen (1896). This method and all the variations on it are based on the assumption that we can estimate the population of animals in a universe using the ratio of marked, previously released animals, to unmarked animals in each sample. Where

N = number of animals in universe
n = number of animals in sample
M = number of marked animals previously introduced into universe
m = number of recaptured (marked) animals in sample

then
$$\frac{N}{M} = \frac{n}{m} \quad \text{or} \quad N = \frac{Mn}{m}$$

This equation assumes that all the marked individuals are released at one time, and recaptured at one time. One reason for the numerous variants of the basic model is that in practice, marking and recapturing occur at a series of times. One of the most widespread methods for dealing with this problem comes from Schnabel (1938), who found that where the number of marked individuals is negligible relative to the total population, and we sum over all times from $t = 1$ to $t = T$ (the last time), N can be calculated from

$$N = \frac{\sum\limits_{t=1}^{T} n_t M_t}{\sum\limits_{t=1}^{T} m_t}$$

where n_t = total sample collected at tth time

M_t = number of marked individuals in population just prior to taking sample at tth time

m_t = number of marked individuals collected in sample at tth time

A great many additional extensions of the mark-recapture method are reviewed by Jones (1964), whose paper constitutes an excellent comprehensive introduction to the burgeoning literature in this field. Jolly (1965) also provides a key to the relevant new mathematical literature.

We can simulate use of the Petersen and Schnabel estimating procedures as follows.

Fill a plastic baby-bathtub with about 400 to 600 blue cloth balls (pompons). This will represent a lake containing a fish population of unknown size, or a forest containing a rodent population of unknown size, or a field containing a grasshopper population of unknown size. Now introduce 50 red cloth balls, and stir all the balls thoroughly with a paddle or large spoon. This is to simulate release of 50 marked individuals which scatter and mix with the resident population after being released. Now remove 50 of the balls at random. This can be done with some sampling device, such as a spoon, or by counting out 50 balls while blindfolded. The important point is to *make sure that the probability of a ball being included in the sample is independent of the color of the ball.* It is amazing how difficult it is to achieve this apparently easy task. Records will be kept as in Table 9.2. We record the total number of red balls in the sample in column 4. Now we assume that an individual cannot stand the constant handling involved in being marked, recaptured, released, recaptured, and so on indefinitely, such handling being of course quite traumatic for a wild animal. On the assumption that such constant handling would introduce supranormal probability of mortality, we will permanently remove a recaptured marked animal, to avoid the possibility of having it recaptured repeatedly. Thus, if our first sample contains 50 individuals (column 3) and 6 of these are red balls (marked individuals), we will discard these 6, and mark and return only 44 of the 50 in the sample. We simulate this step by replacing the 44 blue balls by 44 red balls, which we add to the tub of balls. The worksheet is made up as in Table 9.2, and the balls are once again mixed thoroughly before the next sample is withdrawn. The whole process is repeated until the Schnabel estimate stabilizes. In a classroom, of course, there would be division of labor to gain time: three students manage each tub, one to mix, draw samples, and count; one to keep records; and one to push a slide rule.

After the Schnabel estimate has stabilized, the following problems should be considered.

1. The Schnabel estimate may not in fact stabilize, but may show a systematic drift. Why? Are the balls of one color tending to stick to balls of the same color? What will this do to the estimates? Would some phenomenon analogous to this be expected to occur with wild animals?

2. Which is more satisfactory, the Petersen estimate or the Schnabel estimate? Why?

3. Are there any advantages to using either of these methods to estimate populations of wild animals? Does either method have any disadvantages peculiar to the method?

Method 2   This is the basic catch-per-unit-effort method introduced by DeLury (1947, 1951) and Mottley (1949). The situation contemplated in this, the basic version of the method, is that in which we keep track of the catch per unit effort and the cumulative total catch for a particular age group of a population within a relatively short period of time, during which we can ignore the effect of changing extrinsic variables, such as weather and natural mortality in the exploited population. (Method 6 will extend this method to where we follow particular year classes from year to year, and sort out the effects of harvesting and natural mortality.) The basic theory of method 2 is as follows. We require the following symbols:

t = tth interval

$C(t)$ = catch per unit effort during tth interval (expressed as numbers per rod-hour; tons of fish per standardized trawler-day, or any such unit)

$K(t)$ = total catch to beginning of tth interval

$N(t)$ = size of population at beginning of tth interval

TABLE 9.2

Sample computations for simulated application of Petersen and Schnabel mark-recapture methods

1	2	3	4	5	6	7	8	9	10
	No. of red balls intro- duced into population	Total no. of red and blue balls in sample	No. of red balls dis- covered in sample	Total no. of marked individuals (red balls) in universe at end of interval $t - 1$		Petersen estimate			Schnabel estimate
Time	R_t	n_t	m_t	$M_g = \sum\limits_{1}^{t-1} R_t - m_{t-1}$	$n_t M_t$	$\dfrac{n_t M_t}{m_t}$	$\Sigma n_t M_t$	Σm_t	$\dfrac{\Sigma n_t M_t}{m_t}$
t									
1	50								
2	44	50	6	50	2,500	417	2,500	6	417
3	38	50	12	88	4,400	367	6,900	18	383

N = size of population at beginning of first interval

k = catchability during all intervals (we assume k is constant from one time to another, on the average, and that there is no mortality or recruitment during period N is computed); k thus = proportion of N caught by 1 unit of effort

Then,
$$C(t) = kN(t) \qquad (9.20)$$

$$N(t) = N - K(t) \qquad (9.21)$$

Substituting for $N(t)$ in (9.20),
$$C(t) = kN - kK(t) \qquad (9.22)$$

If we plot $C(t)$ values against $K(t)$ values (or do a linear regression analysis), we can get an estimate of the slope, k, and from this and the intercept on the $C(t)$ axis,

$$N = \frac{kN}{k} \qquad (9.23)$$

This method can be simulated as follows. Start with a plastic baby-bathtub containing "water," (about 1,000 blue cloth balls). Into the "water" introduce about 200 red balls ("fish"). Mix the "fish" and the "water" well. Then, by taking samples of about 100 balls at a time, try to estimate the number of "fish" in the "water," using the foregoing theory. It will probably be necessary to take 10 samples of 100 balls each, to get a stable estimate. This in itself is a revealing commentary on the difficulties associated with the method, because when 1,000 balls have been sampled, about 10/12 of the 200 red balls will have been removed. (Note that in accord with the theory, after each sample, the blue balls are returned to the tub but the red balls are not. We assume that we are running a commercial fishery and our fish are sold and eaten, rather than returned to the water.)

After conducting this exercise, the following questions should be considered.

1. What difficulties might arise in applying this method to estimation of natural populations of fish, mammals, birds, insects, or other organisms?

2. What advantages do we gain in extending this method to changes in catch per unit effort of a year class from one year to another, as in method 6, rather than analyzing data on changes in catch per unit effort in a year class within a season, as in this method?

Method 3 This method is based on the fact that for several kinds of exploited populations, one subgroup within the population is harvested more intensively than other subgroups. For instance, the bulk of the hunting pressure against big game and certain birds is directed against the males. This selective mortality is exploited to give us information on the change in composition of the stock, from which the initial population size is calculated. The method was first conceived by Kelker (1940, 1942) for use with deer and other wildlife populations, and has subsequently been applied to pheasants (Selleck and Hart, 1957) and fish (Lander, 1962). The method has been put on a rigorous mathematical basis by Chapman (1955). The theory of this estimation procedure is as follows. We require the following symbols (subscripts 1 and 2 represent samples taken prior to and after hunting, respectively):

N_1 = total population prior to hunting
F_1 = female population prior to hunting
M_2 = male population after hunting
M_H = number of males killed by hunting
f_1 = number of females in sample prior to hunting
m_1 = number of males in sample prior to hunting

Assumptions: Other forms of mortality are very small relative to hunting mortality during the hunting season, and can be ignored.

$$\frac{f_1}{m_1} = \frac{F_1}{M_1} \qquad \frac{f_2}{m_2} = \frac{F_2}{M_2}$$

Derivation of formulae: Then the ratio of males to females, before and after hunting, respectively, is given by

$$r_1 = \frac{m_1}{f_1} \qquad \text{and} \qquad r_2 = \frac{m_2}{f_2}$$

It follows that given the assumptions, and where only males are hunted,

$$\frac{M_1}{r_1} = \frac{M_H}{r_1 - r_2}$$

or

$$M_1 = \frac{r_1 M_H}{r_1 - r_2}$$

$$F_1 = \frac{f_1}{m_1}\left(\frac{r_1 M_H}{r_1 - r_2}\right) = \frac{M_H}{r_1 - r_2}$$

Suppose, now, that instead of killing all one sex and none of the other there is a different proportion of each group killed.

F_1 could be computed from the following relationship.

$$\frac{m_2(F_1 - F_H)}{f_2} = M_2 = \frac{m_1 F_1}{f_1} - M_H$$

or
$$F_1 - F_H = \frac{f_2}{m_2}\left(\frac{m_1 F_1}{f_1} - M_H\right)$$

or
$$F_1 = F_H + \frac{f_2}{m_2}\left(\frac{m_1 F_1}{f_1} - M_H\right)$$

Similarly, M_1 can be computed.

This procedure can be simulated as follows. Take a plastic tub, representing a game management area, and place in it cloth balls of two colors, representing male and female deer (or pheasants, if this is more appealing). First, simulate the situation in which legislation forbids the hunting of females. The object will be to calculate absolute abundance of both sexes prior to hunting, by using sex ratios based on samples taken before and after "hunting," and a count of the "male animals" killed in the hunting season. In the second part of the exercise, we will assume that both sexes are hunted, but one is hunted far more heavily than the other. The detailed procedure is as follows.

1. Take repeated samples of "male" and "female" deer until the ratio m_1/f_1 stabilizes.
2. "Hunt," by removing 100 "male deer."
3. Take repeated samples of "male" and "female" deer until the ratio m_2/f_2 stabilizes.
4. Compute M_1 and F_1.
5. Count all the "deer" in the "management area," recording data on males and females separately, to see if the M_1 and F_1 estimates were correct.
6. Repeat this exercise, only this time remove 100 "males" and 50 "females."

After performing this exercise, consider the following problems.

1. Derive the formula for M_1, in the case where both sexes are harvested.
2. Do you feel this method is accurate and precise? Why?
3. Can you see any problems involved in applying either of these techniques?
4. Which of the three categories of techniques, mark-recapture, catch per unit effort, or differential mortality, is best for population estimation and why?
5. Explain a statistical procedure to compute the precision (repeatability) of the estimates of F_1 and M_1.

6. What happens to this method as the proportion of each sex removed by hunting approximates more closely the proportion of that sex prior to hunting?

Method 4   The fourth method is based on the idea that animals leave various signs behind them. If we know the number of signs left per animal per unit area per unit time, we can measure the number of signs left, and from this calculate the number of animals that must be in the area. This method could be applied to the number of leaves browsed per unit time by large-game defoliators, or the number of carcasses left by large carnivores, or the number of pellets or dung piles left by mammals, or the number of grass pellets dropped by insects. The method seems to be replete with potential sources of error. Nevertheless, in particular circumstances, it might provide a useful independent check on other methods of population estimation. For example, a series of sample plots could be cleared of deer pellets. Then we could keep a daily record of the number of pellet groups dropped within the same plots, and remove the pellets found at the time of sampling each day. From this, if we know the size of the sample plot, and the normal number of pellets dropped per deer per day, we can compute the deer population density. It is an entertaining exercise for the reader to make a list of the possible sources of bias in this method. Some of the sources are discussed by Van Etten and Bennett (1965).

Method 5   This method was named the "virtual" population estimation procedure by Fry (1949) who introduced the concept into the modern literature. While there are serious difficulties involved in applying the method, as Bishop (1959) has pointed out, it has two great advantages. First, this is the only method that can be relied on absolutely to give us a rock-bottom minimum estimate of the population size; any method yielding a population estimate smaller than that given by virtual population technique *must be wrong*. Second, as Paloheimo (1958) has demonstrated, it is possible to combine the basic idea of the virtual population technique with certain ideas from other techniques and arrive at a procedure that is very powerful indeed.

The basic idea is simplicity itself, in the light of the wisdom which invariably accompanies hindsight (but it took until 1949 before someone thought of it). Suppose a year class of animals is born in 1939, and is first harvested as 4-year-olds in 1943, as 5-year-olds in 1944, 6-year-olds in 1945, and so on. Now, if we are dealing with a closed population (that is, one into which there is no immigration, and out of which there is no emigration), we can make the following statement with absolute certainty of being correct. The absolute minimum number of 4-year-olds that could possibly have been present at the beginning of 1943 is the sum of the 4-year-olds caught in 1943, the 5-year-olds caught in 1944, the 6-year-olds caught in 1945, and so on. In other words, the minimum number of animals of a particular age that could possibly have been present in a certain place at a certain time is the total number of animals of that age group that were ever caught and killed.

Method 6 The final basic method is an extension of method 2. The underlying idea is that for a sequence of pairs of years, we can compute ratios of the following form:

$$\frac{\text{Catch per unit effort of 6-year-olds in year } y + 1}{\text{Catch per unit effort of 5-year-olds in year } y}$$

and can establish a regression of these ratios on the effort in the first of the pair of years. From the slope and intercept of the regression line, we can compute, for example, the fishing mortality coefficient and natural mortality coefficient, respectively. The method was developed originally by Beverton and Holt (1957) and independently by Widrig (1954); important subsequent papers were published by Paloheimo (1958) and Murphy (1965). While some of the new refinements are more powerful, it is perhaps easiest to understand the basic ideas underlying this method in terms of the original derivation of Beverton and Holt; hence that is presented here.

Where Z = instantaneous total mortality coefficient
F = instantaneous fishing mortality coefficient
M = instantaneous natural mortality coefficient
$_aN_t$ = abundance of animals of age a at time t

then $$Z = F + M \qquad \text{and} \qquad \frac{dN}{dt} = -ZN$$

and $$_{a+1}N_{t+1} = {_aN_t}e^{-(F+M)} \qquad (9.24)$$

Now let f represent the fishing intensity (number of units of gear), and c represent the instantaneous mortality caused by one unit of gear, then we may assume that

$$_{a+1}N_{t+1} = {_aN_t}e^{-(cf+M)} \qquad (9.25)$$

Beverton and Holt noted that where $_aN_t$ represents the abundance of a-aged animals at the beginning of a year, the mean abundance throughout the year is given by

$$_a\bar{N}_t = \int_0^1 \frac{{_aN_t}e^{-(cf+M)}dt}{dt} = \frac{{_aN_t}}{cf + M}\left[1 - e^{-(cf+M)}\right] \qquad (9.26)$$

Now let us consider the same year class of fish in two successive seasons. In the first season, the year class will be a years old, and will have the fishing and natural mortality coefficients c_a and M_a. (We are assuming that such coefficients vary from age to age, but, for a given age, do not change in a systematic fashion from year to year. That is, a continued unidirectional change in M_a caused by an environmental change will invalidate the following theory.) The fishing intensity exerted in year t will be indicated by f_t. The mean abundance of the year class a years of age in year t is given by

$$_a\bar{N}_t = \frac{_aN_t}{c_a f_t + M_a}\left[1 - e^{-(c_a f_t + M_a)}\right] \tag{9.27}$$

From (9.25) and (9.26), the mean abundance of $(a + 1)$-year-olds during year $t + 1$, expressed in terms of the a-year-olds at the beginning of year t, is given by

$$_{a+1}\bar{N}_{t+1} = \frac{_aN_t e^{-(c_a f_t + M_a)}\left[1 - e^{-(c_{a+1} f_{t+1} + M_{a+1})}\right]}{c_{a+1} f_{t+1} + M_{a+1}} \tag{9.28}$$

From (9.27) and (9.28), and the relation

$$\frac{1}{e^{-(cf + M)}} = e^{(cf + M)}$$

we may write

$$\frac{_a\bar{N}_t}{_{a+1}\bar{N}_{t+1}} = \left(\frac{c_{a+1} f_{t+1} + M_{a+1}}{c_a f_t + M_a}\right) e^{(c_a f_t + M_a)}\left(\frac{1 - e^{-(c_a f_t + M_a)}}{1 - e^{-(c_{a+1} f_{t+1} + M_{a+1})}}\right) \tag{9.29}$$

Taking logarithms across this expression, to a first approximation, we get

$$\log\left(\frac{_a\bar{N}_t}{_{a+1}\bar{N}_{t+1}}\right) = M_a + c_a f_t \tag{9.30}$$

This is an old friend, the linear regression equation $Y = b_0 + b_1 X$, and hence we can obtain population estimates by means of an iterative process as follows. First, we obtain catch-per-unit-effort indices of abundance of the population for several ages in each of several years. Then, from a regression analysis of

$$\log\left(\frac{_a\bar{N}_t}{_{a+1}\bar{N}_{t+1}}\right)$$

against effort, f_{t+1}, in each of the years, we obtain first-trial values of M_{a+1}

and c_{a+1}. We can repeat for animals 1-year younger, and obtain M_a and c_a. This gives us the values we need to repeat the regression analysis, this time adding a correction factor to the dependent variable, thus

$$\log\left(\frac{{}_aN_t}{{}_{a+1}N_{t+1}}\right) + \log\left[\frac{(c_af_t + M_a)(1 - e^{-(c_{a+1}f_{t+1}+M_{a+1})})}{(c_{a+1}f_{t+1} + M_{a+1})(1 - e^{-(c_af_t+M_a)})}\right] = M_a + c_af_t \quad (9.31)$$

This iterative process is repeated until M_a and c_a stabilize. As Widrig (1954) has pointed out, we now need to convert the instantaneous-fishing-mortality rate, $F = cf$, into an annual rate of exploitation, μ, in order to compute N, the extant population, from C, the catch. We proceed as follows. We obtain a, the annual total death rate from

$$a = 1 - e^{-(F+M)} \quad (9.32)$$

Then μ, the rate of exploitation, is given by

$$\frac{\mu}{a} = \frac{F}{Z} \quad \text{or} \quad \mu = \frac{ac_af_t}{Z} \quad (9.33)$$

where Z is the total instantaneous mortality. By definition, $\mu N = C$, or

$$N = \frac{C}{\mu} \quad (9.34)$$

Thus from (9.32), (9.33), and (9.34), we can obtain N. This is the population size in the middle of a season (${}_a\bar{N}_t$). The population size at the beginning of the season may be obtained by transposing (9.27) to yield

$$_aN_t = {}_a\bar{N}_t \frac{c_af_t + M_a}{1 - e^{-(c_af_t+M_a)}} \quad (9.35)$$

This completes a discussion of the six basic methods of indirect population estimation. As the reader might guess, many additional methods can be generated by using combinations of the basic methods. Methods 5 and 6 have been combined (Paloheimo, 1958; Watt, 1959); combinations of 1 and 6 have been suggested (Murphy, 1965); and many other combinations are possible and would be very useful. However, a more important issue is that of bias, and methods of dealing with it.

Bias in indirect population estimation originates in one or both of two broad categories of factors. We will first explain what we mean by each of these, then illustrate the various types of bias, and explain and give examples of how they can be dealt with.

The first group of bias-producing factors arises because of the interactions between population behavior and the specific methods and type of gear used to harvest the population. For example, as animals age, they tend to move about more, thus the unit of area represented by animals caught by different equipment is different. In general, only a segment of any population is vulnerable to being sampled by any specific type of gear, and one of the greatest problems in indirect population estimation is determining how large this segment is relative to the total population. A related problem is that of determining the area to which a given population estimate must be related. Different kinds of gear have different degrees of effectiveness; therefore we must devise methods that weigh each type by its true effectiveness per unit relative to some standard unit. The estimating techniques discussed previously have assumed a rectilinear relationship between catch per unit effort and abundance; if, in fact, catch per unit effort is a diminishing proportion of abundance as abundance increases, we need some method of dealing with this phenomenon. Certain individuals within a population may be trap happy or trap shy; this can introduce tremendous bias because our inferences about the population may be based on a tiny proportion of its members who are sampled repeatedly, and are atypical in every salient respect. Bias is also introduced in mark-recapture studies if an individual becomes unusual in any respect by virtue of being marked. For example, if the tag, fin clip, or toe cut increases the probability of mortality due to disease, or if a bright-red tag increases the probability that a fish will be attacked by other fish, the probability of recovering the marked individual is clearly lowered.

A second, and more complicated, group of bias-producing factors are products of three-factor interactions among environmental factors, the population being estimated, and the characteristics of the gear and the procedures for operating it. Weather can affect "catchability," and hence changes in catch per unit effort may only in part reflect changes in absolute abundance. Changing weather from year to year may produce significant increases or decreases in the size of animals of a given age. Since size is an important factor in determining the interaction between sampling gear and an animal, such weather changes may have the ultimate effect of changing catch per unit effort where there has been no change in abundance. For example, if cold weather during the third year of life causes an entire year class to consist of very much undersized 4-year-olds the following season, they may be small enough to slip between the meshes of gill nets that would normally be small enough to enmesh 4-year-olds. Finally, changing environmental conditions may affect the way in which a population distributes itself in space at given times of the year. Certain weather conditions may tend to keep a population away from where it is normally encountered and caught in June; again, catch per unit effort will be a biased indicator of absolute population size.

We will now indicate some techniques for circumventing these types of bias. In Table 9.3 we indicate one possible way of correcting population estimates for increased movement of older animals. Mark-recapture estimates of fish caught in traps or pound nets are used to estimate the population of small fish within a few score yards of the net (for fish 2 years old or less); in the case of older fish, net-recaptures are used to estimate a population roaming over a considerable distance. Where numbered tags are placed on individual animals (the time and place of release and recapture are recorded for each animal separately), we can calculate the mean minimum distance in miles per day that must have been traveled by animals of different ages, if we use one of the available techniques to age the animals. From calculations used in Table 9.3, we obtain the probability of capturing an individual of a given age, relative to the probability of capturing an individual of the age group moving around the most. That is, we assume that if 2-year-olds travel x miles per day, and 7-year-olds travel y miles per day, then from the formula for the area of a circle, $A = \pi r^2$, we are only sampling x^2/y^2 as much of the area for 2-year-olds as we are for 7-year-olds. The process is illustrated in Table 9.3. Correction factors by which mark-recapture estimates (based on trap nets) must be multiplied are obtained in the final column. We assume that the 7-year-olds travel enough so that our nets in effect sample the entire population of that age. Then we obtain the correction factors for the other age groups by dividing the relative area covered by the 7-year-olds by the relative area covered by the other age groups. As a check on the validity of this reasoning, in Table 9.4 we compare the correction factors so obtained with those obtained by a quite different method. Anglers' captures of fish released from nets are based on a sample that presumably had ample opportunity to become thoroughly mixed with the rest of the population, because anglers fish from a large number of small boats scattered all over a body of water (that is, the randomizing procedure should be more thorough than in the case of a small number of fixed nets). The ratio of mark-recapture estimates to that from anglers should be a measure, age group by age group, of the mixing involved in net recaptures relative to that involved in anglers' captures. Putting it another way, the ratio is a measure of the extent to which we are sampling the entire population by nets. The reciprocals of these ratios, expressed as a multiple of the reciprocal of the ratio for 7-year-olds, gives us a measure of the correction factor needed for each age group. The data are inadequate to give us a good match in the last two columns of Table 9.4, but the two columns show the same general pattern.

We have mentioned that use of catch-per-unit-effort statistics depends on the assumption that catch per unit effort is a rectilinear function of absolute

TABLE 9.3

Relative area swept by nets for individual South Bay smallmouth bass of different ages (Watt, 1959)

Age, in years	Number of individuals used in calculations	Total distance traveled by all individuals, in miles	Total time taken by all individuals, in days	Mean minimum distance, in miles per day (r)	Relative area covered, in square miles (r^2)	Correction factor by which estimates must be multiplied to produce comparable data for all age groups
2	6	0.6	44	0.014	0.0002	101.00
3	96	46.8	943	0.050	0.0025	8.09
4	66	48.1	639	0.075	0.0056	3.61
5	60	47.8	543	0.088	0.0078	2.59
6	54	75.7	721	0.105	0.0110	1.84
7	49	82.3	578	0.142	0.0202	1.00

TABLE 9.4

Comparison of two methods of obtaining factors for correcting population estimates, based on catches of South Bay smallmouth bass (Watt, 1959)

Age, in years	Schnabel estimates from net recaptures, during 1952, 1953, 1954, 1955 (total)	Petersen estimates from anglers' recaptures during 1952, 1953, 1954, 1955 (total)	Schnabel total / Petersen total	Correction factors	
				Table 9.4	Table 9.3
4	791	7,400	0.107	4.38	3.61
5	3,094	9,100	0.340	1.38	2.59
6	1,868	6,800	0.275	1.69	1.84
7	186	400	0.465	1.00	1.00

TABLE 9.5

Analysis to determine nature of relationship between catch per unit effort, d, and virtual abundance, V, (Watt, 1959)

Age, in years	Statistics on equations $d = a + bV$ and $V = g + hd$					Statistics on equations $\ln d = a + b \ln V$ and $\ln V = g + h \ln d$				
	a	b	g	h	Correlation coefficient r	a	b	g	h	Correlation coefficient r
4	0.0517	0.000315	−50.9	2,902.	0.957	−9.326	1.243	7.725	0.839	0.971
5	0.0511	0.000296	−131.1	3,258.	0.982	−8.000	1.013	7.825	0.940	0.976
6	0.0186	0.000433	17.2	2,087.	0.951	−7.917	1.036	7.588	0.934	0.984
7	−0.0037	0.000497	24.6	1,815.	0.950	−8.160	1.092	7.365	0.877	0.979
8	−0.0084	0.000636	23.8	1,260.	0.895	−10.110	1.566	6.322	0.603	0.972
Data on ages 4–9 pooled	0.0298	0.000329	−58.52	2,872.	0.973	−7.706	0.989	7.761	0.999	0.994

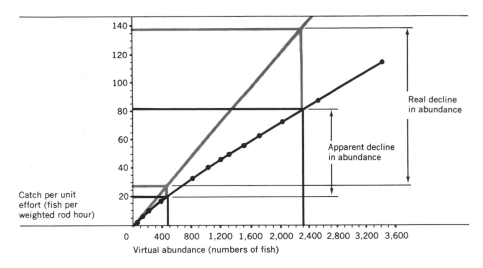

FIGURE 9.1

Relation between catch per unit effort and abundance (Watt, 1959).

abundance. To check this assumption, Watt (1959), using data on a population of smallmouth bass that had been studied intensively for a decade, did statistical analysis of the relation between catch-per-unit-effort measures of abundance and virtual-population measures of abundance (Table 9.5). Figure 9.1 illustrates the relationship between catch per unit effort and virtual abundance, which is slightly curved. Figures on virtual population of fish of age a in year t had to be used instead of the absolute population sizes $_aN_t$, because no reliable measure of the latter was available. However, if enough years and ages of virtual population estimates, $_aV_t$, are available, the effects of temporal changes in effort, vulnerability to gear, and natural mortality rate, which alter $_aV_t$, are minimized. It is clear from Table 9.5 that the relationship between catch per unit effort and virtual population for any given age is not in fact linear, but is best described by the equation

$$\ln {_ad_t} = a + b \ln {_aV_t}$$

where $_ad_t$ represents catch per unit effort of a-aged fish in year t. It is clear from Table 9.5 that the parameters a and b are functions of age. There are several possible explanations for this relationship. With increased population densities, fish may tend increasingly to move about in large, dense schools; also, as Brock and Riffenburgh (1960) have pointed out, schooling is a possible factor in reducing predation in species of prey fish. In the case of smallmouth bass, the anglers in small boats constitute the predators, and the higher the degree of schooling, the lower the proportion of fishing mortality will be. Figure 9.1 depicts the same type of phenomenon as Fig. 6.1: An increase in prey density produces a lowered kill rate per exposed prey. As explained in Secs. 6.1 and 11.6, this occurs because the time taken to attack one prey imposes an upper asymptote on the attack rate per predator, and, as prey density increases, the curve for numbers attacked bends over to meet this asymptote. For example, the fisherman is affected by the time spent handling prey in both angling and netting, because time is withdrawn from active fishing whenever the line is being reeled in with a fish, or fish are removed from a net. The greater the fish (prey) density, the greater will be the proportion of the time spent in such non-fishing activities, and the lower the predation rate per unit prey density will be.

Figure 9.1 indicates the implications of this curvilinear relationship. The dotted line would occur if the relationship were rectilinear. The curved line is the one actually found. Now suppose that the virtual abundance declines 80 percent from one year to the next, as shown. Instead of declining 80 percent also, the apparent decline in catch per unit effort will only be 75 percent, as shown. In order to correct for this effect, two alternative procedures are available (Watt, 1959). The simpler is to use virtual-population estimates instead of catch-per-unit-effort estimates as measures of abundance. This is what Paloheimo (1958) has also done. The other uses the regression relations in Table 9.5 to correct the catch-per-unit-effort measures for the curvature in Fig. 9.1. Thus, we assume that

$$\frac{_a\bar{N}_t}{_{a+1}\bar{N}_{t+1}} = \frac{\theta_a V_t}{_{a+1}V_{t+1}} = \frac{\exp(_a g + {}_a h \ln {}_a d)}{\exp(_{a+1} g + {}_{a+1} h \ln {}_{a+1} d)}$$

The ratios of virtual populations calculated from the above equation are the input for the Beverton-Holt type of analysis previously discussed.

To illustrate the incomplete sampling that occurs when we use mark-recapture methods, Watt (1959) discovered that the estimation procedure outlined above gave estimates up to 3.4 times those from mark-recapture studies, in the case of 5-year-old fish.

Our most complete knowledge of how animal behavior produces biased population estimates comes from mark-recapture work with small mammals. Obviously, all methods for estimating populations depend on the assumption that the first group caught—and all subsequent groups caught—represents a random sample of the population. If, for example, some individuals become trap happy, the probability of catching them repeatedly is too high, and we are not estimating the size of the extant population, but only the size of the trap happy component of the extant population. Young, Neess, and Emlen

(1952) have shown that this is precisely what happens. They trapped, released, and recaptured rodents and kept track of the number of times that each individual was recaptured. We can predict from the Poisson distribution the number of individuals that should be recaptured no, one, two, three times, etc., knowing only the mean number of times all individuals were recaptured, m. Where x takes the values 0, 1, 2, 3, 4, 5, etc., and represents the number of times an animal was recaptured, and $p(x)$ represents the probability of being recaptured x times, then from the Poisson distribution,

$$p(x) = \frac{e^{-m} m^x}{x!}$$

Comparing the observed frequency distribution of recaptures with that predicted by the Poisson, Young, Neess, and Emlen found that there were too many animals caught once and never again (trap shy) and too many were recaptured several times (trap happy). The explanation for this phenomenon may lie in the work of Calhoun (1963). He studied Norway rats and found that the tendency to enter traps is dependent on the psychological history of individual rats. If a rat has been subjected to unusual social stress, this will be reflected in reduced growth for a given age. Calhoun measured the degree of inhibition of growth by a "maturity index." This measures how old an individual is for its size. A high maturity index means a small individual is unusually old for its size, and hence has had inhibited growth due, on the average, to increased social stress during life. Calhoun's data show a positive correlation between the maturity index and a trap-proneness index. This means that conditions of social stress that inhibit growth reduce the tendency of rats to avoid noxious stimuli or situations deleterious to them. The implication is serious: If we try to deduce the characteristics of a population from trapped individuals, we not only underestimate population size, but we also underestimate size of individuals for a given age, and perhaps also underestimate the growth rate.

We can compensate for such difficulties by means of stratification. A simple example will illustrate the effect.

Case 1   We release 5,000 individuals, and do not record any characteristics of the individuals. We recapture 500 in a new collection of 1,000; so we estimate our population size as

$$\frac{1,000}{500} \times 5,000 = 10,000$$

Case 2   We release 5,000 individuals, but note that 4,000 of these are stunted, and 1,000 are "normal" size for their age. We capture 1,000 individuals of which 500 are recaptures, as before, but note that 600 individuals are stunted, including 480 recaptures, and 400 are "normal," including 20 recaptures. Thus, our population estimate is:

$$Stunted: \quad \frac{600}{480} \times 4,000 = \quad 5,000$$

$$Normal: \quad \frac{400}{20} \times 1,000 = 20,000$$

$$Total: \qquad\qquad\qquad \overline{25,000}$$

A fundamental difficulty with all methods of population estimation is that we need to know the area to which a given population estimate is applicable. Brant (1962) recognized that the problem of relating population estimates to areas was central in estimation. He also realized that the type of vegetation had a profound effect on population density. He used an elaborate grid layout of traps, and computed the distance traveled between successive captures. From this computation, he tried to calculate the size and shape of the area covered by the small mammals captured on a trap line.

This shape is calculated on the assumption that the average distance between successive captures/unit time (92 feet) could be used to define the size of this area. He assumed that the area covered by the trap line was that enclosed by a line parallel to the trap line and 92 feet on either side of it, and 92 feet out from the end traps. Then the catch on the trap line could be divided by the area covered to get a population density.

Brant found that these features affect rodent densities:

1. *Microtus* becomes most abundant in areas where grass is the most common cover.

2. Areas of highest population peaks can have the lowest populations at a later date. In areas of less favorable habitat, populations were more stable, and did not reach such marked peaks.

Note that in *all* population estimation research, we should try to get the same estimate by using several techniques. Brant's use of the average distance is evidently reasonably satisfactory, from the following figures on *Peromyscus* abundance.

Day of study	Density calculated from line trap and average distance	Density estimated from grid live trapping
166–168	2.67	1.75
313–315	0.76	0.88
351–353	0.66	0.71
454–456	0.17	0.29

Different kinds of traps also have different catchabilities. In this study, buried cans caught 24 percent more rodents per unit time than did the small Sherman traps.

We will now consider the second class of factors that produce bias in indirect population estimation: factors produced by interactions among environmental factors, the population being estimated, and the gear. If weather fluctuations are causing significant differences from year to year in the size of animals of a given age, it may be necessary to correct population estimates because certain age groups of the population have a subnormal chance of being caught because they are too small. The way to deal with this situation is to make use of any available technique for reconstructing the size distribution of, say, 4-year-olds for all the years in which we are interested. Then we can compare the actual size distribution with the size distribution of 4-year-olds caught in each year, and use a planimeter to determine the proportion of the frequency distribution of 4-year-olds actually vulnerable to capture in each year. Reconstruction of the actual frequency distribution (versus size) curve can be accomplished by using a computer and making use of the well-known relationships between the length and weight of animals at each year of life, and the annuli radii laid down in hard parts of the animal, such as fish scales (Watt, 1959).

An even more complex type of interaction is illustrated in Table 9.6. The raw data are in lines 1 to 4. We could get an estimate of the pounds of cisco per lift during June and July by dividing the total poundage taken by the total number of lifts in June and July. The simple means which result (line 7) suggest that, on the basis of this one net, there were 85 percent as many fish present in 1954 as in 1953. However, these indices are misleading (Watt, 1956).

TABLE 9.6

Cisco landings in pound net 5 in South Bay (Watt, 1956)

Lines	Catch	1953	1954
1	Pounds taken during June	6,003	4,669
2	Pounds taken during July	351	366
3	Lifts during June	8	6
4	Lifts during July	6	7
5	Pounds taken in June and July	6,345	5,035
6	Lifts during June and July	14	13
7	Pounds per lift during June and July	454	387
8	Pounds per lift in June	750	778
9	Pounds per lift in July	59	52

Lines 8 and 9 give the pounds per lift in June and July, respectively. In both years the availability of fish was much less in July than in June, presumably due to some behavioral factor. In this case the decreased availability is caused by migration into deeper water in response to increasing temperature in the upper layer. Obviously a shift from year to year in the bulk of the fishing effort from June to July would result in a decreased annual catch per unit effort. We can correct for this by treating the June and July catches per unit effort separately each year. The resulting annual indices of catch per unit effort give a very different picture from the unweighted means in line 7. The figures in line 8 show that the June abundance as measured by this net actually increased by 4 percent from 1953 to 1954.

All of the foregoing problems are made even more complex by two additional problems. First, in the case of fish, catch per unit effort may be related to the interaction of the interval between clearing nets, and the population density (Kennedy, 1951). Presumably this principle may operate for all types of nets and traps.

Second, we may have to weight all catch-per-unit-effort data relative to some standard unit because, for example, cotton, linen, multifilament and monofilament nylon gill nets all have different catchabilities, fishing power depends on the size of the trawler, and angling efficiency is a diminishing function of the number of anglers per boat (Table 9.7).

TABLE 9.7

Catch per unit effort of smallmouth bass by anglers as a function of number of anglers per boat (Watt, 1959)

Number of men in boat	Total fish caught	Total hours fishing	Fish per man per hour	Weight for effort
1	182	172	1.058	1.00
2	2,633	1,853	0.710	0.67
3	2,429	1,555	0.521	0.49
4	2,129	1,115	0.477	0.45
5	657	389	0.338	0.32
6	279	163	0.285	0.27

REFERENCES

Beverton, R. J. H., and S. J. Holt: On the Dynamics of Exploited Fish Populations, *Fish. Invest., London*, ser. 2, 1–533 (1957).

Bishop, Y. M. M.: Errors in Estimates of Mortality Obtained from Virtual Populations, *J. Fish. Res. Bd. Can.*, **16**:73–90 (1959).

Brant, D. H.: Measures of the Movements and Population Densities of Small Rodents, *Univ. Calif. Publ. Zool.*, **62**:105–184 (1962).

Brock, V. E., and R. H. Riffenburgh: Fish Schooling: A Possible Factor in Reducing Predation, *Extrait du J. Conseil Perm. Intern. Exploration Mer*, **25**:307–317 (1960).

Calhoun, J. B.: The Ecology and Sociology of the Norway Rat, *U.S. Dept. Health, Education and Welfare Public Health Serv. Publ.*, Bethesda, Md., no. 1008 (1963).

Cameron, J. M.: Use of Variance Components in Preparing Schedules for the Sampling of Baled Wool, *Biometrics*, **7**:83–96 (1951).

Chapman, D. G.: Population Estimation Based on Change of Composition Caused by a Selective Removal, *Biometrika*, **42**:279–290 (1955).

Cochran, W. G.: "Sampling Techniques," 2d ed., John Wiley & Sons, Inc., New York, 1963.

DeLury, D. B.: On the Estimation of Biological Populations, *Biometrics*, **3**:145–167 (1947).

————: On the Planning of Experiments for the Estimation of Fish Populations, *J. Fish. Res. Bd. Can.*, **8**:281–307 (1951).

Fry, F. E. J.: Statistics of a Lake Trout Fishery, *Biometrics*, **5**:27–67 (1949).

Jolly, G. M.: Explicit Estimates from Capture-Recapture Data with Both Death and Immigration-Stochastic Model, *Biometrika*, **52**:225–257 (1965).

Jones, R.: A Review of Methods of Estimating Population Size from Marking Experiments, *Rapports et Proces Verbaux, J. Conseil Perm. Intern. Exploration Mer*, **155**:202–209 (1964).

Kelker, G. H.: Estimating Deer Populations by a Differential Hunting Loss in the Sexes, *Proc. Utah Acad. Sci., Arts and Letters*, **17**:65–69 (1940).

————: Sex-Ratio Equations and Formulas for Determining Wildlife Populations, *Proc. Utah Acad. Sci., Arts and Letters*, **19**:189–198 (1942).

Kennedy, W. A.: The Relationship of Fishing Effort by Gill Nets to the Interval between Lifts, *J. Fish. Res. Bd. Can.*, **8**:264–274 (1951).

Lander, R. H.: A Method of Estimating Mortality Rates from Change in Composition, *J. Fish. Res. Bd. Can.*, **19**:159–168 (1962).

Morris, R. F.: The Development of Sampling Techniques for Forest Insect Defoliators, with Particular Reference to the Spruce Budworm, *Can. J. Zool.*, **33**:225–294 (1955).

Mottley, C. M.: The Statistical Analysis of Creel-census Data, *Trans. Am. Fish. Soc.*, **76**:290–300 (1949).

Murphy, G. I.: A Solution of the Catch Equation, *J. Fish. Res. Bd. Can.*, **22**:191–202 (1965).

Neyman, J.: On the Two Different Aspects of the Representative Method: The Method of Stratified Sampling and the Method of Purposive Selection, *J. Roy. Statist. Soc.*, **97**:558–606 (1934).

Paloheimo, J. E.: A Method of Estimating Natural and Fishing Mortalities, *J. Fish. Res. Bd. Can.*, **15**:749–758 (1958).

Petersen, C. G. J.: The Yearly Immigration of Young Plaice in the Limfjord from the German Sea, *Rep. Danish Biol. Sta.*, **6**:1–48 (1896).

Schnabel, Z. E.: The Estimation of the Total Fish Population of a Lake, *Am. Math. Monthly*, **45**:348–352 (1938).

Selleck, D. M., and C. M. Hart: Calculating the Percentage of Kill from Sex and Age Ratios, *Calif. Fish Game*, **43**:309–316, 1957.

Siniff, D. B., and R. O. Skoog: Aerial Censusing of Caribou Using Stratified Random Sampling, *J. Wildlife Mgt.*, **28**:391–401 (1964).

Van Etten, R. C., and C. L. Bennett, Jr.: Some Sources of Error in Using Pellet-Group Counts for Censusing Deer, *J. Wildlife Mgt.*, **29**:723–729 (1965).

Watt, K. E. F.: The Choice and Solution of Mathematical Models for Predicting and Maximizing the Yield of a Fishery, *J. Fish. Res. Bd. Can.*, **13**:613–645 (1956).

————: Studies on Population Productivity II. Factors Governing Productivity in a Population of Smallmouth Bass, *Ecol. Monogr.*, **29**:367–392 (1959).

Widrig, T. M.: Method of Estimating Fish Populations, with Application to Pacific Sardine, *U.S. Fish Wildlife Serv. Fishery Bull.*, **56** (94):141–166 (1954).

Young, H., J. Neess, and J. T. Emlen, Jr.: Heterogeneity of Trap Response in a Population of House Mice, *J. Wildlife Mgt.*, **16**:169–180 (1952).

10 | SYSTEMS ANALYSIS

10.1 STRAIGHTFORWARD, STEPWISE, AND MIXED-MODE MULTIPLE LINEAR REGRESSION ANALYSIS

Chapter 9 discussed measurement of data required to build a systems model for showing how best to manage a resource. The next problem is to determine which of the measured variables are sufficiently important to be worthy of inclusion in the systems model. We determine whether a variable is important or not by determining how great a contribution it makes to the variance in the dependent variable under study, using multiple analysis of variance in conjunction with multiple linear regression analysis, or multiple iterative regression analysis. In mathematical terms, using an unrealistically simple model for purposes of illustration, suppose Y represents the productivity of deer in a management unit, and $X_1, X_2, \ldots, X_7$ are seven factors suspected to be important in regulating deer productivity. We have measured Y and each of the seven X's in each of 10 study plots for each of 10 years. Now we wish to determine which of the seven variables should be included in a model of deer productivity. The simplest model we can possibly postulate is

$$Y = b_o + \sum_{i=1}^{7} b_i X_i \tag{10.1}$$

We test the importance of each of the seven variables by multiple regression analysis, using the 100 sets of eight variate values, and using ANOVA to test the significance of the contribution made to variance in Y by each of the X values in turn. A simple linear model is not needed for multiple regression analysis, since equations such as

$$Y = b_o X_1^{b_1} X_2^{b_2} X_3^{b_3} \qquad (10.2)$$

or

$$Y = b_o e^{b_1 X_1 + b_2 X_2 + b_3 X_3} \qquad (10.3)$$

can be transformed to

$$\ln Y = \ln b_o + b_1 \ln X_1 + b_2 \ln X_2 + b_3 \ln X_3$$

and

$$\ln Y = \ln b_o + b_1 X_1 + b_2 X_2 + b_3 X_3$$

respectively, by taking logarithms.

This writer's experience is that few biologists have an intuitive understanding of the difference in meaning between the principal statistics used in connection with regression analysis; therefore we begin by explaining these in terms of the simple relation between two variables:

$$Y = a + bX \qquad (10.4)$$

where n will be the number of sets of data in which Y and X are measured. As we shall show below, a high value for one statistic may not occur with a high value of another statistic, so it is very important that we be absolutely clear what each statistic means.

The first statistic of interest is the variance about the fitted line. This quantity, sometimes referred to as the *standard error of estimate*, is of concern since it is a measure of the scatter about the line used for prediction. The statistic can be derived as follows. (Note that we use the subscripts o and c to distinguish observed and calculated values.)

$$s_{yx}^2 = \frac{\Sigma(Y_o - Y_c)^2}{n - 2} \qquad (10.5)$$

The denominator is $n - 2$ because we lose 2 degrees of freedom for the estimates of a and b. A little algebra produces a form of (10.5) that can be computed much more quickly.

$$\Sigma(Y_o - Y_c)^2 = \Sigma(Y_o - a - bX_o)^2$$
$$= \Sigma Y_o(Y_o - a - bX_o) - a\Sigma(Y_o - a - bX_o)$$
$$- b\Sigma X_o(Y_o - a - bX_o) \quad (10.6)$$

We can see how to simplify (10.6) by considering the derivation of a and b estimates. If $Y = a + bX$, and we seek the a and b values best fitting the data, then from the relation

$$\Sigma(Y_o - Y_c)^2 = \Sigma(Y_o - a - bX_o)^2$$

we obtain a and b estimates by minimizing the sum of squared deviations by solving the pair of equations

$$\frac{\partial \Sigma(Y_o - Y_c)^2}{\partial a} = -2\Sigma(Y_o - a - bX_o) = 0 \tag{10.7}$$

$$\frac{\partial \Sigma(Y_o - Y_c)^2}{\partial b} = -2\Sigma X_o(Y_o - a - bX_o) = 0 \tag{10.8}$$

Since we set $\Sigma(Y_o - a - bX_o)$ and $\Sigma X_o(Y_o - a - bX_o)$ both equal to zero to obtain a and b estimates, the last two terms in (10.6) are zero, and

$$\Sigma(Y_o - Y_c)^2 = \Sigma Y_o(Y_o - a - bX_o) = \Sigma Y_o^2 - a\Sigma Y_o - b\Sigma X_o Y_o \tag{10.9}$$

This can be computed readily once we have solved the normal equations to obtain a and b, because (10.7) and (10.8) become, after transposing,

$$an + b\Sigma X_o = \Sigma Y_o \tag{10.10}$$

$$a\Sigma X_o + b\Sigma X_o^2 = \Sigma X_o Y_o \tag{10.11}$$

so after calculating the sums we need to solve these, and solving for a and b, we have four of the five values needed to solve (10.9). Only ΣY_o^2 is still needed.

The second important statistic in regression analysis is the slope of a regression line. In the simple linear case, this is obtained by solving (10.10) and (10.11) to obtain

$$b = \frac{n\Sigma XY - \Sigma X\Sigma Y}{n\Sigma X^2 - (\Sigma X)^2}$$

Another common statistic is the correlation coefficient, r, given by

$$r = \frac{s_{xy}}{s_x s_y} = \frac{n\Sigma XY - \Sigma X\Sigma Y}{\sqrt{[n\Sigma X^2 - (\Sigma X)^2][n\Sigma Y^2 - (\Sigma Y)^2]}}$$

Squaring r, we obtain

$$\frac{(n\Sigma XY - \Sigma X\Sigma Y)^2}{[n\Sigma X^2 - (\Sigma X)^2][n\Sigma Y^2 - (\Sigma Y)^2]} \tag{10.12}$$

and this is the same as the proportion of the variance in Y accounted for by the regression,

$$\frac{\Sigma[b(X - \bar{X})]^2}{\Sigma(Y - \bar{Y})^2} = \frac{b^2[n\Sigma X^2 - (\Sigma X)^2]}{n\Sigma Y^2 - (\Sigma Y)^2} \tag{10.13}$$

$$\frac{\Sigma[b(X - \bar{X})]^2}{\Sigma(Y - \bar{Y})^2} = \left[\frac{n\Sigma XY - \Sigma X\Sigma Y}{n\Sigma X^2 - (\Sigma X)^2}\right]^2 \frac{n\Sigma X^2 - (\Sigma X)^2}{n\Sigma Y^2 - (\Sigma Y)^2}$$

$$= \frac{(n\Sigma XY - \Sigma X\Sigma Y)^2}{[n\Sigma X^2 - (\Sigma X)^2][n\Sigma Y^2 - (\Sigma Y)^2]} \qquad (10.14)$$

which is identical to (10.12).

We now need to consider the difference between b, r, and s_{yx}^2. The regression coefficient, b, is merely a slope, or measure of rise per unit run. That is, if b is 2, this means that a unit increase in X produces a 2-unit increase in Y. The square of the correlation coefficient r is a measure of the proportion of the variation accounted for by the regression, and the standard error of estimate reflects the average squared deviation, per observation, about the fitted line. Note, however, that the standard error of estimate is not exactly equal to the average squared deviation, because the denominator in (10.9) is $n - 2$, not n.

We may gain deeper insight into the meaning of these variables by noting that (10.14) can be written as

$$\left[\frac{n\Sigma XY - \Sigma X\Sigma Y}{n\Sigma X^2 - (\Sigma X)^2}\right]\left[\frac{n\Sigma XY - \Sigma X\Sigma Y}{n\Sigma X^2 - (\Sigma X)^2}\right]\left[\frac{n\Sigma X^2 - (\Sigma X)^2}{n\Sigma Y^2 - (\Sigma Y)^2}\right] \qquad (10.15)$$

Note that the first term in brackets is b, the regression coefficient, and the denominator of the second term in brackets cancels the numerator of the third term in brackets. Thus (10.15) reduces to

$$b\left[\frac{n\Sigma XY - \Sigma X\Sigma Y}{n\Sigma Y^2 - (\Sigma Y)^2}\right] \qquad (10.16)$$

The correlation coefficient r is thus the square root of (10.16), which may be thought of as being three components, two in the numerator: the slope of the regression line, and the covariance (or variance in each variable related to the variance in the other); and one in the denominator, the variance in Y. Since the denominator contains a term for variance in Y, the greater the scatter in Y values for a given X value, the lower the correlation coefficient will be. Perhaps the matter can be clarified further by considering Figs. 10.1, 10.2, 10.3, and 10.4.

In Figs. 10.1 and 10.2, the slope is greater than in Figs. 10.3 and 10.4, therefore b is greater in the former two. In Figs. 10.1 and 10.3, the scatter about the lines of best fit is very small; therfore s_{yx} is small; in the other two figures, the reverse is true. The correlation coefficient is high in (10.1), where the slope is great and scatter is slight, and is low in the other three cases, unless s_{yx} is very small in (10.3).

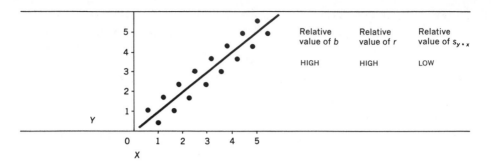

FIGURE 10.1

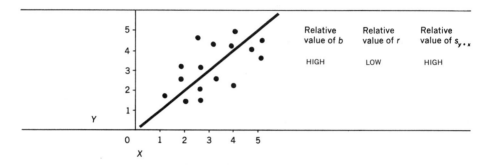

FIGURE 10.2

The statistics b, r, and s_{yx} are all based on sampling, and are themselves all subject to sampling error. Therefore, we might wish to test any of these null hypotheses.

$b = 0 =$ no regression
$r = 0 =$ no regression
$s_{yx}^2 =$ variance within sets of Y values for given X values. If s_{yx}^2 is greater than within-sets variance, regression is nonlinear

In practice by far the most useful test is to test the hypothesis $b = 0$, because if $b = 0$, $r = 0$ in any case, and testing the linearity of the regression model is unnecessary if graphical testing was done at the appropriate step in the analysis. By this, we mean that Y should have been plotted against X *before* any statistical tests were performed, to ensure that we did in fact have a straight line. Alternatively, we should have performed a graphical test to ensure that the transformation used produced a straight line.

In order to explain the test of the hypothesis that $b = 0$, we must introduce the concept of partitioning variance.

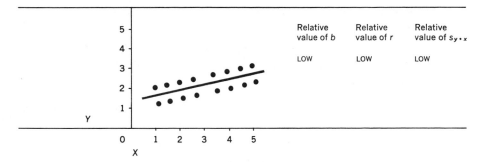

FIGURE 10.3

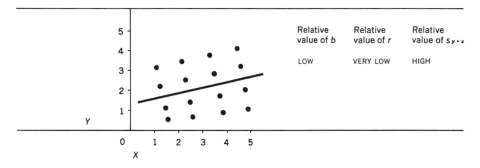

FIGURE 10.4

The total variation about a fitted regression line has two major components: that explained by regression, and that not explained by regression. By using the analysis of variance format, we can construct the table:

Source of variation	Degrees of freedom	Sums of squares	Mean squares	Expected mean squares
Regression (slope of line)	1	$b^2 \Sigma (X - \bar{X})^2 = \Sigma (Y_c - \bar{Y})^2$	s_r^2	$\sigma^2 + \beta^2 \Sigma (X - \bar{X})^2$
Deviations of observed values about estimated line	$n - 2$	$\Sigma (Y_o - \bar{Y})^2 - b^2 \Sigma (X - \bar{X})^2$ $= \Sigma (Y_o - Y_c)^2$	s^2	σ^2
Total	$n - 1$	$\Sigma (Y_o - \bar{Y})^2$		

We test the hypothesis that $s_r^2/s^2 = 1$, by entering a table of F values with 1 and $n - 2$ degrees of freedom.

One last point needs mentioning before proceeding to consideration of multiple linear regression. Least squares regression testing techniques are based on two assumptions:

1. The distribution of Y values for any X value is normal (described by a normal distribution).

2. The variances in Y around the regression line are the same for all X values (that is, we have homoscedasticity).

Where these assumptions are violated, we can either use some kind of transformation, or obtain weighted sums of squares and cross products, weighting reciprocals of the variances in values of Y for each value of X. That is, for the ith X value, if there is no homoscedasticity, we assign weights $w = 1/\sigma_i^2$, where σ_i^2 is the variance among the Y values at $X = X_i$. In this case, our normal equations become, instead of (10.10) and (10.11),

$$a \sum_i w_i + b \sum_i w_i X_i = \sum_i w_i Y_i \qquad a \sum_i w_i X_i + b \sum_i w_i X_i^2 = \sum_i w_i X_i Y_i$$

Multiple regression analysis

From this point on, explanation of regression techniques can be clarified and facilitated enormously by recourse to matrix algebra. Otherwise the reader will become so drowned in algebraic detail that it will be all too easy to lose sight of the central thread of the arguments. Furthermore, since the regression and curve-fitting techniques in the remainder of this chapter are likely to be performed on a computer in any case, it seems sensible to explain them in a language closely related to Fortran (that is, matrix algebra). We will digress for several pages for a brief introduction of vectors and matrices, before we resume discussion of multiple regression analysis.

A vector is an ordered array of numbers in which the order is significant. The numbers may be arranged in the form of a column or a row. The symbol for a column vector is a boldface letter without superscript, and for a row vector, it is a boldface letter with superscript. Normally, lower case letters are used to denote vectors:

$$\mathbf{v} = \begin{bmatrix} v_1 \\ v_2 \\ v_3 \\ v_4 \end{bmatrix} \qquad \text{or} \qquad \mathbf{v}' = [v_1, v_2, v_3, v_4]$$

A matrix is an ordered rectangular array of numbers in which the arrangement is significant. Matrices are denoted in most modern literature by italicized capitals. A matrix has m rows and n columns, and the elements, or individual entries, have two subscripts, the first designating the row and the second the column in which the element is found:

$$
A = \begin{bmatrix}
a_{11} & a_{12} & \cdots & a_{1n} \\
a_{21} & a_{22} & \cdots & a_{2n} \\
\cdots & \cdots & \cdots & \cdots \\
a_{m1} & a_{m2} & \cdots & a_{mn}
\end{bmatrix}
$$

which can also be written as

$$
A = [a_{ij}]
$$

(where $i = 1, 2, \ldots, m$; $j = 1, 2, \ldots, n$). If $m = 1$, we have a row vector; if $n = 1$, we have a column vector. Of particular significance are square matrices, in which $m = n$.

Matrices are added by adding corresponding elements in the two or more addend matrices. Thus,

$$
C = A + B = \begin{bmatrix}
a_{11} + b_{11} & a_{12} + b_{12} & a_{13} + b_{13} \\
a_{21} + b_{21} & a_{22} + b_{22} & a_{23} + b_{23} \\
a_{31} + b_{31} & a_{32} + b_{32} & a_{33} + b_{33} \\
a_{41} + b_{41} & a_{42} + b_{42} & a_{43} + b_{43} \\
a_{51} + b_{51} & a_{52} + b_{52} & a_{53} + b_{53}
\end{bmatrix}
$$

Matrix multiplication involves addition and multiplication. It can only be performed where the number of columns of matrix A equals the number of rows of matrix B. Where this prerequisite holds, an element of the product, $C = AB$, is obtained as follows. The element c_{ij} is the sum of the products of the elements in the ith row of A with the corresponding elements of the jth column of B. Thus

$$
c_{ij} = a_{i1}b_{1j} + a_{i2}b_{2j} + \cdots + a_{in}b_{nj} = \sum_{k=1}^{n} a_{ik}b_{kj}
$$

Of special importance in matrix operations are identity, or unit, matrices, designated by I. These are square matrices in which only the elements along the principal, or leading, diagonal are different from zero, and all the elements along the diagonal are equal to 1. Thus

$$I = \begin{bmatrix} 1 & 0 & \cdots & 0 \\ 0 & 1 & \cdots & 0 \\ \multicolumn{4}{c}{\cdots\cdots\cdots} \\ 0 & 0 & \cdots & 0 \end{bmatrix}$$

For any matrix the following identity holds: $AI = A$. Also very important in matrix operations are transposed matrices, or transposes. The transpose of A, denoted as A^T, is obtained by interchanging rows and columns of A. Thus, if

$$A = \begin{bmatrix} a_{11} & a_{12} & \cdots & a_{1n} \\ a_{21} & a_{22} & \cdots & a_{2n} \\ \multicolumn{4}{c}{\cdots\cdots\cdots\cdots} \\ a_{m1} & a_{m2} & \cdots & a_{mn} \end{bmatrix}$$

$$A^T = \begin{bmatrix} a_{11} & a_{21} & \cdots & a_{m1} \\ a_{12} & a_{22} & \cdots & a_{m2} \\ \multicolumn{4}{c}{\cdots\cdots\cdots\cdots} \\ a_{1n} & a_{2n} & \cdots & a_{mn} \end{bmatrix}$$

The final concept we need from matrix algebra in order to apply the notions from this field to regression procedures is that of the inverse matrix, or inverse. Where A is a square matrix and B is another square matrix such that

$$BA = I$$

then we say that B is the inverse of A. Where A actually has an inverse B, we denote this inverse as A^{-1}. The two following relations hold.

$$AA^{-1} = I \quad \text{and} \quad A^{-1}A = I$$

Any square matrix can have only one inverse. This can be proved as follows. Suppose that the matrix A has a second inverse, B, which is different from A^{-1}. Then

$$B = BI = B(AA^{-1}) = (BA)A^{-1}$$

The last-named equality holds because the associative law applies to matrix multiplication. Thus $(AB)C = A(BC)$. Note, however, that the commutative law does not apply, and therefore $AB \neq BA$.

$$(BA)A^{-1} = IA^{-1} = A^{-1}$$

Therefore $B = A^{-1}$ and the matrix A has only one inverse. We shall now consider, first, the condition under which A has an inverse, and, second, the method of calculating the inverse.

Any square matrix A can only have an inverse if the determinant of the matrix is not equal to zero (that is, if the matrix is nonsingular). The determinant, $|A|$, of a square matrix A, is calculated according to the following rule. Where we have the square matrix

$$A = \begin{bmatrix} a_{11} & a_{12} & \cdots & a_{1n} \\ a_{21} & a_{22} & \cdots & a_{2n} \\ \cdots & \cdots & \cdots & \cdots \\ a_{n1} & a_{n2} & \cdots & a_{nn} \end{bmatrix}$$

the determinant $A = \det [a_{ik}]$ is the sum of the $n!$ terms $(-1)^r a_{1k_1} a_{2k_2} \cdots a_{nk_n}$, each corresponding to one of the $n!$ ordered sets $k_1, k_2, \ldots, k_n$, obtained by r permutations of elements from the set $1, 2, \ldots, n$. The number n is the order of the determinant. To illustrate the application of this rule, consider second- and third-order determinants.

$$\begin{vmatrix} a_{11} & a_{12} \\ a_{21} & a_{22} \end{vmatrix} = a_{11}a_{22} - a_{21}a_{12}$$

$$\begin{vmatrix} a_{11} & a_{12} & a_{13} \\ a_{21} & a_{22} & a_{23} \\ a_{31} & a_{32} & a_{33} \end{vmatrix} = \begin{array}{l} a_{11}a_{22}a_{33} - a_{11}a_{23}a_{32} + a_{12}a_{23}a_{31} \\ - a_{13}a_{22}a_{31} + a_{13}a_{21}a_{32} - a_{12}a_{21}a_{33} \end{array}$$

We require only one more term to complete our discussion of the inverse. The cofactor A_{ik} of the element a_{ik} is also the coefficient of a_{ik} in the expansion of the determinant of A. A determinant D may be represented in terms of the elements and cofactors of any one row or column as follows.

$$D = \det [a_{ik}] = \sum_{i=1}^{n} a_{ij}A_{ij} = \sum_{k=1}^{n} a_{jk}A_{jk}$$

The adjoint, or adjugate matrix of any square matrix is given by

$$
C = \begin{bmatrix}
A_{11} & A_{21} & \cdots & A_{n1} \\
A_{12} & A_{22} & \cdots & A_{n2} \\
\cdots & \cdots & \cdots & \cdots \\
A_{1n} & A_{2n} & \cdots & A_{nn}
\end{bmatrix}
$$

Note that C is the transpose of the matrix of cofactors of A. It can be shown that

$$AC = |A| I \tag{10.17}$$

Now suppose A has the inverse B. Then

$$AB = I \tag{10.18}$$

It follows that

$$A^{-1} = B = \frac{1}{|A|} C \tag{10.19}$$

because

$$AB = A \frac{1}{|A|} C = \frac{1}{|A|} AC = I$$

Therefore the inverse of any matrix A is calculated from the determinant A, and from C, the transpose of the matrix of cofactors. We can illustrate the process using a 2×2 matrix and its inverse. From (10.18), we have

$$
\begin{bmatrix} a_{11} & a_{12} \\ a_{21} & a_{22} \end{bmatrix}
\begin{bmatrix} b_{11} & b_{12} \\ b_{21} & b_{22} \end{bmatrix}
= \begin{bmatrix} 1 & 0 \\ 0 & 1 \end{bmatrix}
= \begin{bmatrix} a_{11}b_{11} + a_{12}b_{21} & a_{11}b_{12} + a_{12}b_{22} \\ a_{21}b_{11} + a_{22}b_{21} & a_{21}b_{12} + a_{22}b_{22} \end{bmatrix}
$$

or

$$a_{11}b_{11} + a_{12}b_{21} = 1 \tag{10.20}$$

$$a_{11}b_{12} + a_{12}b_{22} = 0 \tag{10.21}$$

$$a_{21}b_{11} + a_{22}b_{21} = 0 \tag{10.22}$$

$$a_{21}b_{12} + a_{22}b_{22} = 1 \tag{10.23}$$

We wish to solve for the values of b_{ij} in the inverse matrix. If we multiply (10.20) by a_{22}, and (10.22) by a_{12}, we get

$$a_{11}a_{22}b_{11} + a_{12}a_{22}b_{21} = a_{22} \tag{10.24}$$

$$a_{12}a_{21}b_{11} + a_{12}a_{22}b_{21} = 0 \tag{10.25}$$

If we subtract (10.25) from (10.24), we get

$$b_{11}(a_{11}a_{22} - a_{12}a_{21}) = a_{22}$$

or
$$b_{11} = \frac{a_{22}}{a_{11}a_{22} - a_{12}a_{21}} \qquad (10.26)$$

Similarly,
$$b_{12} = \frac{-a_{12}}{a_{11}a_{22} - a_{12}a_{21}} \qquad (10.27)$$

$$b_{21} = \frac{-a_{21}}{a_{11}a_{22} - a_{12}a_{21}} \qquad (10.28)$$

$$b_{22} = \frac{a_{11}}{a_{11}a_{22} - a_{12}a_{21}} \qquad (10.29)$$

Note that the denominator of each term is the determinant of A. The matrix of cofactors in the determinant of A is

$$\begin{vmatrix} a_{22} & -a_{21} \\ -a_{12} & a_{11} \end{vmatrix}$$

and the transpose of this matrix, or the adjoint matrix, is

$$C = \begin{vmatrix} a_{22} & -a_{12} \\ -a_{21} & a_{11} \end{vmatrix}$$

and these are the four numerators in the four values.

Enough of the basic notions of matrix algebra have now been presented to explore the way in which this mode of expression can be used as a short-hand notation to describe regression operations.

Suppose that the model postulated to describe some resource system can be expressed in the form

$$Y = b_o + \sum_{i=1}^{n} b_i X_i \qquad (10.30)$$

Then we proceed as in the derivation of (10.7) and (10.8), obtaining the array of equations:

$$\frac{\partial \Sigma(Y_o - Y_c)^2}{\partial b_0} = 2\Sigma(-Y_o + b_o + b_1 X_1 + b_2 X_2 + b_3 X_3 + \cdots + b_n X_n)$$

$$\frac{\partial \Sigma(Y_o - Y_c)^2}{\partial b_1} = 2\Sigma(-X_1 Y_o + b_o X_1 + b_1 X_1^2 + b_2 X_1 X_2 \\ + b_3 X_1 X_3 + \cdots + b_n X_1 X_n)$$

$$\frac{\partial \Sigma(Y_o - Y_c)^2}{\partial b_n} = 2\Sigma(-X_n Y_o + b_o X_n + b_1 X_1 X_n + b_n X_2 X_n \\ + b_3 X_3 X_n + \cdots + b_n X_n^2)$$

After setting each of these equations equal to zero, to obtain the values of b_o and b_i, $i = 1 \cdots n$, which minimizes $\Sigma(Y_o - Y_c)^2$, we obtain, where N is the number of observations, the set

$$b_oN + b_1\Sigma X_1 + b_2\Sigma X_2 + \cdots + b_n\Sigma X_n = \Sigma Y_o$$

$$b_o\Sigma X_1 + b_1\Sigma X_1^2 + b_2\Sigma X_1 X_2 + \cdots + b_n\Sigma X_1 X_n = \Sigma X_1 Y_o$$

$$b_o\Sigma X_n + b_1\Sigma X_n X_1 + b_2\Sigma X_n X_2 + \cdots + b_n\Sigma X_n^2 = \Sigma X_n Y_o$$

Now, from the first equation of this set, we obtain the relation

$$b_o = \frac{\Sigma Y_o - \sum_{i=1}^{n} b_i \Sigma X_i}{N}$$

Substituting for b_o in the remaining equations of the set, and making use of the notation

$$x_1 x_2 = \Sigma X_1 X_2 - \frac{\Sigma X_1 \Sigma X_2}{N}$$

$$x_1 y = \Sigma X_1 Y_o - \frac{\Sigma X_1 \Sigma Y_o}{N}$$

and so on, the normal equations appear as

$$b_1 x_1^2 + b_2 x_1 x_2 + \cdots + b_n x_1 x_n = x_1 y$$

$$b_1 x_1 x_2 + b_2 x_2^2 + \cdots + b_n x_2 x_n = x_2 y$$

$$b_1 x_n x_1 + b_2 x_n x_2 + \cdots + b_n x_n^2 = x_n y$$

Notice that we now have a symmetrical family of equations in which, on the left-hand side, there is a diagonal consisting of terms of the form $b_i x_i^2$ with the other terms in pairs, and in which the ith row and the jth column are identical to the jth row and ith column.

Thus, in terms of the matrix theory already presented, the system of equations that must be solved for the values of b_i, $i = 1, \ldots, n$ may be written as

$$a_{11}b_1 + a_{12}b_2 + \cdots + a_{1n}b_n = c_1$$

$$a_{21}b_1 + a_{22}b_2 + \cdots + a_{2n}b_n = c_2$$

$$\cdots \cdots \cdots \cdots \cdots \cdots \cdots \cdots \cdots$$

$$a_{n1}b_1 + a_{n2}b_2 + \cdots + a_{nn}b_n = c_n$$

where
$$a_{11} = \Sigma X_1^2 - \frac{(\Sigma X_1)^2}{N}$$

$$a_{12} = a_{21} = \Sigma X_1 X_2 - \frac{\Sigma X_1 \Sigma X_2}{N}$$

$$c_1 = \Sigma X_1 Y_o - \frac{\Sigma X_1 \Sigma Y_o}{N}$$

and so on, where N is the number of observations. From the rule for matrix multiplication, it appears that the foregoing system of equations is the product of multiplication of a square matrix by a column vector, and hence may be expressed as

$$
\begin{bmatrix}
a_{11} & a_{12} & \cdots & a_{1n} \\
a_{21} & a_{22} & \cdots & a_{2n} \\
\cdots & \cdots & \cdots & \cdots \\
a_{n1} & a_{n2} & \cdots & a_{nn}
\end{bmatrix}
\begin{bmatrix}
b_1 \\
b_2 \\
\cdot \\
b_n
\end{bmatrix}
=
\begin{bmatrix}
c_1 \\
c_2 \\
\cdot \\
c_n
\end{bmatrix}
$$

or, more succinctly, as

$$A\mathbf{b} = \mathbf{c} \tag{10.31}$$

There is thus available an extremely powerful, succinct shorthand for expressing multiple regression analysis. The straightforward solution of multiple regression normal equations can be expressed as a matrix operation. For example, to solve systems such as (10.31) in matrix terms, if A is nonsingular, we premultiply both sides of the equation by A^{-1}, obtaining

$$A^{-1}A\mathbf{b} = A^{-1}\mathbf{c}$$

$$I\mathbf{b} = \frac{1}{|A|} C\mathbf{c}$$

or

$$\mathbf{b} = \frac{1}{|A|} C\mathbf{c}$$

where $|A|$ is the determinant of A, and C is the matrix of cofactors. Solving regression equations is then the same as finding inverses of matrices.

There are numerous methods for finding such inverses in regression analysis. The important point to note is that these fall into two classes. One approach is to solve for all the b values, on the assumption that these have some meaning, even if most of them make no significant contribution to the variance in Y. The other approach is referred to as the *stepwise approach*.

The basic idea behind this approach is that variables are added to the multiple regression equation only if it can be shown that they are making a significant contribution to Y. The contribution of the ith independent variable by itself to variance in Y_o is given by

$$\sum^{N} [b_i(X_i - \bar{X}_i)]^2 = b_i^2 \sum^{N} (X_i - \bar{X}_i)^2 = \frac{\left(\Sigma X_i Y_o - \dfrac{\Sigma X_i \Sigma Y_o}{N}\right)^2}{\Sigma X_i^2 - \dfrac{(\Sigma X_i)^2}{N}}$$

in which the sums of squares and cross products have been adjusted. Thus the computer calculates the vector of terms of the form c_i^2/a_{ii}, $i = 1, \ldots, n$. The variable for which this term is largest, at each iteration, is removed from the matrix, and regression coefficients for this variable and all variables already removed are calculated. After a variable has been added to the regression equation, the square matrix A is reduced by one row and one column, and this matrix and the column vector $\mathbf{c}$ are corrected for the removal of the variable. The process is repeated until we no longer find a variable making a significant contribution to the variance as measured by an F test.

Another possible approach to regression analysis is to combine straight-forward regression, in which we solve for all the b values, with stepwise regression. For example, suppose we have 23 independent variables. We may decide that on the basis of prior field and experimental knowledge, variables 3, 17, and 22 must be important, even if our regression analysis cannot confirm this by F tests. We program the computer so that these three variables are removed first, and the others are removed stepwise, beginning with the variable that makes the greatest contribution to the variance in the system, and proceeding to the variable that makes the least contribution. This procedure has been referred to as *mixed-mode* regression analysis.

Many computer programs are now available, either from computer manufacturers, users' groups, or most computer centers, that perform all these calculations automatically.

Because of the nature of most complex systems, the two basic approaches to regression may yield rather different results. In many cases, the independent variables are related to each other. They generally fall into several groups of variables; within the groups there is a high degree of association between the variables. Weather factors, for example, are often related, humidity being inversely related to temperature, and directly related to rainfall. What happens when we use straightforward regression on such a complex of factors is that the variance due to the complex is attributed to several factors, none of which can appear significant precisely because of the intercorrelations within the group of factors. If we use stepwise regression, only the most important factor in each complex will typically show up as significant. This is because, for example, removing humidity removes some of the variance due to rainfall and temperature. Therefore, once the effect of humidity has been accounted for, there is little variance left to be accounted for by other variables in the weather complex of factors.

This difficulty suggests that there may be something basically unrealistic in the use of multiple linear regression analysis to describe highly complex biological systems. What we really need is some means of describing the effect of a natural group of factors that expresses how they *actually interact*, rather than having to force our mathematical description into a linear additive model. One helpful solution is to treat all possible pairs of interactions between independent variables as if they were additional independent variables (Mott, 1967).

There are a great many difficulties with the use of multiple regression analysis. The volume of computation involved will typically be great enough to require electronic computation. The great speed with which these machines process large-scale computations makes it likely that research workers will attempt to fit a multiple regression without having any reason for including some of the independent variables except that they were embarking on a "fishing trip," hoping that some independent variables will turn out to be significant, and particularly, that some factor hitherto unsuspected of much importance may turn out to be surprisingly important.

At best, this procedure will yield the researcher a rough idea as to which variables are accounting for a significant proportion of the variance in the system. At worst, blind application of multiple regression analysis leads to erroneous conclusions. A significant regression coefficient may not mean that a factor is important; rather, the factor may be highly correlated with some important factor that has not been included in the regression analysis. Furthermore, a factor may not appear to be important when in fact it is, because the wrong model may have been postulated.

In general, while empirical models relate causes to effects, they yield little insight into the mechanics of the causal pathway. Such models really tell us little about how to manage complex systems.

A linear multiple regression equation will be a woefully bad predictor if the true mode of action of relevant variables is markedly nonlinear. The relationships between many variables in the real world cannot be crammed into a linear additive model. For example, most phenomena have a maximum rate of action which cannot be exceeded no matter how much an independent variable is increased. This implies the existence of asymptotes, which cannot be fitted into a linear model. For example,

$$Y = K(1 - e^{-aX})$$

Excessive dependence on linear multiple regression as a research tool can have an insidious effect on the development of knowledge in a biological field. This statistical tool was not developed because linear models describe the behavior of complex systems in the real world. Rather, a linear additive model is theoretically straightforward to deal with, and leads to a simple, if tedious, computational procedure. Trying to cram research findings into a model that is best from a statistical point of view—but poorest from a biological or management point of view—is a procedure that can only result in preventing the development of a suitable theoretical substructure for biology.

To conclude, multiple regression analysis is only a tool to be used as a guide in preliminary work on construction of more realistic models based on differential equations or other structured models, as in physics. Where such models have been developed, we can use the regression methods described in the next two sections.

10.2 ITERATIVE REGRESSION AND CURVE FITTING

As pointed out, many equations can be fitted using the methods of multiple linear regression analysis, after they are subjected to suitable transformation. However, in many cases, particularly in biology, no transformation will convert the equation into a form that can be handled this way. This is not a great problem because a variety of other curve fitting and regression methods are available. The only difficulty with most of these other methods is that they are iterative, and therefore require a prodigious amount of computation. These methods have only been routinely used since the arrival of computers.

In principle, we need a process for estimating parameter values which:

1. should be a trial-and-error method so that analysis of the error (or residual) left after each try (or iteration) shows us how to improve the next try. In other words, we seek a cyclical process of estimation that converges to the best values of the parameters with each iteration.

2. should have a straightforward mathematical basis and a computing procedure (algorithm) that is simple in principle and easy to routine for a computer, even if it is tedious and massive. Ideally, then, we wish to get our nonlinear estimation problem into some form that allows the use of a set of linear additive equations. This is achieved as follows.

Suppose the nonlinear model to be fitted to the data is

$$E(y) = f(x_1, x_2, \ldots, x_m; \beta_1, \beta_2, \ldots, \beta_k) = f(\mathbf{x}, \boldsymbol{\beta})$$

where the x's are independent variables and the β's are population values of k parameters, and $E(y)$ is the expected (average) value of the dependent variable y. Let

$$Y_i, X_{1i}, X_{2i}, \ldots, X_{mi} \qquad i = 1, 2, \ldots, n$$

represent the data points. We wish to compute the estimates of β's that minimize the residual sum of squares

$$\phi = \sum_{i=1}^{n} (Y_i - \hat{Y}_i)^2$$

where $\hat{Y}_i$ is the value of Y predicted by the ith set of data. When the residuals between n observed and calculated values of Y are denoted by R, we can write these equations

$$R_i = f(X_{1i}, X_{2i}, \ldots, X_{mi}, \beta_1, \beta_2, \ldots, \beta_k) - Y_i \qquad i = 1, 2, \ldots, n$$

$$(10.32)$$

Assume that by using some graphical or algebraic procedure, we can obtain approximate values, $\mathbf{b}$, for the true population parameters $\boldsymbol{\beta}$. We wish to correct the $\mathbf{b}$ at each iteration by incremental amounts δ. Then we have the vector equation

$$\boldsymbol{\beta} = \mathbf{b} + \boldsymbol{\delta} \qquad (10.33)$$

Now, if we substitute the values from (10.33) into (10.32), we have n equations of form

$$R_i + Y_i = f(X_{1i}, X_{2i}, \ldots, X_{mi}; b_1 + \delta_1, b_2 + \delta_2, \ldots, b_k + \delta_k)$$

$$i = 1, \ldots, n \quad (10.34)$$

By using Taylor's theorem for a function of several variables, and ignoring derivatives of the second, and higher orders, (10.34) can be expanded to

$$R_i + Y_i = f(X_{1i}, X_{2i}, \ldots, X_{mi}; b_1, b_2, \ldots, b_k) + \delta_1 \left(\frac{\partial f_i}{\partial b_1} \right)$$

$$+ \delta_2 \left(\frac{\partial f_i}{\partial b_2} \right) + \cdots + \delta_k \left(\frac{\partial f_i}{\partial b_k} \right) \qquad i = 1, \ldots, n \quad (10.35)$$

In (10.35), the partial derivatives of form $\partial f_i / \partial b_i$ refer to the value of the partial derivative evaluated at the two vectors of values $X_{1i}, X_{2i}, \ldots, X_{mi}$ and $\mathbf{b}$. Now the set of n equations in (10.35) can be rewritten as

$$R_i = \sum_{j=1}^{k} \delta_j \left(\frac{\partial f_i}{\partial b_j}\right) + f(\mathbf{X}_i, \mathbf{b}) - Y_i \qquad i = 1, \ldots, n \qquad (10.36)$$

These equations are the analog of (10.30), only now, instead of having to find appropriate parameter values, we must find appropriate corrections to parameter values. The vector $\mathbf{b}$ in (10.30) corresponds to the vector δ in (10.36). The sought-for vector, δ, is obtained by a least squares method (at each iteration we seek the vector of values, δ, which minimizes the sum of squares of residuals, R_i). Thus, we have

$$\Sigma R_i^2 = \Sigma \left[\sum_{j=1}^{k} \delta_j \left(\frac{\partial f_i}{\partial b_j}\right) + f(\mathbf{X}_i, \mathbf{b}) - Y_i \right]^2$$

From now on, to obtain a more succinct notation, we will denote $f(\mathbf{X}_i, \mathbf{b}) - Y_i$ by r_i. The format of the set of partial differential equations of form $\partial \Sigma R_i^2 / \partial \delta_j$ can perhaps be grasped most readily at first if we set $K = 3$. Then we have

$$\frac{\partial \Sigma R_i^2}{\partial \delta_1} = 2 \left[\delta_1 \Sigma \left(\frac{\partial f_i}{\partial b_1}\right)^2 + \delta_2 \Sigma \left(\frac{\partial f_i}{\partial b_1}\right)\left(\frac{\partial f_i}{\partial b_2}\right) + \delta_3 \Sigma \left(\frac{\partial f_i}{\partial b_1}\right)\left(\frac{\partial f_i}{\partial b_3}\right) + \Sigma \left(\frac{\partial f_i}{\partial b_1}\right) r_i \right]$$

$$\frac{\partial \Sigma R_i^2}{\partial \delta_2} = 2 \left[\delta_1 \Sigma \left(\frac{\partial f_i}{\partial b_1}\right)\left(\frac{\partial f_i}{\partial b_2}\right) + \delta_2 \Sigma \left(\frac{\partial f_i}{\partial b_2}\right)^2 + \delta_3 \Sigma \left(\frac{\partial f_i}{\partial b_2}\right)\left(\frac{\partial f_i}{\partial b_3}\right) + \Sigma \left(\frac{\partial f_i}{\partial b_2}\right) r_i \right]$$

$$\frac{\partial \Sigma R_i^2}{\partial \delta_3} = 2 \left[\delta_1 \Sigma \left(\frac{\partial f_i}{\partial b_1}\right)\left(\frac{\partial f_i}{\partial b_3}\right) + \delta_2 \Sigma \left(\frac{\partial f_i}{\partial b_2}\right)\left(\frac{\partial f_i}{\partial b_3}\right) + \delta_3 \Sigma \left(\frac{\partial f_i}{\partial b_3}\right)^2 + \Sigma \left(\frac{\partial f_i}{\partial b_3}\right) r_i \right]$$

Setting each of these equations equal to zero, and dividing across by 2, we obtain the set

$$\delta_1 \Sigma \left(\frac{\partial f_i}{\partial b_1}\right)^2 + \delta_2 \Sigma \left(\frac{\partial f_i}{\partial b_1}\right)\left(\frac{\partial f_i}{\partial b_2}\right) + \delta_3 \Sigma \left(\frac{\partial f_i}{\partial b_1}\right)\left(\frac{\partial f_i}{\partial b_3}\right) = -\Sigma \left(\frac{\partial f_i}{\partial b_1}\right) r_i$$

$$\delta_1 \Sigma \left(\frac{\partial f_i}{\partial b_1}\right)\left(\frac{\partial f_i}{\partial b_2}\right) + \delta_2 \Sigma \left(\frac{\partial f_i}{\partial b_2}\right)^2 + \delta_3 \Sigma \left(\frac{\partial f_i}{\partial b_2}\right)\left(\frac{\partial f_i}{\partial b_3}\right) = -\Sigma \left(\frac{\partial f_i}{\partial b_2}\right) r_i \qquad (10.37)$$

$$\delta_1 \Sigma \left(\frac{\partial f_i}{\partial b_1}\right)\left(\frac{\partial f_i}{\partial b_3}\right) + \delta_2 \Sigma \left(\frac{\partial f_i}{\partial b_2}\right)\left(\frac{\partial f_i}{\partial b_3}\right) + \delta_3 \Sigma \left(\frac{\partial f_i}{\partial b_3}\right)^2 = -\Sigma \left(\frac{\partial f_i}{\partial b_3}\right) r_i$$

A means of expediting the computational procedure for forming such arrays can be pointed out by examining matrix algebra again. Consider the product of a matrix and its transpose.

$$A = \begin{bmatrix} a_{11} & a_{12} & a_{13} \\ a_{21} & a_{22} & a_{23} \\ a_{31} & a_{32} & a_{33} \end{bmatrix} \quad \text{and} \quad A^T = \begin{bmatrix} a_{11} & a_{21} & a_{31} \\ a_{12} & a_{22} & a_{32} \\ a_{13} & a_{23} & a_{33} \end{bmatrix}$$

$$AA^T =$$

$$\begin{bmatrix} a_{11}a_{11} + a_{12}a_{12} + a_{13}a_{13} & a_{11}a_{21} + a_{12}a_{22} + a_{13}a_{23} & a_{11}a_{31} + a_{12}a_{32} + a_{13}a_{33} \\ a_{21}a_{11} + a_{22}a_{12} + a_{23}a_{13} & a_{21}a_{21} + a_{22}a_{22} + a_{23}a_{23} & a_{21}a_{31} + a_{22}a_{32} + a_{23}a_{33} \\ a_{31}a_{11} + a_{32}a_{12} + a_{33}a_{13} & a_{31}a_{21} + a_{32}a_{22} + a_{33}a_{23} & a_{31}a_{31} + a_{32}a_{32} + a_{33}a_{33} \end{bmatrix}$$

This can be written as

$$AA^T = \begin{bmatrix} \sum_{j=1}^{3} a_{1j}^2 & \sum_{j=1}^{3} a_{1j}a_{2j} & \sum_{j=1}^{3} a_{1j}a_{3j} \\ \sum_{j=1}^{3} a_{1j}a_{2j} & \sum_{j=1}^{3} a_{2j}^2 & \sum_{j=1}^{3} a_{2j}a_{3j} \\ \sum_{j=1}^{3} a_{1j}a_{3j} & \sum_{j=1}^{3} a_{2j}a_{3j} & \sum_{j=1}^{3} a_{3j}^2 \end{bmatrix}$$

which obviously corresponds to (10.37). The problem of iterative regression can then be translated in terms of matrix operations into the following succinct language.

$$\langle Y(\mathbf{X}_i, \mathbf{b} + \boldsymbol{\delta}_t) \rangle = f(\mathbf{X}_i, \mathbf{b}) + \sum_{j=1}^{k} \left(\frac{\partial f_i}{\partial b_j} \right) (\delta_t)_j$$

or
$$\langle \mathbf{Y} \rangle = \mathbf{f}_o + P\boldsymbol{\delta}_t \tag{10.38}$$

In (10.38), $\mathbf{b}$ represents the vector of least squares estimated values of the parameters, $\boldsymbol{\beta}$, toward which the iterative process converges; the vector $\boldsymbol{\delta}_t$ represents the small Taylor series corrections at each step; and the brackets $\langle \rangle$ are used to distinguish predictions based on the linearized model from those based on the actual nonlinear model. Thus, what we are actually seeking to do is minimize the quantity

$$\langle \phi \rangle = \sum_{i=1}^{n} (Y_i - \langle Y_i \rangle)^2$$

We find the vector $\boldsymbol{\delta}_t$ by solving

$$A\boldsymbol{\delta}_t = \mathbf{g} \tag{10.39}$$

the equivalent of (10.37), by obtaining

$$\boldsymbol{\delta}_t = A^{-1}\mathbf{g}$$

where

$$A^{[k \times k]} = P^T P$$

(superscripts in square brackets indicate the number of rows and columns in the matrix),

$$P^{[n \times k]} = \frac{\partial f_i}{\partial b_j} \quad \begin{cases} i = 1, 2, \ldots, n \\ j = 1, 2, \ldots, k \end{cases}$$

$$\mathbf{g}^{[k \times l]} = \left(\sum_{i=1}^{n} (Y_i - f_i) \frac{\partial f_i}{\partial b_j} \right)$$

$$= P^T(\mathbf{Y} - \mathbf{f}_o) \quad j = 1, 2, \ldots, k$$

An approximate estimate of the contribution of each variable to variance in Y can be computed in an analogous fashion to that in the previous section.

The vector of trial parameter values is found by using a combination of algebraic and graphical tricks developed ad hoc for each type of nonlinear equation. For example, suppose we wished trial values for a, b, and K in

$$N_a = PK(1 - e^{-aN_o P^{1-b}}) \tag{10.40}$$

A trial value for K is obtained by making a plot of N_a versus N_o, and dividing the asymptote, PK, by P. Now (10.40) can be rearranged as follows.

$$1 - \frac{N_a}{PK} = e^{-aN_o P^{1-b}}$$

$$\ln\left(\frac{PK}{PK - N_a}\right) = aN_o P^{1-b}$$

$$\frac{\ln\left(\dfrac{PK}{PK - N_a}\right)}{N_o P} = aP^{-b}$$

$$\ln\left[\frac{\ln\left(\dfrac{PK}{PK - N_a}\right)}{N_o P}\right] = \ln a - b \ln P$$

Thus we can plot

$$\frac{\ln\left(\dfrac{PK}{PK - N_a}\right)}{N_o P}$$

against P on log-log graph paper, and find a and b from the intercept and slope.

Another kind of iterative computer procedure that is necessary for resource management simulation studies involves solving equations. Suppose we wish to solve the following equation for searching time, TS.

$$\pi VRAKRGM^2 \left[(HK^2 - 2HKHTS)\, TS + \frac{2HK(HO - HK)}{AD}(1 - e^{-ADTS}) \right.$$

$$+ \frac{2HK(HTO - HTS)}{AF}(e^{-AFTS} - 1) + HTS^2 TS$$

$$- \frac{2HTS(HO - HK)}{AD}(1 - e^{-ADTS}) - \frac{2HTS(HTO - HTS)}{AF}(e^{-AFTS} - 1)$$

$$+ \frac{(HO - HK)^2}{2AD}(1 - e^{-2ADTS})$$

$$+ \frac{(HO - HK)(HTO - HTS)}{AD + AF}(e^{-(AD+AF)TS} - 1)$$

$$\left. + \frac{(HTO - HTS)^2}{2AF}(1 - e^{-2AFTS}) \right] + \frac{8\pi}{6}[AKRGM (HO - HTO)]^3$$

$$- \frac{1}{NOSPSS} = 0 \quad (10.41)$$

in which since we know the value for all variables except TS, (10.41) takes the form

$$f(TS) = 0$$

and thus lends itself to being solved for TS using the Bolzano, or bisection, method (McCormick and Salvadori, 1964). This very flexible and powerful method for solving equations lends itself to processing on large computers. We begin by arbitrarily assigning TS a trial value that we know must be smaller than the true value of TS. Let us call this value TS_1. The computer then calculates $f(TS_1)$. Then we increase TS_1 by an increment h and calculate $f(TS_1 + h)$. The computer then tests to see if $f(TS_1)$ and $f(TS_1 + h)$ have opposite signs [that is, it determines if the value of $f(TS_1)$ multipled by $f(TS_1 + h)$ is less than zero]. If not, it continues calculating terms of form $f(TS_i + h)$, successively incrementing TS_i by h, until it discovers that $f(TS_i + h)$ and $f(TS_{i+1} + h)$ have opposite signs. At this point, the correct value of TS must fall between $TS_i + h$ and $TS_{i+1} + h$. The step size h is now halved at each iteration. When $f(TS_i + h)$ and $f(TS_i + h/2)$ have opposite signs, the next values compared are $f(TS_i + h)$ and $f(TS_i + h/4)$; when the two criterion functions have the same sign, the next values compared are $f(TS_i + h/2)$ and $f(TS_i + 3h/4)$. The correct value of TS_i is thus approached by going from small trial values to larger trial values; at any iteration, the width of the interval at the previous stage gives the upper bound of the error. When the width of the error interval is equal to or less than a predetermined tolerance, the process ceases. In 10 iterations, the width of the error interval is reduced to $1/2^{10} = 1/1,024$.

Note that in both types of iterative processes discussed, the closer the first trial value is to the correct answer, the smaller will be the number of iterations required to obtain the estimated parameter value, or values, with predetermined accuracy. Thus a high premium is placed on ingenious devices for obtaining the best possible initial value.

Methods of speeding up convergence are discussed in the next section; the methods discussed in Sec. 13.4 will also be useful in iterative curve fitting.

10.3   MAXIMIZING RATE OF CONVERGENCE ON BEST PARAMETER VALUES

While the iterative regression method of the previous section is theoretically sound, in practice it often produces estimates of parameter values that oscillate from iteration to iteration and converge extremely slowly (Box, 1960), or actually diverge (Marquardt, 1963). The reason for these difficulties is that the contour surface of the residual sum of squares is greatly attenuated in some directions and elongated in others so that the minimum lies at the bottom of a long curving trough (Marquardt, 1963). As might be expected in view of the great importance of iterative methods, the difficulties that arise in application have led to a great deal of research in numerical analysis to discover means of forcing iterative processes to converge, and converge rapidly. Much of this work has involved new research in matrix algebra (for example, Bodewig, 1956, Faddeeva, 1959, Marquardt, 1963, and Varga, 1962).

Mathematicians have a number of tricks available to them for producing convergence, and for increasing the speed of convergence in iterative regression, because finding the parameter values that minimize a residual sum of squares is merely a special case of the general problem of finding the values of independent variables that minimize (or maximize) a dependent variable. The iterative regression problem is merely the extremum problem of mathematics presented in a somewhat different guise. Thus all the methods of Chap. 13, plus other extremum methods not discussed there (Spang, 1962; Wilde, 1964), are available to us. Another possibility that has received considerable attention from mathematicians and statisticians is the combination of different methods. Of particular interest for computers has been the development of iterative regression methods that combine the Taylor series technique with the gradient method explained in Sec. 13.4 (Marquardt, 1963). Consideration of the special properties of these two different methods suggests how techniques could be constructed that improve on either of them.

The strong point of the gradient methods is their ability to converge toward the best parameter values even when the initial trial values are far removed from the correct values; the weakness of the gradient methods is their inability to converge on the best values rapidly, once a correct region for detailed exploration has been located by the computer (Marquardt, 1963; Spang, 1962, p. 347).

The Taylor series iterative method has the opposite characteristics: if the initial trial values are too far removed from the correct values, it may not converge at all; however, once the vicinity of the correct values has been discovered, the method converges rapidly (Marquardt, 1963).

The situation suggests three courses of action, to ensure convergence and speed in the convergence.

1. The matrices of values used in the Taylor series can be subjected to one of a variety of operations that guarantee faster approach to the correct values (Faddeeva, 1959, pp. 211–219; "overrelaxation" iterative methods are considered by Varga, 1962, Chaps. 3 and 4).

2. Gradient methods can be used for enough iterations to locate the general area of the best parameter values, then the Taylor series method can be used for detailed exploration of the promising area thus located.

3. A method that employs features from the gradient and Taylor series methods at each iteration (Marquardt, 1963) can be used. In the previous two procedures, we decide on the direction of the correction vector, *then* the step size. In the method outlined hereinafter, both are determined simultaneously.

I have selected the third approach for a complete exposition for two reasons. First, a computer program for applying this method is readily available as IBM Share Program No. 1428. Second, Marquardt's procedure is clearly the outcome of a great deal of actual computer experience in fitting nonlinear functions.

To explain the Marquardt algorithm (computing procedure) in succinct symbolism, it is necessary for us to refer back to equation (10.39):

$$A\boldsymbol{\delta}_t = \mathbf{g}$$

Marquardt shows, using methods beyond the scope of this book, that a simple modification of the procedure explained in Sec. 10.2 will cause great improvement in convergence. However, since the new method combines the features of gradient methods and Taylor series methods, we must scale the **b** space because properties of the gradient methods are not scale invariant, and all authors writing on gradient methods stress the need for scaling (for example, Marquardt, 1963; Wilde, 1964). Marquardt suggests scaling in terms of the standard deviations of the derivatives $\partial f_i/\partial b_j$, taken over the sample points $i = 1$, $2, \ldots, n$. The current trial values of b_j are used as necessary in the evaluation of the derivatives. Then, using this scale, the A matrix is transformed into the matrix of simple correlation coefficients among $\partial f_i/\partial b_j$. This step defines a scaled matrix A^* and a scaled vector $\mathbf{g}^*$. The analog of (10.39) is now

$$A^*\boldsymbol{\delta}_t^* = \mathbf{g}^*$$

The algorithm for obtaining the parameter values proceeds as follows. At the rth iteration we construct the equation

$$(A^{*(r)} + \lambda^{(r)}I)\,\boldsymbol{\delta}^{*(r)} = \mathbf{g}^{*(r)}$$

(λ is a Lagrange multiplier or "slack variable" which will be defined shortly.) The next step is to obtain a descaled $\boldsymbol{\delta}^{(r)}$. The new trial vector

$$\mathbf{b}^{(r+1)} = \mathbf{b}^{(r)} + \boldsymbol{\delta}^{(r)}$$

leads to a new sum of squares $\Phi^{(r+1)}$. The trick in the computational procedure is to select $\lambda^{(r)}$ so that

$$\Phi^{(r+1)} < \Phi^{(r)} \tag{10.42}$$

Marquardt demonstrates that a sufficiently large $\lambda^{(r)}$ always exists such that (10.42) will be true, unless $\mathbf{b}^{(r)}$ is such that Φ is already a minimum. Trial and error is required to find the value of λ that maximizes the rate of convergence. The appropriate strategy for selecting λ is to make it large in early iterations, where the unmodified Taylor series method wouldn't converge fast, if at all, and small in late iterations, where the unmodified Taylor series method converges fast. Accordingly, Marquardt proposes the following strategy.

Let $v > 1$.

Let $\lambda^{(r-1)}$ denote the value of λ from the previous iteration.

Initially let $\lambda^{(0)} = 10^{-2}$, say. Compute $\Phi(\lambda^{(r-1)})$ and $\Phi(\lambda^{(r-1)}/v)$.

1. If $\Phi(\lambda^{(r-1)}/v) \leqslant \Phi^{(r)}$, let $\lambda^{(r)} = \lambda^{(r-1)}/v$.
2. If $\Phi(\lambda^{(r-1)}/v) > \Phi^{(r)}$, and $\Phi(\lambda^{(r-1)}) \leqslant \Phi^{(r)}$, let $\lambda^{(r)} = \lambda^{(r-1)}$.
3. If $\Phi(\lambda^{(r-1)}/v) > \Phi^{(r)}$, and $\Phi(\lambda^{(r-1)}) > \Phi^{(r)}$, increase λ by successive

multiplication by v until for some smallest w,

$$\Phi(\lambda^{(r-1)}v^{w}) \leqslant \Phi^{(r)} \qquad \text{let} \qquad \lambda^{(r)} = \lambda^{(r-1)}v^{w}$$

In (1), we have fast convergence, in (2) slow convergence, and in (3) very slow convergence. More details are contained in the Marquardt paper, and in IBM Share Program No. 1428.

All that remains now is the problem of determining which variables are making the most important contributions to the total sum of squares, once the process has converged to a predetermined stability (as the computer will determine by test at each iteration). The analysis of variance calculations is performed in an analogous fashion to those outlined in Sec. 10.1.

REFERENCES

Bodewig, E.: "Matrix Calculus," North Holland Publishing Company, Amsterdam, 1956.

Box, G. E. P.: Fitting Empirical Data, *Ann. N.Y. Acad. Sci.*, **86**:792–816 (1960).

Faddeeva, V. N.: "Computational Methods of Linear Algebra," Dover Publications, Inc., New York, 1959.

McCormick, J. M., and M. G. Salvadori: "Numerical Methods in FORTRAN," Prentice-Hall, Inc., Englewood Cliffs, N.J., 1964.

Marquardt, D. W.: An Algorithm for Least-Squares Estimation of Non-Linear Parameters, *J. Soc. Ind. App. Math.*, **11**:431–441 (1963).

Mott, D. G.: The Analysis of Determination in Population Systems, in K. E. F. Watt (ed.), "Systems Analysis in Ecology," Academic Press Inc., New York, 1967.

Spang, H. A., III: A Review of Minimization Techniques for Non-Linear Functions, *SIAM Review*, **4**:343–365 (1962).

Varga, R. S.: "Matrix Iterative Analysis," Prentice-Hall, Inc., Englewood Cliffs, N.J., 1962.

Wilde, D. J.: "Optimum Seeking Methods," Prentice-Hall, Inc., Englewood Cliffs, N.J., 1964.

11 | SYSTEMS DESCRIPTION

11.1 TYPES OF MODELS FOR DESCRIBING POPULATION SYSTEMS

All the models proposed for describing populations in resource management problems fall into four categories, whether one considers pest control, fisheries management, wildlife, agriculture, or forestry. It is necessary to have a clear understanding of the fundamental differences in the different types of models, because it is the model type chosen that determines the whole character of the data-collection program in a particular instance. If we choose the model that will be used for data interpretation before starting data collection, which is the rational procedure, we will collect precisely the type and amount of data required as input by the model, and thereby have an efficient research program. Suppose, on the other hand, we collect the data first, then decide what type of model will be used to describe and interpret the data. The research program will likely be woefully inefficient because it is easily possible to collect more data than are required to serve as input for one of the simple models, and to collect less data than are required to construct one of the complicated types of models.

   In view of the great importance of the choice of model type for the design of research programs on problems of natural resource management, it is necessary to understand precisely the strengths and weaknesses of the different types of models. The models differ with respect to their information requirements, the assumptions on which they are based, their verisimilitude, their predictive reliability, their analytic usefulness, the possibility that they will yield surprising new results through serendipity, and the cost of collecting the information required to build them. The four classes of models are:

1. Models that attempt to explain changes in the size of a population on the basis of the relationship between the size of the reproductive segment of the population, and the size of the resultant offspring population.

2. Models that use regression methods to relate the stock in each age in each year to the stock in one or more age groups the previous year.

3. Models that attempt to explain changes in populations only in terms of factors intrinsic to the population (complicated "steady-state" models). Factors extrinsic to the population, such as weather, or changing temperature, velocity, or direction of ocean currents, are assumed to remain constant. The model considers such factors as change in growth of individual animals, harvesting intensity, age at which year classes are first harvested, and natural mortality.

4. Models that are complicated but not steady-state models. This category is "open-ended" in that there is no limit to the degree of complexity that can be built into the model. As many environmental factors as required can be built into the model, in addition to a great deal of detail on competitor species, parasites, predators, diseases, dispersal, and the results of a variety of strategies imposed by man.

The reader should note carefully that there is no question of some of these models being good, in some sense, and others being bad. Rather, some of the models have smaller information demands, and yield less information; others have higher information demands, and yield more information. In cases where it is simply impossible to meet the information demands of the more elaborate types of models, the simpler types perform an important function in showing how to maximize the interpretive value of such information as we do have.

Type 1 models

There is an enormous literature on models for interpreting population changes in terms of the relationship between the size of the reproductive stock and the size of the progeny population produced by this reproductive stock (see the last three paragraphs of Sec. 3.1). Ricker's (1954) monograph explores the relationship between stock-recruitment curves and temporal trends in the population. Chapters 3 and 4 and Sec. 5.1 in this book deal at length with the stock-recruitment curve, and it is also the basis for much important recent work in entomology (Morris, 1959; Neilson and Morris, 1964).

The main advantage of using this method to interpret and predict population behavior is that the only data requirement is the size of the reproductive stock and the size of the progeny population produced by this stock in a consecutive sequence of years. The main disadvantage of the method is that it only works for prediction or explanation where the bulk of the variance in a population system is accounted for by the size of the parental stock (that is, where there is very tight density-dependent regulation). This will not always be the case, as we have already shown (see especially, Secs. 5.3 and 5.4). Furthermore, even if the population is largely under density-dependent control, the type 1 model ignores the fact that all functions of an animal change with age (Sec. 3.1). Serious error may be introduced by treating the parental stock as a large homogeneous mass of individuals, all of whom have the same probability of survival, tendency to cannibalism or less direct competition, fecundity, and tendency to disperse. Shifts in the age structure of the parental stock from time to time will greatly diminish the ability of type 1 models to account for the bulk of the variance in the temporal trend of the population.

However, as Schaefer (1954) has pointed out, it is difficult in the extreme to obtain all the data required for one of the following three types of models for many sea fishes (and also other animal populations), and hence we must often be content to obtain what information we can from this, the simplest of the four types of models.

Type 2 models

The second type of model is one step more sophisticated than the first, in that it utilizes data on the age structure of the catch, or that segment of the population we can sample. An example of application of this approach is given by Royce and Schuck (1954) who used it to predict catches of Georges Bank haddock. Where

$$C_N = \text{number (or weight) of year-class } i \text{ caught in year } N$$
$$C_{N-1} = \text{corresponding number or weight for same year class caught previous year}$$
$$E_N = \text{total effort expended to catch or collect all year classes in year } N$$
$$E_{N-1} = \text{total effort expended in year } N - 1$$

the Royce-Schuck formulation can be represented by a family of regression equations, one for each age group of animals, of the form

$$C_N = a + bC_{N-1} + cE_{N-1} + dE_N$$

where a, b, c, and d are regression coefficients obtained through regression analysis of the data for a consecutive series of years, and the four parameters presumably take different values for each of the age groups on which the analysis is performed.

As with the type 1 model, the type 2 model only works well where the environment is quite stable. Royce and Schuck found that 83 percent of the year-to-year variability in catch was accounted for by their model, so that in a reasonably constant environment, this approach may have great predictive usefulness. However, where the environment is unstable, this method breaks down. First, as in the case of the smallmouth bass discussed in Sec. 5.4, wide variations in environmental conditions from year to year may cause great variability in the strength of the youngest year class entering the vulnerable array of ages for a resource, from one year to another. Also, there may be wide variations in the size of the animals of a given age, causing different vulnerability to the gear for animals of the same age, from one year to another. Marked changes in environmental conditions from one year to another may also result in great variation in catch per unit effort which does not reflect corresponding changes in population abundance.

Since the type 2 model does not consider factors governing recruitment, it does not show us how to maximize recruitment, with respect to biomass or numbers. Since recruitment may be one of the most important aspects of productivity, the model has, in principle, limited usefulness.

However, as with the type 1 model, this approach is important in that it shows us how to make the best use of the information we have, when only limited information is available. The big advantage of this method is that for a population in a stable environment, it gives high predictive reliability in exchange for minimal information that can be obtained cheaply and readily.

Type 3 models

Considerably more ambitious and information-hungry are models that attempt to describe population processes in some detail, while still assuming a constant extrinsic environment. Such models consider not only the numbers of animals present at each age, but also their growth rate. The most elaborate, widely applied, and discussed model of this type was proposed by Beverton and Holt (1956, 1957) for use in the understanding and management of commercially exploited stocks of oceanic fisheries. This model was developed as follows.

Consider the rate of change of biomass yield, Y_w, with respect to time t. The average weight of fish of age t is given by w_t, and F denotes the instantaneous fishing (or harvesting) mortality coefficient, where N_t is the number of t-aged fish still alive in the typical year class, then

$$\frac{dY_w}{dt} = FN_t w_t \qquad (11.1)$$

gives the rate of change of biomass yield for the typical year class with respect to its age. The problem is to evaluate N_t and w_t. Let t_p represent the age at which fish first enter the area where they are liable to be caught, and let t'_p represent a subsequent age, at which fish are large enough to be caught by the particular type of gear in use. Between these two ages, only natural mortality is operating. Where M is the instantaneous natural mortality rate and C is the constant of integration, we have

$$\frac{dN}{dt} = -MN$$

or
$$\ln N_t = -Mt + C \qquad (11.2)$$

The number of individuals at age t_p is R, the number of recruits. Therefore, when $N_t = R$ and $t = t_p$, we have

$$\ln R = -Mt_p + C$$

or
$$C = \ln R + Mt_p$$

Substituting in (11.2), we obtain

$$\ln N_t = -Mt + \ln R + Mt_p = \ln R - M(t - t_p)$$

or
$$N_t = Re^{-M(t-t_p)} \qquad (11.3)$$

The number of young fish surviving to the age t'_p at which they are first liable to capture by the gear is given by

$$N'_{t_p} = Re^{-M(t'_p - t_p)} \qquad (11.4)$$

After this age, the year class is subject to fishing mortality as well as natural mortality, so we have

$$\frac{dN}{dt} = -(F + M)N \qquad (11.5)$$

If we write (11.4) as

$$R' = Re^{-Mp} \qquad (11.6)$$

we can combine (11.5) and (11.6) to produce

$$N_t = R'e^{-(F+M)(t-t'_p)} \qquad (11.7)$$

We now need an expression for w_t. Beverton and Holt use a theory of Von Bertalanffy (1938, 1949) for this purpose. Let

$$\frac{dw}{dt} = \text{rate of change in weight of animal}$$

$$H = \text{coefficient of anabolism}$$

$$D = \text{coefficient of catabolism}$$

$$s = \text{magnitude of "resorbing surfaces" of animal}$$

$$w = \text{body weight}$$

Von Bertalanffy assumes that

$$\frac{dw}{dt} = Hs - Dw \tag{11.8}$$

If we assume that where l represents the length of the animal, $s = pl^2$, and $w = ql^3$, we have

$$\frac{dw}{dt} = \frac{d(ql^3)}{dt} = \frac{dw}{dl}\frac{dl}{dt} = 3ql^2\frac{dl}{dt} \tag{11.9}$$

Hence, from (11.8) and (11.9),

$$\frac{dl}{dt} = \frac{1}{3ql^2}\frac{dw}{dt} = \frac{1}{3ql^2}(Hs - Dw) = \frac{1}{3ql^2}(Hpl^2 - Dql^3) = \frac{Hp}{3q} - \frac{Dl}{3} \tag{11.10}$$

Now, if we set

$$\frac{Hp}{3q} = E \quad \text{and} \quad \frac{D}{3} = K$$

(11.10) reduces to

$$\frac{dl}{dt} = E - Kl$$

or

$$\int \frac{dl}{E - Kl} = \int_0^t dt$$

or

$$-\frac{1}{K}\ln(E - Kl) = t - t_0$$

and

$$l = \frac{E}{K} - \frac{1}{K}e^{-K(t-t_0)} \tag{11.11}$$

Now E/K is the upper asymptote to which the length of the animal grows. Setting $L_\infty = E/K'$ (where K' is a constant), (11.11) can be rewritten as

$$l = L_\infty(1 - e^{-K'(t - t_0)}) \tag{11.12}$$

Now $w = ql^3$, so we can convert (11.12) to an expression for the weight of an animal at any age, and obtain

$$w_t = W_\infty(1 - e^{-K'(t - t_0)})^3 \tag{11.13}$$

If we expand the cubic term, it may be rewritten as

$$w_t = W_\infty \sum_{n=0}^{3} \Omega_n e^{-nK'(t - t_0)} \tag{11.14}$$

where

$$\Omega_0 = +1 \qquad \Omega_2 = +3$$
$$\Omega_1 = -3 \qquad \Omega_3 = -1$$

Now we can substitute from (11.7) for N_t, and from (11.14) for w_t in (11.1) to obtain

$$\frac{dY_w}{dt} = FR'W_\infty e^{-(F + M)(t - t'_p)} \sum_{n=0}^{3} \Omega_n e^{-nK'(t - t_0)} \tag{11.15}$$

The yield to a fishery from the entire life span of a year class is found by integrating (11.15) with respect to t, between the limits t'_p and t_λ, where t_λ is the greatest age attained by any members of the year class. Also, from (11.6),

$$R' = Re^{-Mp}$$

so that (11.15) becomes, using the rule that

$$\int e^{-at} dt = -\frac{1}{a} e^{-at}$$

$$Y_w = FRe^{-Mp} W_\infty \sum_{n=0}^{3} \frac{\Omega_n e^{-nK(t'_p - t_0)}}{F + M + nK'} (1 - e^{-(F + M + nK)t_\lambda}) \tag{11.16}$$

This particular conceptual framework allows us to maximize the yield per recruit, Y_w/R, by manipulating F and t'_p. This model clearly yields a great deal more insight into the dynamics of a resource, and the interaction of exploitation with the resource, than did the previous two models. However, it also has a number of drawbacks. First, it assumes constant environmental conditions, whereas several of the populations discussed in this book are almost entirely regulated by very variable environmental conditions. Second, as Beverton and Holt have recognized in extensions of this basic model, it is necessary to build into (11.16) the ability to deal with density-dependence of R on spawning-stock size at previous points in time. Also, the effects of dispersal, interspecific competition, predation, parasitism, disease, and numerous other factors must be built into (11.16) if it is to be universally applicable.

Type 4 models

Because of these difficulties, general and flexible types of models have been proposed (Watt, 1956, 1961, 1962b), and applied (Watt, 1959a, Morris, 1963). For example, let

N_{ij} = number of individuals of ith year class surviving to jth birthday
S_{in} = probability of survival of individual in ith year class from $n - 1$ to n
N_{i0} = number of individuals of ith year class hatched, spawned, born, or germinated

Then
$$N_{ij} = N_{i0}S_{i1}S_{i2} \cdots S_{ij} = N_{i0} \prod_{n=1}^{j} S_{in} \qquad (11.17)$$

(S_{in} represents functions of all the relevant environmental factors).

Similarly, where G_{ij} represents the average weight growth of individuals of the ith year class during their jth year of life (that is, G_{ij} is a multiple of weight at the $j - i$th birthday), we may write, for weight at the jth birthday:

$$w_{ij} = w_{i0}G_{i1}G_{i2} \cdots G_{ij} = w_{i0} \prod_{n=1}^{j} G_{ij} \qquad (11.18)$$

The biomass of year-class i present at age j is then given by

$$P_{in} = N_{in}w_{in} = \left(N_{i0} \prod_{n=1}^{j} S_{in} \right)\left(w_{i0} \prod_{n=1}^{j} G_{in} \right) \qquad (11.19)$$

It is understood that all the values $S_{in}, n = 0, 1, \ldots, j, w_{i0}$ and $G_{in}, n = 1, 2, \ldots, j$ are functions of all relevant independent variables. The analysis methods used to determine the appropriate mathematical form for these functions are explained in the following chapter, and the methods used to determine the best values for the parameters in the functions are explained in Chap. 10.

Once the models have been derived, and the parameters have been estimated, we can use the methods outlined in Chaps. 12 and 13 to find out how to maximize the productivity of a useful species, or to minimize the productivity of a pest.

11.2 METHODS OF DEVELOPING LARGE-SCALE SYSTEMS MODELS

When we construct a mathematical model for a complex process, instead of merely selecting an empirical formula useful for interpolation, as in regression analysis, we try to write one or more differential, difference, or difference-differential equations based on insight into the mechanics of the process under study.

Before discussing the methods for building large models, the difference between the three kinds of equations mentioned will be explained.

Differential equations

Differential equations arise whenever a rate of change of some variable with respect to another can be expressed in terms of a continuous set of variate values of one or more other variables. Consider, for example, the simple attack equation (Watt, 1959b) in which

$$\frac{dN_A}{dN_0} = aP^{1-b}(PK - N_A)$$

where N_A = numbers attacked
$\quad N_0$ = numbers vulnerable to attack
$\quad P$ = numbers of attackers
$\quad a, b, K$ = constants; all numbers have been measured in the same universe

This is a differential equation because the variables P and N_A are assumed to be able to vary continuously. The notion of a differential equation can in fact be generalized considerably, to include accelerations and higher-order derivatives, squares of derivatives, or higher degrees. Partial differential equations, with several independent variables, are possible, as in

$$\frac{\partial^2 Z}{\partial x^2} + \frac{\partial^2 Z}{\partial y^2} = x^2 + y$$

The common element in all differential equations is the notion that the variables can take any value on a continuous scale (including noninteger values). A moment's reflection indicates that such equations do not give an accurate picture in many biological systems, unless we handle them in a special way. Consider an animal laying eggs. It lays 8 eggs, 1 egg, 233 eggs, or 119,236 eggs, but never 6.237 eggs. Consider an area of ground, divided into squares 10 meters by 10 meters. If there are nine mice living in one such square, emigration from the square will occur one mouse at a time, not 1.23 mice at a time. Many biological processes have this character, in which variables can only take discrete values. Difference equations are often more appropriate than differential equations, or at least differential equations must often be treated as difference equations.

Difference equations

Difference equations are used wherever the variables take only discrete values. For example, the differential equation for exponential growth is

$$\frac{dY}{dt} = rY \tag{11.20}$$

In contradistinction, the difference equation for exponential growth is

$$\frac{Y_{k+1} - Y_k}{t_{k+1} - t_k} = \frac{\Delta Y}{\Delta t} = rY_k \tag{11.21}$$

Equation (11.20) states that at any instant of time, the rate of change of growth in Y will be proportional to the value of Y at that instant. Equation (11.21) states that the difference in magnitude of Y from time t_k to time t_{k+1} will be r times the value of Y at t_k. The essential difference is that Y and t are assumed to vary only by discrete (that is, steplike or integer) values. Goldberg (1958) has written an excellent introduction to this field for behavioral scientists.

Difference-differential equations

Suppose, now, that we wished our dependent variable to be a derivative, as in differential calculus, but our independent variables are to be treated as discrete values at some prior point in time. Here the exponential growth equation would take the form

$$\frac{dY}{dt} = Y_{h-1}(t) \tag{11.22}$$

Such equations are treated at length by Bellman and Cooke (1963).

No matter what type of equation is most suitable for describing a complex system, the problem faced in all cases is that of determining the particular structure for the model which gives the most realistic description of the system.

If large complex problems have many interaction terms, the most convenient way to approach model construction is by trying to split the problem into bits, so the bits can be dealt with individually. For example, suppose we know that Y, the volume of timber grown in a forest per unit time, is a function of 15 other variables, so we could write

$$Y = f(X_1, X_2, \ldots, X_{15}) \tag{11.23}$$

The first step is to determine how to split (11.23). Do we have

$$Y = \frac{f(X_1, X_2, X_3)}{f(X_4)} + f(X_5, \ldots, X_{15})$$

or

$$Y = [f(X_1, X_2, X_3)][f(X_4, X_5, X_6)][f(X_7, \ldots, X_{15})]$$

How to combine such component terms, or submodels, to make a large model will be apparent from the following two probability theorems.

The addition theorem The probability that one out of m events, any two of which are mutually exclusive, occurs is equal to the sum of the probabilities of the occurrence of each event separately.

The multiplication theorem The probability that two stochastically independent events occur together is equal to the product of the probabilities of the occurrence of each event separately.

Given that the researcher has determined how to split his problem into pieces, how does he write equations for a submodel? Suppose that the whole model is $Y = A \cdot B \cdot C \cdot D$, and the submodel we are concerned with is

$$A = f(X_1, X_2, X_3) \tag{11.24}$$

To determine the particular form of (11.24), the analyst will sort the data, using cards, to separate out the effects of X_1, X_2, and X_3, obtaining tabulations as in Table 11.1. After making such a table, he will plot families of

TABLE 11.1

Result from sorting and tabulating data to reveal form of functional relationship

Number of cards in subdeck	A		X_1		X_2		X_3	
	Total	*Mean*	*Total*	*Mean*	*Total*	*Mean*	*Total*	*Mean*
5				2		2		40
5				2		2		80
5				2		2		120
5				2		4		40
5				2		4		80
5				2		4		120
5				2		8		40
				2		8		80
				2		8		120
								and so on

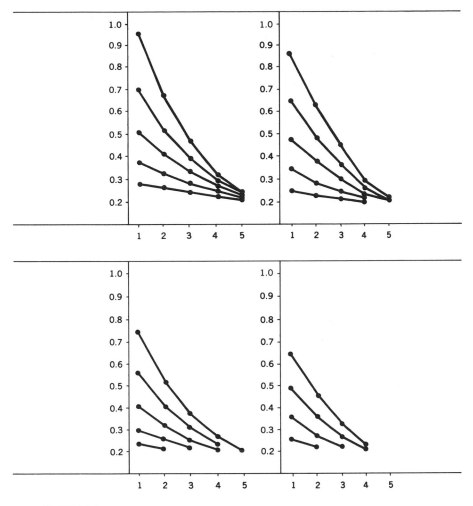

FIGURE 11.1

Plot of row means from table as in Table 11.1, to determine form of functional relationship $A = f(X_1, X_2, X_3)$. In this figure, A is plotted on the y axis, X_1 is plotted on the x axis, different values of X_2 are represented by the different lines in each panel, and different values of X_3 are represented by the four different panels (Watt, 1961).

graphs, as in Fig. 11.1. The order in which he then develops sub-submodels to express the effects of X_1, X_2, and X_3 on A will depend on the proportion of the variance in A accounted for by each of the X's. He should begin by modeling the effect of that independent variable for which the graph of A on X has steepest slope and least scatter.

The researcher will then make various assumptions about the mode of operation of, say, X_3 on A, on the basis of his understanding of the phenomenon, and check these by examining the graphs. It is best to check assumptions systematically by asking a series of questions, the answers to which form a logical branched tree. For example, using only a small list of questions, we form a logical tree as in Fig. 11.2.

The following list of questions used in Fig. 11.2 could be expanded to make a tree from which all known equations could be derived.

1. Is dA/dX_3 proportional to X?
2. Is dA/dX_3 proportional to A? (That is, are we dealing with some form of compound growth law?)
3. Is dA/dX_3 inversely proportional to X_3?
4. Does dA/dX_3 approach zero as A approaches some upper asymptote A_{max}?
5. Does dA/dX_3 approach infinity as X_3 approaches some lower limit X_{3min}?

Integral forms of some of the most commonly encountered equations are given below.

$$A - 32 \qquad Y = a + bX$$

$$A - 28 \qquad Y = a + b \ln X$$

$$A - 24 \qquad \ln Y = a + bX \qquad \text{or} \qquad Y_x = Y_0 e^{bX}$$

$$A - 16 \qquad Y = a + bX - c'X^2$$

$$A - 20 \qquad Y = aX^b$$

$$A - 22 \qquad Y = \frac{Y_{max}}{1 + e^c - b'X} \qquad \text{logistic}$$

$$A - 30 \qquad Y = Y_{max}(1 + e^{-bX})$$

Once an elementary equation has been chosen to describe the effect of X_3 on A, this equation is integrated (using tables of integrals if necessary), and transformed into a form suitable for testing. Graphical testing of the validity of the equation is done as follows. Suppose we decide that $A - 30$ describes the effect of X_3 on A, while X_1 and X_2 are constant. That is

$$\frac{dA}{dX_3} = b(A_{max} - A)$$

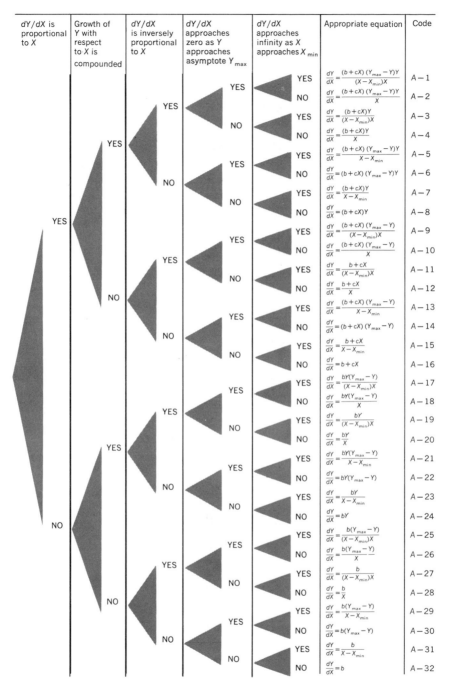

dY/dX is proportional to X	Growth of Y with respect to X is compounded	dY/dX is inversely proportional to X	dY/dX approaches zero as Y approaches asymptote Y_{max}	dY/dX approaches infinity as X approaches X_{min}	Appropriate equation	Code
YES	YES	YES	YES	YES	$\frac{dY}{dX} = \frac{(b+cX)(Y_{max}-Y)Y}{(X-X_{min})X}$	A-1
				NO	$\frac{dY}{dX} = \frac{(b+cX)(Y_{max}-Y)Y}{X}$	A-2
			NO	YES	$\frac{dY}{dX} = \frac{(b+cX)Y}{(X-X_{min})X}$	A-3
				NO	$\frac{dY}{dX} = \frac{(b+cX)Y}{X}$	A-4
		NO	YES	YES	$\frac{dY}{dX} = \frac{(b+cX)(Y_{max}-Y)Y}{X-X_{min}}$	A-5
				NO	$\frac{dY}{dX} = (b+cX)(Y_{max}-Y)Y$	A-6
			NO	YES	$\frac{dY}{dX} = \frac{(b+cX)Y}{X-X_{min}}$	A-7
				NO	$\frac{dY}{dX} = (b+cX)Y$	A-8
	NO	YES	YES	YES	$\frac{dY}{dX} = \frac{(b+cX)(Y_{max}-Y)}{(X-X_{min})X}$	A-9
				NO	$\frac{dY}{dX} = \frac{(b+cX)(Y_{max}-Y)}{X}$	A-10
			NO	YES	$\frac{dY}{dX} = \frac{b+cX}{(X-X_{min})X}$	A-11
				NO	$\frac{dY}{dX} = \frac{b+cX}{X}$	A-12
		NO	YES	YES	$\frac{dY}{dX} = \frac{(b+cX)(Y_{max}-Y)}{X-X_{min}}$	A-13
				NO	$\frac{dY}{dX} = (b+cX)(Y_{max}-Y)$	A-14
			NO	YES	$\frac{dY}{dX} = \frac{b+cX}{X-X_{min}}$	A-15
				NO	$\frac{dY}{dX} = b+cX$	A-16
NO	YES	YES	YES	YES	$\frac{dY}{dX} = \frac{bY(Y_{max}-Y)}{(X-X_{min})X}$	A-17
				NO	$\frac{dY}{dX} = \frac{bY(Y_{max}-Y)}{X}$	A-18
			NO	YES	$\frac{dY}{dX} = \frac{bY}{(X-X_{min})X}$	A-19
				NO	$\frac{dY}{dX} = \frac{bY}{X}$	A-20
		NO	YES	YES	$\frac{dY}{dX} = \frac{bY(Y_{max}-Y)}{X-X_{min}}$	A-21
				NO	$\frac{dY}{dX} = bY(Y_{max}-Y)$	A-22
			NO	YES	$\frac{dY}{dX} = \frac{bY}{X-X_{min}}$	A-23
				NO	$\frac{dY}{dX} = bY$	A-24
	NO	YES	YES	YES	$\frac{dY}{dX} = \frac{b(Y_{max}-Y)}{(X-X_{min})X}$	A-25
				NO	$\frac{dY}{dX} = \frac{b(Y_{max}-Y)}{X}$	A-26
			NO	YES	$\frac{dY}{dX} = \frac{b}{(X-X_{min})X}$	A-27
				NO	$\frac{dY}{dX} = \frac{b}{X}$	A-28
		NO	YES	YES	$\frac{dY}{dX} = \frac{b(Y_{max}-Y)}{X-X_{min}}$	A-29
				NO	$\frac{dY}{dX} = b(Y_{max}-Y)$	A-30
			NO	YES	$\frac{dY}{dX} = \frac{b}{X-X_{min}}$	A-31
				NO	$\frac{dY}{dX} = b$	A-32

FIGURE 11.2

Logical branching tree for obtaining appropriate differential equation to describe a set of data (Watt, 1961).

and integrating gives

$$- \ln (A_{max} - A) = bX_3 - \ln A_{max}$$

and since we wish first to see if b is indeed constant, this can be rearranged to yield

$$\ln \left(\frac{A_{max}}{A_{max} - A} \right) = bX_3 \qquad (11.25)$$

If we plot the transformed data on semilog graph paper, it will yield a straight line if (11.25) describes the relation between A and X_3, X_1 and X_2 being held constant. If a straight line is not obtained, some other elementary integral form will be needed in place of $A - 30$. If there is rectilinearity, we proceed to the next step.

The next step involves plotting the data in terms of some other independent variable, say X_2. We plot, against X_2,

$$Z = \frac{\ln \left(\dfrac{A_{max}}{A_{max} - A} \right)}{X_3}$$

If plotting this transformation against X_2 yields a straight line parallel to the X_2 axis, X_2 has no significant effect on A. Otherwise, we know that b is not a constant, but is some function of X_2. If so, we decide the appropriate form of the function for X_2, and test this guess by plotting the appropriate transformation to see if we get a straight line. For example, if we decide that

$$\frac{dZ}{dX_2} = \frac{cZ}{X_2}$$

then

$$Z = gX_2^c$$

which we test by plotting Z against X_2 on log-log graph paper.

If we get a straight line, we know that

$$\ln \left[\frac{\ln \left(\dfrac{A_{max}}{A_{max} - A} \right)}{X_3} \right] = \ln g + c \ln X_2$$

In this case, we can proceed to determine the effect of X_1 on A. We do this by plotting, against X_1,

$$\frac{\ln \left(\dfrac{A_{max}}{A_{max} - A} \right)}{X_3 X_2^c}$$

When this step is complete, we solve the resultant form for A. The other submodels are handled in the same way, and finally the whole model is put together.

Now that the basic process of cyclical model building and testing has been explained, there are many questions on detail that need to be considered.

First, where we have postulated that A is a function of X_1, X_2, and X_3, how do we determine the order in which we will test the significance of X_1, X_2, and X_3? We proceed as in stepwise multiple regression (Chap. 10), and examine the effects of independent variables starting with the most important, and proceeding to the least important. As explained in Chap. 10, the reason for proceeding in this sequence has to do with the logic of statistical testing: by leaving the least significant variables until last, we have stronger error terms with which to test the significance of the first variables removed, using the variance ratio, or F test.

Second, how do we decide the fundamental structure of the whole systems model into which we incorporate terms for submodels, into which we have in turn built sub-submodels? One way of splitting up a model for a system is to structure it in terms of time. Consider, for example, an equation to account for the change in numbers of an insect population from one generation to another. We will need the following symbols. Let

N_t = density of adult insects present immediately prior to oviposition in year t

N_{t+1} = density of adult insects present at corresponding time in year $t+1$

$T_{t:t+1}$ = N_{t+1}/N_t, that is, trend index of population from t to $t+1$

P_t = proportion of N_t consisting of females that oviposit at t

F_t = mean fecundity of $N_t P_t$ females

S_E = proportion of eggs surviving to eclosion

$S_I, S_{II}, \ldots, S_{VI}$ = proportion surviving of first instar larvae, second instar larvae, . . . , sixth instar larvae

S_p, S_A = proportion surviving of pupae and adults, respectively; adult populations are measured just before oviposition in year $t+1$

We have, then, the model

$$T_{t:t+1} = \frac{N_{t+1}}{N_t} = P_t F_t S_E S_I \cdots S_{VI} S_p S_A$$

This equation structures our system in a commonsense sequence, and is called a model. It is built up out of terms such as

$$S_E = f(X_1, X_2, X_3, X_4, \ldots, X_n)$$

which are called submodels, and these in turn may be built up out of sub-submodels, which, for example, relate the variate value of X_4 to factors that govern X_4.

Another matter that needs to be considered is the list of five questions which produced the 32 (2^5) equations in Fig. 11.2. The list is by no means complete; it was presented to show that most of the simple, commonly encountered differential equations can be derived from a simple logical branching tree. Biologists will be able to think of a variety of other questions that might be asked to expand the logical tree enormously. It is, however, much easier to think of a differential equation than to solve it, and the biologist will need a powerful aid if he does much of this type of model building. Apart from the usual handbooks, the most useful such aid is the compendium of methods of solution by Murphy (1960).

A logical branching tree is not the only way to arrive at the equation that best describes a particular process (Turner, Monroe, and Lucas, 1961; Turner, Monroe, and Homer, 1963). Another approach is to make use of general, underlying equations from which a great variety of commonly used models may be derived as special cases. By using such formulas, we can employ curve-fitting techniques to determine the values of various parameters that provide best fit of the general equation to a particular body of data. The particular values the parameters take indicate which of the special cases we are dealing with. To illustrate, consider the differential equation used by Turner and his associates.

$$\frac{d\eta}{d\xi} = \frac{\delta(\eta - \alpha)}{\xi - \gamma\delta}$$

This equation integrates to produce

$$\eta = \alpha + \beta(\xi - \gamma\delta)^\delta \tag{11.26}$$

The constant, δ, determines which special case we have represented in our data. If δ is a positive integer, we have a polynomial process, which is nonlinear if δ exceeds 1. As δ approaches either positive or negative infinity, (11.26) approaches the exponential model

$$\eta = \alpha + \beta l^{-\xi/\gamma}$$

The following values of $2/\delta$ produce the given simple models

$2/\delta$	model
-4	inverse square root law
-3	inverse two-thirds law
-2	rectangular hyperbola
-1	inverse square law
0	exponential
1	parabola
2	straight line

When $2/\delta$ is negative, the curves have two asymptotes; when $2/\delta$ is zero, they have one; and when $2/\delta$ is positive, they have no asymptotes.

Another such basic equation has been proposed by Grosenbaugh (1965). Consider

$$Y = H + A[e^{(N^2 - 1)U} - NU]^{NM + 1} \qquad (11.27)$$

where U represents a one- or two-parameter function such as

$$U = e^{-B(X - G)} \qquad (11.28)$$

Equation (11.27), with variants (11.28) inserted, takes the form of almost all well-known mathematical functions for two variables, depending on the values assigned to N and M. The idea behind Grosenbaugh's research is to facilitate the computations for iterative regression, by developing tables of the partial derivatives used in equations (10.37) for each of the elementary functions corresponding to a particular N and M pair of values. This suggests a large-scale computer routine for doing iterative regression, in which the computer starts with only the raw data. In the first part of the program, the computer determines the values of N and M and the form of U. Then, using these three pieces of information, it selects from memory the appropriate derivatives for calculating the values in the set of equations (10.37). The second part of the program is the iterative regression routine.

It should be noted that both the logical branching tree method and the general equation method suggest means of programming a computer so that it decides which model is best to describe a given body of data.

Model-building techniques of the type we have described can obviously lead to extremely complicated models. The question naturally arises as to how one can perform logically valid statistical or other tests on the model, to determine how accurately it describes reality. One approach is to test the whole model statistically piece by piece (submodel, or sub-submodels separately), using iterative regression where necessary, because parameters enter submodels nonlinearly. Unfortunately, this approach has a logical weakness, in that we lose a degree of freedom every time a new parameter enters an equation, and our error terms in statistical tests lose strength. This means that as our models become more and more complex, it becomes progressively more difficult to demonstrate statistically that they do not in fact describe the data. In other words, *any* sufficiently complicated model is a highly flexible interpolatory formula, even though its structure bears no relationship to the structure of the process we are trying to describe. The reader can satisfy himself on this point by noting that in tables of F values, the smaller the number of numerator degrees of freedom, the higher the F value must be to produce significance at a given probability level. To circumvent this degrees-of-freedom logical impasse, the following procedure is suggested. Obtain data from one set of surveys, or experiments on components of whole systems, then build a systems model. The model is tested by seeing how it can predict the outcome from another set of surveys or experiments on whole systems, or systems of a different type. No degrees of freedom are lost in such a test. For example, we might build a systems model describing a historical process, using data from experiments on particular components of the whole process. Then, having used these submodels to build the whole model, the model is tested against data on the history of the whole system. Such an approach has been used by Holling (1966) and Watt (1955).

This concludes a general discussion on methods of developing models. A typical situation that arises subsequent to following the steps outlined is for the researchers to find that their systems model only accounts for 40 percent of the variance in the system under study. It often develops that a major proportion of the entire model-building effort is expended on efforts to refine the initial systems model. A number of procedures are available to help with this step.

1. Using the model, obtain values of Y calculated for each case on which data were obtained. Then make a large graph in which "Y calculated" values are plotted against "Y observed" values. Every point on the graph should be labeled as to the characteristics of the sample that produced the datum. When all points are plotted, we study the graph to observe if particular sets of sample values deviate consistently above or below the 45° line on which all points would lie in an error-free situation, in an accurate model. Such sets of values may provide clues about factors, omitted from the analysis, which should have been measured.

2. Another method for discovering structural weakness in the model

is to plot Y observed and Y calculated values against each of the independent variables in turn. Systematic departures from a straight line parallel to the x axis indicate that a term in the model does not mimic nature accurately.

3. A third source of difficulty in a systems model occurs if one or more variables have been incorrectly measured, defined, or coded. Suppose we know from field experience that torrential downpours or high-velocity wind gusts can drive delicate insects off their food plants and onto the ground where they may starve, or be drowned. It is not adequate to express the force of such factors in standard meteorological units, such as inches of rain collected in a rain guage, or miles of wind per hour moving past a stationary point. Rather, we should record and enter into our equations the biologically relevant variables: maximum drop velocity within a 24-hour period, or maximum gust velocity.

4. If some of the independent variables that entered into multiple regression analyses were of the form $X_1 X_2$, then computer output might reveal that cross product or other interaction terms are needed in the systems model.

The experimental components approach

The preceding discussion outlined mathematical techniques by which one could develop a systems model for a system on which a large body of data had already been collected. However, such a procedure is only useful as a "macro" approach; in order to obtain the kind of insight into the mechanics of a system that will allow us to manipulate it profitably, more detailed information about the quantitative nature of processes will be required than those available from field studies. Furthermore, the macro approach, typical of field studies, limits us to the ranges of variate values provided by nature. A "micro" approach is called for in addition to the macro approach. Holling (1961, 1965) has presented an elaborate exposition of the logic and methodology of the micro approach, which he calls *experimental components analysis*; the following discussion is based in part on his publications.

The essence of the experimental components approach is that the processes of experimentation and mathematical model construction are conducted as two interlocking parts of an integrated program. A systems model grows out of a sequence of steps, which proceed as follows.

The process we wish to study (for example, predation, parasitism, dispersal, reproduction) is conceived of as being describable by a systems model constructed out of component processes, or constituent fragments. We determine by experiment and observation what these fragments are for the process in question, then through a priori considerations sort them into two groups: basic and subsidiary. A *basic component* of a process is a constituent factor that invariably operates where that process occurs. For example, prey density and predator density must always operate where predation occurs. On the other hand, predator speed is a *subsidiary component*, because it is not relevant in the case of an ambush predator, or a filter-feeder, or a Portuguese man-of-war. In short, we assume that for any process, such as predation, there is a basic model that explains the underlying processes common to all species pairs of predators and prey. We further assume that the great diversity of different predator-prey processes is caused by additions to this basic model that occur in various situations as subsidiary components. A sequence of experiments is conducted, beginning with experiments on a small group of basic components of a process. The experiments are designed to analyze the operation of these components in sufficient depth that a mathematical model can be built to describe the operation of the components. The model is chosen to meet two desiderata: it must give a statistically satisfactory fit to the data and it must incorporate real insight into the phenomenon.

Additional experiments are conducted until all the basic components of the process have been analyzed and modeled and can be incorporated into the systems model. Then the whole process is repeated for any subsidiary components that can be found in the various situations where the type of process under study occurs. Finally, a systems model is constructed to simulate the process on a computer. Any particular instance of the process (representing a particular variant of the model) can be simulated merely by informing the computer, by means of control and parameter cards, which terms are to be added to the basic model, and what values the relevant parameters are to take.

The preceding is a very general description of what is in practice a highly complex experimental and mathematical procedure. In order to give any real insight into the kind of reasoning used in the experimental components approach, it is necessary to describe a particular application. Holling's program on predation will be explained to highlight the sequence of steps in his reasoning. There is no need to discuss his model in detail here since it is explained in Sec. 11.6.

Holling first decided, on the basis of observations, experiments on many predator-prey systems, and study of the literature, that five groups of variables affect predation:

Density of the prey species
Density of the predator species
Characteristics of the prey species
Characteristics of the predator species
Characteristics of the environment

The first two variables must operate in every predator-prey situation: they are basic variables. There are situations in which the last three groups of variables do not affect predation: they are subsidiary variables. The two basic factors can each have their effect through numerous causal pathways. This means there are four components of the response to prey density by predators, each of which can be subdivided as follows.

1. Searching rate:
   a. Predator speed relative to prey speed
   b. The maximum distance from which a predator notices and attacks a prey, and the effect of hunger on this distance
   c. Proportion of attacks resulting in successful capture

2. Time predators are exposed to prey:
   a. Time predators spend in nonfeeding activities
   b. Time predators spend in feeding activities

3. Time spent handling each prey:
   a. Time spent in pursuit for prey
   b. Time spent eating
   c. Digestive pause, while predator is not hungry enough to attack

4. Hunger:
   a. Rate of digestion and assimilation
   b. Maximum food capacity of the gut

Experiments are designed to obtain the form and the parameter values, in mathematical equations, for each of these components and subcomponents. As such equations are developed, they are incorporated into a systems model which is carefully checked against experimental data on the whole system. Finally, the systems model, when complete, serves as the basis for computer simulation and optimization studies. These give rise to new insight about the most important criteria for evaluating biological-control agents, and the most effective parameter values for those criteria.

A similar components analysis can be conducted for any type of complex process in the ecology-behavior domain. It should be emphasized that the key concept in this whole approach is the intimate reciprocal feedback between the experimental program and computer analysis and simulation, with the model arising from one type of data and being subjected to test against other data at each stage in its development. Whenever the simulation studies predict a result not corroborated by experiment, the model clearly needs reexamination.

11.3 WEATHER

In developing mathematical descriptions for the effect of weather on living organisms, we are faced with the following six problems.

1. Which mathematical model for the effect of constant temperature on the rate of biological reactions gives the most satisfactory description of the effects of fluctuating temperatures for the data under consideration?

2. How is this model modified to account for the fact that measured temperatures in nature are fluctuating, not constant?

3. How do simultaneously varying temperature and relative humidity affect living organisms?

4. How do we express the effect of hot or cold lethal temperatures on the probability of survival?

5. How do we discover which meteorological indices constitute the appropriate independent variable for our model?

6. Suppose meteorological stations are some distance away from the site at which the biological data were collected. How far away can the station be and still supply data useful in interpreting the biological information? What topographic features of the landscape determine this distance?

Many thousands of papers have reported the effect of temperature and humidity on various organisms. A smaller, but imposing body of literature attempts interpretation of these data in terms of several mathematical models. Nevertheless, for a variety of reasons, this monumental amount of work has not yet produced a temperature-humidity model that is completely satisfactory for use in large mathematical population models.

Contributions have been made from a wide range of specialties among which there is incomplete exchange of data and theories. A cursory examination of reviews on temperature-humidity effects reveals the exceedingly diverse viewpoints from which precisely the same phenomenon can be studied: insect physiology (Chauvin, 1956; Wigglesworth, 1950); general physiology (Fry, 1947, 1958; Precht et al., 1955); insect ecology (Andrewartha and Birch, 1954; Messenger and Flitters, 1959); cell physiology (Belehradek, 1957a,b); mathematical physiology (Hearon, 1952); and reaction-rate kinetics (Johnson, Eyring, and Polissar, 1954). It is difficult for any one individual to comprehend all this literature. The result, for example, is that no insect ecologist has yet attempted to apply modern reaction-rate theory to interpretation of the effect of temperature on the whole insect. Conversely, the reaction-rate theoreticians do not seem to have generalized their theories to cope with facts and theories about fluctuating temperatures (as have Messenger and

Flitters, 1959), or the combined effect of temperature and humidity (as has Da Fonseca, 1958). Insufficient attention seems to have been paid to Fry's (1947) important point that temperature-activity curves may reflect the difference between the effect of temperature on active and basal metabolism. If basal metabolism rises with increasing temperature, and active metabolism rises and then falls with increasing temperature, activity (determined by active metabolism less standard metabolism) must rise and then fall with increasing temperature. Fry presents this notion as an alternative to that of destruction of enzymes at high temperatures as an explanation for the descending limbs of temperature-activity curves.

Without exception, every mathematical model ever put forth to account for the effect of temperature on living systems has been or can be criticized for either or both of two reasons: the theoretical basis presented for the model is invalid, or the model simply doesn't fit the type of data it is intended to describe.

Another cause of confusion in this literature is that few studies have ever been presented that compare the fits of two or more models to several sets of data. Janisch (1932) showed that a number of early models did not fit data on embryonic development of *Ephestia kuhniella*, which were fitted reasonably well by a catenary. Satomura (1950) compared the fits of a number of early formulas, including those of Van't Hoff, Arrhenius, and Janisch, and found that none of them fitted his temperature–development-rate data very well. Most authors who have attempted to describe data on the effect of temperature on development rate by a mathematical model have not attempted to demonstrate statistically that the model they used fitted their data better than any of the other well-known models. The result has been that different models are used in different countries to describe precisely the same type of data. For example, the catenary proposed by Janisch (1925, 1927, 1928, 1932) has been used in German journals (Andersen, 1938; Noll, 1942) and in the United States (Huffaker, 1944; Allen and Smith, 1958); the Australians, following Davidson (1942, 1944) have used the logistic (Birch, 1944; Andrewartha and Birch, 1954); and Pradhan's formula (1945, 1946a,b) has been used in Japan (Matsazawa et al., 1957).

There is evidently need to reexamine this problem from two points of view. First, we should evaluate the theoretical justification for various models and construct one or more new models if this seems necessary. Second, it is obviously worthwhile to compare the fits obtained when the various models are tested against typical, well-collected bodies of data.

Our discussion here will progress from consideration of the simplest phenomenon, the effect of constant temperatures on development rate when humidity is held constant, to treatment of increasingly complex situations.

The effect of constant temperatures on development rate

I will not review the literature on mathematical models for the effect of temperature on development rate since so many excellent reviews are available (for example, Chauvin, 1956; Wigglesworth, 1950). The models I shall mention are those that have attracted serious attention in the past five decades.

Crozier's (1924–1925) and Bliss' (1926) interpretation of temperature–reaction-rate data as a sequence of Arrhenius plots joined at angles has been criticized for a number of reasons. Buchanan and Fulmer (1930) pointed out that graphs of a series of joined straight lines could be made to appear like a smooth curve if the scale was changed. Kistiakowsky and Lumry (1949) showed that sharply bending straight lines in Arrhenius plots were theoretically impossible, and supported their position by experimental results that showed gradually curving lines. Hearon (1952) and Johnson, Eyring, and Polissar (1954) summarized the modern theory of reaction rates from two points of view, both of which indicated that Crozier's interpretation almost certainly could not be correct.

Janisch (1925) proposed symmetrical and asymmetrical catenaries to describe the effect of temperature on rate of insect development. The justification was only that this curve gave a good fit to the data. However, Martini (1928) questioned that catenaries gave better fits than even the simple hyperbola, which was used for such data decades ago.

Davidson's (1942) application of the logistic to data on temperature versus rate of development does not seem justified either for theoretical reasons or as giving a good fit to such data. This equation has typically been fitted to temperature-rate data by leaving off points from one or both ends of the empirically obtained curve (Birch, 1944; Browning, 1952). Perhaps one could argue that the logistic is useful for ecological description because only a small proportion survive of the individuals subjected to high temperatures at which the temperature-rate curve is descending. But this is not a valid argument. In Birch's (1944) own work, as many as 81 percent of the *Calandra oryzae* L. eggs survived heating at temperatures 3°C above the temperature that produced most rapid development. Browning (1952) showed that the logistic did not give a very satisfactory fit to temperature and rate of development data.

Pradhan (1945, 1946a,b) has published a new approach to analysis of this type of data. The equation he proposed, like those of Davidson and Janisch, was essentially an empirical formula, in that it gave a reasonable fit to the data, but did not have a physicochemical basis. However, his treatment was very thorough. He considered fluctuating temperatures (1945) and showed how to compute their effects, considered new temperature-recording instruments to facilitate application of his model (1946a), and discussed fitting of his equation in some detail (1946b). His equation may be written as

$$y = y_o e^{-ax^2} \qquad (11.29)$$

where y_o = highest value of developmental index
$\quad y$ = developmental index at temperature t
$\quad x = T - t$
$\quad T$ = temperature corresponding to y_o

One immediately obvious drawback to (11.29) is that it describes a symmetrical curve. Cursory examination of any temperature–development-rate data (for example, Birch, 1944) shows that the descent from the maximum development rate is steeper than the ascent to it. We shall evaluate Pradhan's formula as an interpolatory device in the next section, along with the other formulas under discussion.

The next equation is based on modern work in reaction-rate kinetics, quantum mechanics, biochemistry, and cell physiology. It is discussed fully in modern texts on reaction-rate kinetics (for example, Johnson et al., 1954; Bray and White, 1957); so I shall only give enough detail on the derivation to make clear the nature of the assumptions involved.

Where E_n, the native, active form of an enzyme E, is in equilibrium with (E_d), the reversible, denatured, inactive form; and K_1 is the equilibrium constant, we may write

$$E_o = E_n + E_d \qquad (11.30)$$

and
$$(E_d) = K_1(E_n) \qquad (11.31)$$

by substituting for (E_d) in (11.30),

$$(E_o) = (E_n) + K_1(E_n)$$

and
$$(E_n) = \frac{(E_o)}{1 + K_1} \qquad (11.32)$$

Now suppose we assume that the rate, Z, of a particular biological process is primarily limited by the amount and activity of enzyme E; that is, there is virtual saturation of the enzyme by the substrate. In such a case, the overall reaction proceeds as if governed by a single specific reaction-rate constant k':

$$Z = bk'(S)(E_n) \qquad (11.33)$$

in which (S) represents substrate concentration, a constant, and b is a proportionality constant. If we substitute for (E_n) in (11.33), we get

$$Z = \frac{bk'(S)(E_o)}{1 + K_1} \tag{11.34}$$

According to the modern viewpoint in theoretical physical chemistry,

$$k' = (\kappa kT/h)K^{\ddagger} \tag{11.35}$$

where T = absolute temperature

h = Planck's constant

k = Boltzmann's constant

κ = probability that formation of activated complex will lead to reaction

$K^{\ddagger}$ = constant of equilibrium between normal and activated states

Equation (11.35) is based on the assumption that every elementary rate process can be treated as an unstable equilibrium between reactants and an activated complex with a lifetime of about 10^{-13} second.

Now, from thermodynamics, where R is the gas constant, Δ before a property represents the difference between the initial and final values of the property, and $\Delta F^{\ddagger}$, $\Delta H^{\ddagger}$, and $\Delta S^{\ddagger}$, are the standard free energy, heat, and entropy of activation, respectively, we have

$$K^{\ddagger} = e^{-\Delta F^{\ddagger}/RT} = e^{-\Delta H^{\ddagger}/RT} e^{\Delta S^{\ddagger}/R} \tag{11.36}$$

If we substitute for k', $K^{\ddagger}$, and K_1 in (11.34), we get

$$Z = \frac{b(\kappa kT/h)e^{-\Delta H^{\ddagger}/RT}e^{\Delta S^{\ddagger}/R}(S)(E_o)}{1 + e^{-\Delta H_1/RT}e^{\Delta S_1/R}} \tag{11.37}$$

Belehradek (1957a) notes that (11.37) contains 11 constants, and concludes that "the attempt at a mathematical analysis ends in a tangle that can hardly encourage an experimental cytologist to follow the same path in search of sound conclusions." This comment need not concern us too much, however, because (11.37) for practical purposes reduces to

$$Z = \frac{bTe^{-c/T}}{1 + de^{-a/T}} \tag{11.38}$$

which has only four constants. Unfortunately, there are other criticisms of (11.37). Belehradek (1957a) notes that since enzymes in vitro work much faster than in vivo, molecular resistance must be a very important factor in intracellular reactions, and then (11.37) is not sufficiently complex if applied to the molecular kinetics of intracellular processes. He also asserts that (11.37) simply does not fit chemical data.

Belehradek's point about molecular resistance does not seem relevant, because Johnson and his associates specifically note that the theory of the activated complex, as embodied in (11.37), is intended to be applicable to all elementary rate processes, including diffusion and lubrication.

I have not attempted to fit Belehradek's formula to entomological data, since it will not account for decreasing rate when temperature rises beyond the optimum. Also, careful study of Belehradek's figures (1954, 1957a) reveals slight curvature in many of them.

Biochemical studies shed considerable light on the question of the validity of (11.37). For example, Chadwick (1957) concluded that the apparent energy of activation was itself dependent on temperature. Evidently we can ignore irreversible destruction of the reactive enzyme, because this only occurs at temperatures too high to be of ecological relevance.

The most serious objections raised against (11.37) are that some quite different mechanism may account for the facts, or that (11.37) needs to be made very much more complex. For example, we might develop equations to account for individual rate processes connected serially to form catenary reactions (straight or branched) of any conceivable degree of complexity with respect to interconnections among steps in the multistep process (Hearon, 1952; Bray and White, 1957).

Kavanau (1950) has proposed an alternative equation to (11.37), based on the assumption that at low temperatures the formation of intramolecular hydrogen bridges converts reactive enzyme particles to a catalytically inactive condition. There is evidence in support of this assumption since, as Kavanau showed, the heat of activation falls with increasing temperature for simple hydrolysis and prepupal development in *Drosophila*. However, the main body of modern opinion seems to favor the activated complex model of Eyring, and therefore I have chosen the empirical formula version of this model (11.38) for extensive statistical tests. My defense of the use of (11.38) as an empirical formula to describe temperature effects on the whole animal is that, while the theoretical justification for the formula is not beyond question, extensive fitting of (11.38) by ecologists may help to build a bridge between ecology and reaction-rate studies on enzymes, which could then lead to improved models.

Statistical tests of models

Only three of the aforementioned models were chosen for critical testing: those of Pradhan, Davidson (the logistic), and Eyring, equation (11.38). To test the fit of the models in describing rates of biological reactions, we selected, from the enormous mass of data in the biological literature, four sets that seemed particularly suitable for a critical comparison. We will describe each of the four sets briefly, explaining why each is particularly significant.

1. The first set concerns rate of development of *Lucilia caesar* larvae (Peairs, 1927). The work was carefully done, in specially constructed incubators, and the replication was awesome: 96,814 animals were used in the experiment; 2,000 to 17,340 animals were used at each of the nine temperatures tested. The temperatures covered an adequately wide range (5°C to 35°C) and careful records were kept of survival. The results were subjected to thorough statistical analysis, and were all published (table 1, Peairs).

2. Powsner (1935), determining the effect of temperature on development rate in *Drosophila melanogaster*, took great pains to eliminate sources of variability from his experimental procedure. Among other features, his experiments used flies that were the products of 75 generations of pair-mating inbreeding, in order to reduce heterozygosity to a minimum. The resulting variability within his replicate cohorts was very low. We used Powsner's table VIII.

3. Birch (1944) measured the effect of temperature on rate of development of *Calandra oryzae* eggs. This work was also done meticulously, with great attention to potential technical errors. Birch defined moisture effect as the total evaporating power of the air, which he took to be equal to the duration of the experiment multiplied by the saturation deficit. He kept this product very nearly constant in all treatments by placing the eggs in Petri dishes in Fowler jars in which humidity control was achieved with sulfuric acid solutions.

4. Messenger and Flitters (1958) measured the effect of temperature on rate of development in eggs of *Dacus dorsalis*. They used elaborate, high-quality equipment, excellent technique, adequate replication, and a wide range of temperatures (54.5°F to 98°F).

Each of the three models was tested against each of these four sets of data. The logistic was fitted using the Taylor series iterative least squares procedure (Chap. 10) and a linear procedure. Where R is developmental rate and t is temperature, if

$$R = \frac{K}{1 + e^{a + bt}}$$

then

$$1 + e^{a + bt} = \frac{K}{R}$$

and

$$\ln\left(\frac{K - R}{R}\right) = a + bt$$

Thus we can obtain least squares fits of a and b by using trial values of K. This gives values very close to those from the iterative method, and is much easier to apply. Pradhan's formula was fitted by converting

$$R = y_o e^{-\frac{1}{2}at^2}$$

to

$$R = e^{bt - a't^2 + c}$$

or

$$\ln R = c + bt - a't^2$$

in which we can find a', b, and c by standard regression techniques (Chap. 10). We can get reasonably close approximations to the correct values of the four parameters in (11.38) by using a number of ad hoc methods, after which we can get improved estimates by using iterative least squares techniques. Two of the constants can be estimated by plotting $\log (R/T)$ against $1/T$; this would leave us with two equations and two unknowns, for which we can solve by substituting in values of the other parameters and the temperature at any point on the curve. (T represents absolute temperature because Eyring's model is developed in terms of absolute temperature.)

Residual sum of squares

Data	Logistic	Pradhan equation	Eyring model (before iterative correction)
Lucilia	18.12	27.10	10.58
Drosophila	7.45	30.81	29.62
Calandra	5.32	4.37	1.71
Dacus	421.53	101.35	100.97

We conclude that the Eyring model not only is the only model that accounts for temperature-reaction rates in whole organisms and has a generally accepted theoretical foundation, but it also gives us somewhat better fits to the data. Furthermore, it loses only one more degree of freedom in statistical tests than the more generally adopted empirical formulas, because it only has four fitted parameters, whereas they have three. We do not mean to imply that the reactions of whole organisms to changing temperatures can be understood in terms of a single enzyme reaction. Rather, one is more likely to get clues about the mechanism of temperature effect on whole organisms by expressing ecological findings in terms of a model with a well-understood physicochemical meaning. Sustained use of such a model will probably lead to a concept of reaction-rate kinetics like that discussed by Hearon.

Our next major problem concerns description of the effect of regimes of naturally fluctuating temperature on living organisms. Ahmad (1936) made a comprehensive analysis of the effect of alternating temperatures on the development of three different species of insects. He subjected all his results to elaborate statistical analysis. He discovered that there are no general rules that cover all species and cases. For example, pupae of the stable fly *Muscina stabulans* show very little change in development due to alternating temperatures. But development of *Locusta migratoria* eggs shows a complicated pattern of responses determined by the stage of development, the high temperature, the low temperature, and the time at each. Eggs were incubated daily from 5:00 P.M. to 9:00 A.M. at temperatures of 27°, 33°, or 37°C, depending on the treatment (for the rest of the day, the eggs were kept at 5°C). In every case, development is accelerated over the controls (all the time at the high temperature), but increasing the difference between low and high temperatures reduces the acceleration.

Messenger and Flitters (1959) have made a very critical analysis of data from their own experiments, in which they compare results of a constant temperature on rate of development with results when the same temperature was the mean value for a 10°, 20°, and 30°F diurnal range. This procedure was tested for eggs of each of three species of fruit flies, over the range 52° to 90°F. Observed rates of development under fluctuating conditions were compared with rates predicted by summing the effects of the accumulated heat experience under the fluctuating temperature regime, using the development rate versus temperature curve at constant temperatures as the basis for prediction. Messenger and Flitters found that at the extremes of temperatures, development of eggs proceeds at rates greater than would be predicted on the basis of constant temperatures alone. In the medial range of temperatures, the rate of development under a fluctuating-temperature regime is very similar to that predicted from the constant-temperature relationship. Messenger and Flitters suggest that the acceleration of development at extreme temperatures may be due to some development occurring at supposedly subthreshold and supralimiting thermal levels, provided that the organism spends only a brief period at such temperatures.

The most elaborate theoretical attempt of which I am aware to compare the effects of fluctuating and constant temperatures on organisms is by Pradhan (1945). Consider Fig. 11.3 a hypothetical plot of rate of activity (development, feeding, flying, swimming, etc.) versus temperature. Let E_X represent the effect of fluctuations about an average temperature X; these go down to $X - B$ and up to $X + B$. When the temperature goes up to $X + B$, the activity rate will be higher than at X, and when it goes down to $X - B$, the rate will be lower. We may say that where R_X represents the rate of a biological activity under the constant temperature X, the rate of the same

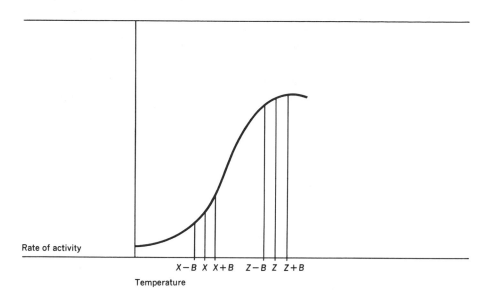

X−B X X+B   Z−B Z Z+B

Temperature

FIGURE 11.3

Effect of fluctuation in temperature on rate of biological activity (Pradhan, 1945).

activity under a fluctuating temperature with mean at X may be thought of as

$$R_{(X-B):(X+B)} = R_X + E_{(X-B):(X+B)}$$

$R_{(X-B):(X+B)}$ represents the rate of activity under a temperature regime fluctuating in the range $X - B$ to $X + B$; and $E_{(X-B):(X+B)}$ represents the correction that must be made to the rate predicted from the constant-temperature effect of the mean of all the fluctuating temperatures. $E_{(X-B):(X+B)}$ is thus the sum of the effects of all the temperature experience in the range $X - B$ to $X + B$, including that at X. Where the curve is rising rapidly, as at X, the accelerative effect of temperatures in the range X to $X + B$ will exceed the retardative effect of temperatures in the range $X - B$ to X, and $E_{(X-B):(X+B)}$ will be a positive number. However, when we consider deviations from a higher temperature, Z, retardative effects in the range $Z - B$ to Z will exceed accelerative effects in the range Z to $Z + B$, and $E_{(Z-B):(Z+B)}$ will be a negative number. Pradhan proposes that we compute $R_{(X-B):(X+B)}$ by obtaining the integral for the temperature effects in the range $X - B$ to $X + B$ that particular organisms have actually experienced. In practice, we would probably compute a sum, rather than an integral.

The conclusion we draw from this discussion about the effect of fluctuating temperatures on animals is, that the fluctuations have no special effect per se. Rather, the effects of fluctuations can be understood by summing the effects at each of the temperatures experienced, as each of the effects would be predicted from a curve for the effect of constant temperature. The only complication, as Messenger and Flitters point out, is that short exposures to lethal temperatures may actually have a positive effect on an organism. This, however, may merely mean that it takes a considerable period to heat up or cool down an organism because of the great specific heat of water, which constitutes a high percentage of the mass of all protoplasm.

Da Fonseca (1958) has proposed a simple and ingenious solution to the old problem of expressing the biological impact of joint variation in temperature and humidity. He suggests that we use a single measure to express the effect of temperature and humidity: the wet bulb temperature. Where the relative humidity is 100 percent, the wet bulb temperature will be the same as the dry bulb temperature; the drier the air, the lower the wet bulb temperature will be relative to the dry bulb temperature. This idea should be tested in a variety of situations, because it makes sense, for certain types of organisms. High humidities are more dangerous to an organism than low humidities at high temperatures, because it is more difficult at high humidites to cool the blood by using the latent heat of vaporization.

Much has been written about lethal temperature effects on animals, but different authors have emphasized different aspects of the problem. Fry (1947) has summarized the viewpoint in which acclimation temperature is held to be an important variable in determining the lethal temperature. Salt (1950) has emphasized the important role of time in determining the probability of death. There are extensive discussions of this subject, by Luyet (1957), Payne (1927), Smith (1958, 1961), and a great many others.

Unfortunately, I cannot give a general mathematical model for the probability of death due to lethal cold (or high temperatures) as a function of the factors that determine this probability. Review of the literature shows that five basic factors are involved in this phenomenon, and no experimenters, to my knowledge, have ever varied and recorded the effects of all five in a single experiment. The five factors are:

1. The temperature to which the organism was acclimated prior to exposure to the lethal temperature

2. The lethal temperature

3. The length of time the organism was exposed to the nonlethal temperature

4. The length of time the organism was exposed to the lethal temperature

5. The length of time taken for the change from the acclimation temperature to the lethal temperature

After we know how to develop suitable mathematical models to express the effects of weather on living organisms, the next problem is that of determining what meteorological data to use as independent variables in the models. In Sec. 6.2, I have argued that in some cases, at least, the explanation for large-scale biological wavelike outbreaks radiating outward from epicenters is to be sought in unusual climatic patterns at certain points on the earth's surface. Some of the questions that arise are: What is unusual about the climatic patterns at epicenters? How do we recognize a particular locality as being an epicenter? How do we recognize areas surrounding an epicenter as not being epicenters? I discovered that, in the case of the spruce budworm, climatic sequences at Green River, New Brunswick, explained the population fluctuations of the budworm in the area surrounding Green River, but the sequence of weather conditions from year to year at Edmundston, New Brunswick, only 30 miles away, did not (Watt, 1964a). The key to budworm population dynamics seemed to be the mean of maximum daily temperatures in June and in the first 15 days of July (Morris, 1963). However, the odd weather patterns required to trigger outbreak are only found at certain locations. Precisely how odd these weather patterns can be at an epicenter is brought out by Fig. 11.4. Here mean maximum daily temperature at Green River for the period June 1 to July 15 is plotted against year from 1925 to 1959. We see that there is an unusual clustering of high temperatures in the years 1944 to 1953; this produced the great budworm outbreak beginning in the late 1940's. In the 6-year period 1944 to 1949, inclusive, not only were June 1 to July 15 temperatures high in five of the 6 years, but in four of the five, the high temperatures were more than one standard deviation above the June 1 to July 15 mean for the 35-year period. If we assume we have a normal distribution of these temperature indices about the mean, only 16 percent of the years should have these indices as high as, or higher than, one standard deviation above the mean. That is, in a randomly selected 6-year interval, we would expect about 6(16/100) or 1 year with a June 1 to July 15 temperature one standard deviation or more above the mean. The odds against getting 4 years or more in the 6-year period are about 6^4 to 1 against, or 1,296 to 1 against, by chance alone. Figure 11.4 then represents a truly unusual situation. The interesting point is that I did not discover this same clustering of unusually warm June 1 to July 15 periods in the data for Edmundston, which is only 30 miles away.

Some readers may question that climate could be so different at two places quite close together. In support of the general picture presented here, I submit Fig. 11.5, from Wellington (1964). Wellington has studied the relation between microclimate and population dynamics in the tent caterpillar *Malaco-*

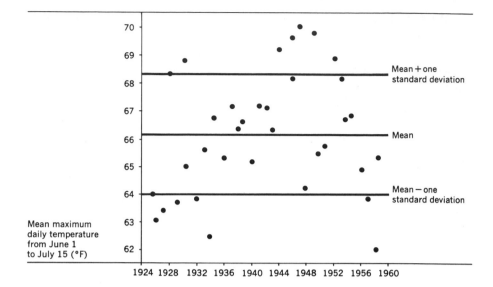

FIGURE 11.4

Temporal trend in June-July mean maximum temperatures at Green River, New Brunswick.

soma pluviale for many years. Figure 11.5 shows that there is a different sequence of events in a valley and an adjacent hill, and this difference is reflected in the biological consequence (length of time required to pass through first-instar larval period). Wellington also gives a detailed explanation of the meteorological events near Victoria, B.C., in the period 1956 to 1962, and their relation to the biological consequences in *Malacosoma* populations. Perhaps the most important finding, which relates to the discussion of waves in Sec. 11.7, is that changing weather not only has an effect on the temporal trends in a population, but also on the spatial distribution. When weather was worst (1958), the mosaic of terrain climates provided some pockets that were tolerable for the most vigorous colonies. As weather improved, the area in which the species could exist spread. This demonstrates the way in which weather can regulate wavelike phenomena. But a great deal of research is needed to discover if weather can trigger an outbreak at a single or small number of epicenters from which consequences radiate out, or if populations are merely allowed to build up at more places than would normally be the case. The latter is called the scattered type of outbreak and the former is called the spreading-out type of outbreak by Miyashita (1963), who believes only the former are due to weather, and that the latter are due to a widespread change in food.

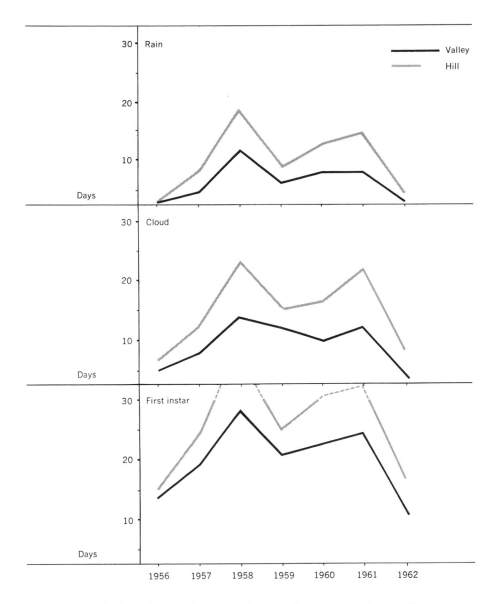

FIGURE 11.5 *Cloudy and rainy days recorded in valley and on adjacent hill during the emergence and development of first-instar Malacosoma pluviale larvae near Victoria, B.C. Broken lines represent years in which there was virtually no survival of first-instar larvae. Period of first-instar development always included part of April, but occasionally began late in March or ended early in May. Length of period varied in response to weather (bottom panel). Note that lines are not parallel. If we were to express weather effect or biological effect at either location as a linear function of weather at a nearby meteorological station, the constant b in the equation Y = a + bX for the hill would differ from that for the adjacent valley (Wellington, 1964).*

Another problem of using weather records to interpret biological data has to do with detecting the particular sequence of weather events that determine biological instability. This problem arises, in part, because one location may have very unstable weather in August, but be stable (relative to other locations) in March. The reverse might be true of a second location. The following data illustrate this point.

Mean monthly temperatures, degrees Fahrenheit, 1945 to 1956

	January range	May range	August range
Lethbridge	−14–29 (43)	49–54 (5)	59–65 (6)
Toronto	11–29 (18)	51–57 (6)	65–73 (8)
Vancouver	21–42 (21)	51–57 (6)	61–64 (3)

Thus, while Lethbridge has the most unstable weather of the three locations in January, it has the most stable in May. Toronto, the most stable in January, is least stable in August. The indices used in statistical analysis must clearly be selected with care and insight into the weather, as well as the biology of each situation.

The final question has to do with the extent that we can use the weather records at one locality to interpret biological events at another. The distance over which we can safely make this extrapolation depends on the terrain. I have found that there is a high degree of correlation between the mean monthly temperatures in Magdeburg and those in Berlin, 80 miles away. I found a poor correlation between those in Green River and those in Edmundston, only 30 miles away. The explanation is that Magdeburg and Berlin lie in a flat plain, whereas Green River is in mountainous country. We demonstrate this same phenomenon for two other pairs of localities in Figs. 11.6 and 11.7, where we examine the correlation between mean September temperatures in two cities. In Fig. 11.6, the two cities are Lethbridge and Summerland, 320 miles apart but with intervening mountain ranges. In Fig. 11.7, the two are Toronto and Montreal, 347 miles apart but in flat country. The correlation is least for the two cities closest together, because of the mountains in between. Hence the extent to which we attempt to relate biological events to weather as recorded at a distant weather station must be determined by the surrounding topography.

11.4 THE EFFECT OF DENSITY ON RATE OF REPRODUCTION

Reproduction is one of the most important component processes in the dynamics of animal populations, and we must learn how best to model this process if we wish to simulate the behavior of natural resources. Many math-

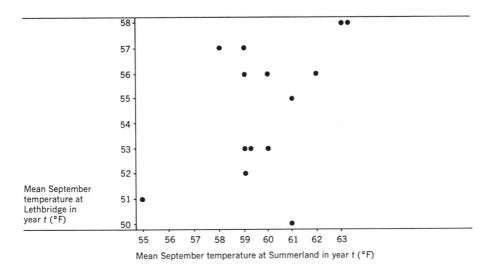

FIGURE 11.6

Correlation between mean September temperatures in Summerland and Lethbridge (data from Canada Department of Transport).

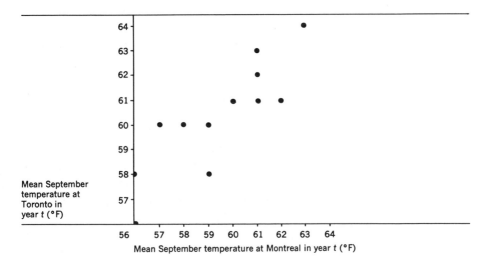

FIGURE 11.7

Correlation between mean September temperatures in Montreal and Toronto (data from Canada Department of Transport).

ematical models for the effect of population density on fecundity have been introduced into the literature, and these models are of fundamentally different types. This group of models is useful to demonstrate the logic and data-manipulation techniques used in determining which of an available array of models is best to describe a given process.

At the outset we should caution the reader that all our models and data are obtained from entomological literature, simply because many entomologists have concerned themselves with quantitative analysis of this problem. Certain features of the following discussion can be generalized to other animals, and certain features cannot. For example, in vertebrate populations it may not be true that lowering the population density to very low levels affects the reproductive rate because social forces may keep all the individuals in a large area together in one group (for example, musk ox). Territorialism may also have an effect on the curve for reproductive rate versus density. The reader should extrapolate from these models to interpretation of data on other types of animals with great care. However, the basic components of the fecundity versus density relation must be found in some form in all kinds of animals. These basic components and their subsidiary components can be tabulated as in Table 11.2.

We have mentioned that the multiplicity of models available for description of this phenomenon allows us to demonstrate how best to select the most appropriate model. It is necessary to lay down, at the outset, the conditions that must be met by a mathematical model in order for it to be a suitable description of a biological process. First, the equation must provide a statistically satisfactory fit to data collected on the phenomenon. Second, it should be no more complex than is necessary to provide such fit. Third, it ought to incorporate insight into the biological mechanism of the phenomenon it describes. Fourth, it ought to be formulated so that it is vulnerable to statistical test (that is, all the variables in the equation can be measured by using a completely repeatable and definable operational procedure). Finally, the equation must be general, in that it can be made to reproduce any set of measurements on the phenomenon (allowing, of course, for sampling error), merely by altering parameter values. In short, the equation should be steeped in, and a product of, the lore and data of biology. The formula specifically should not have been selected for any of the following reasons: purely a priori considerations; ease of mathematical manipulation; it gave a good fit, regardless of its meaning; it provided a convenient analog with mathematical thinking in another field (for example, physics).

Review of empirical findings

When we make graphs of oviposition rate versus population density for the data from various experiments, we find that these graphs fall into three basic categories. These categories have been labeled "*Drosophila* type,"

TABLE 11.2

Causal pathways affecting reproductive rate in animals

Determining factors	Subsidiary components	Basic components
Mean fecundity per female	Mean frequency of copulations per female, and effect of this frequency on fecundity	Population density, and hence probability of contact between prospective mates
	Deleterious effects of interference with copulation, increased fighting between males, etc.	Population density
	Nutritional plane of female, affecting rate of female sterility and reproduction in nonsterile individuals	Competition for food, which depends on food availability and population density
	Activity level of female (activity increases the proportion of incoming energy used to support metabolism, as opposed to reproduction)	Interference from other animals, which in turn depends on population density
	Age of female	Age of female when reproduction occurs depends on competition for reproduction sites and interference, and hence ultimately on population density
Mean availability of sites in which reproduction can occur	Density of sites in medium	Structure of environment
	Number of young already reproduced	Population density

"intermediate type," and "Allee type" by Fujita (1954), and examples, in that order, are given here as Figs. 11.8, 11.9, and 11.10. We must not allow ourselves to be deceived into regarding such a situation as representing the operation of three different laws; such different-appearing graphs can be produced by one basic and underlying law, provided the mathematical formulation of the law is sufficiently comprehensive. In statistical terms, this means that a complex equational form will have enough parameters and terms to be quite flexible as a description of various bodies of data. The only dif-

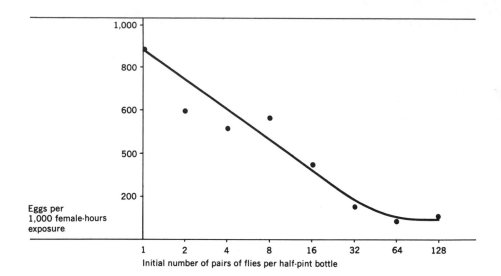

FIGURE 11.8

"Drosophila type" curve: Effect of population density on fecundity in Drosophila. (Pearl, 1932).

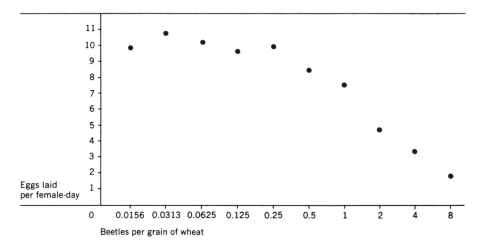

FIGURE 11.9

"Intermediate type" curve: Effect of population density on fecundity in Rhizopertha (Crombie, 1942).

ficulty associated with such flexibility is that for each parameter we add to an equation, the statistical test of the validity of the equation loses a degree of freedom for the denominator of the F test, and it is more difficult to show

292   *The methods of resource management*

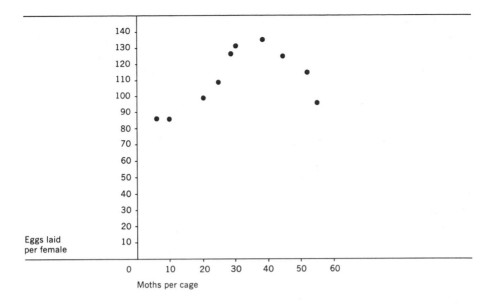

FIGURE 11.10

"Allee type" curve: Effect of population density on fecundity in Anagasta kuhniella (Zell) (Ullyett, 1945).

that the law is wrong. Thus, as an equation becomes more complex, the volume of data required to test it increases (that is, we increase the requirement for degrees of freedom). Fujita (1954) recognized that graphs such as Figs. 11.8, 11.9, and 11.10 were in fact different expressions of the same law. I have noted earlier (1960) the various bodies of data supporting this view. All three types can be found in any species; which type we find depends on universe size, the Allee curve occurring in the largest universe for a given density (Kirchner, 1939), on nutritional level (Robertson and Sang, 1944), and on various environmental factors such as relative humidity (Utida, 1941a). It can be demonstrated mathematically and experimentally that the *Drosophila* type of curve is really a variant of the Allee type, which occurs in situations where the density is not allowed to drop low enough for the effects of extreme low density in depressing fecundity rate to reveal themselves. Therefore, in all situations, there is, in principle, some optimum density for fecundity rate; the fecundity rate falls off on either side of this density.

Whenever we discover a phenomenon that works best at an intermediate value for an independent variable, we are led to suspect the operation of factors that depress efficiency of the phenomenon at high and low variate values for that variable. In this case, we can discover several such factors.

MacLagan and Dunn (1935), working with *Sitophilus oryza* (L.) (Curculionidae), and Utida (1941b), working with the azuki bean weevil *Callosobruchus chinensis* (L.), found that the copulation frequency of females was lower at the lowest densities tested than at intermediate densities. Crombie (1942) found that fecundity and fertility in *Rhyzopertha dominica* (Fab.) were not affected by further copulations after the first. From these three sets of data we know that the effect of density on fecundity via copulation frequency must be expressed by some function that rises gradually toward an asymptote with increasing density.

Park (1933) found in *Tribolium* that the Allee type of curve was produced by the interaction of two opposing groups of factors, one of which, copulation and recopulation, has a beneficial effect with rising densities and the other, interference and competition for oviposition sites, has a deleterious effect with rising densities. The problem is to design critical observational procedures or experiments that allow us to examine separately the effect of interference and competition for oviposition sites. Smirnov and his associates have found clear-cut evidence of the effect of interference.

Smirnov and Wiolovitsh (1934) showed that a *Drosophila* type of curve might be due to the effect of increasing density in raising the incidence of sterility in females. They used the shield bug, *Chionaspis salicis* L. In this instance, almost all of the depression of fecundity with increasing density could be accounted for by sterility. Smirnov and Polejaeff (1934) showed the same kind of situation with the coccid *Lepidosaphes ulmi* L., also in the Diaspididae. While in both these studies, especially in the latter, the results are obscured by sampling error due to the small volume of replication, the fact that the findings were obtained from natural populations makes them of particular interest.

Crombie (1942) showed that increasing the density of males can decrease the total number of copulations by causing increased fighting. Some clarification of the difference between mode of operation of competition for oviposition sites and other types of interference seems necessary at this point. If *N* adults live in a large area that includes only a small area suitable for oviposition, there will be no sterility or failure to copulate because of interference. Nevertheless, fecundity may be low due to competition for oviposition sites. If, on the other hand, *N* adults live in a small area, most of which is suitable for oviposition, there will be appreciable depression of fecundity due to interference-induced phenomena but not much due to competition for available oviposition sites.

Experiments have been performed which clearly demonstrate, by manipulating oviposition area and total area independently, that phenomena other than competition for oviposition sites can depress fecundity. For example, Robertson and Sang (1944) varied the oviposition area available to 10 pairs of *Drosophila* in bottles. A sevenfold decrease in density of adult *Drosophila* per unit of available oviposition area produced only an 11 percent increase in the fecundity per female per day. Robertson and Sang concluded that Pearl's (1932) *Drosophila*-fecundity results could not be explained on the basis of competition for oviposition sites alone. Rather, some other effect that depresses fecundity with increasing densities must be postulated to account for the observed magnitude of Pearl's fecundity-depression. Ito (1955) showed that fecundity of flour beetles, *Tribolium castaneum* (Hbst.), decreased if there was no air space above the flour in the container. Since all other variables were held constant, while removing the air space depressed fecundity, the depression must have been caused by increasing density-dependent competition of some type other than competition for oviposition sites. Ishida (1952) performed an experiment in which number of available oviposition sites per female and volume of the universe per female were the two independent variables in a factorial design. His experiment showed that when competition for oviposition sites, as measured by females per unit oviposition area, was held constant, some other component of competition depressed the fecundity as density increased. This factor is interference.

Evaluation of mathematical models to describe effect of density on fecundity

Lotka (1923) wrote the differential equation

$$\frac{dN_1}{dt} = b_1 N_1 - d_1 N_1 - a_1 N_1 N_2$$

where N_1 = number of unattacked hosts
N_2 = number of adult parasites
$b_1 N_1$ = number of hosts born per unit time
$d_1 N_1$ = number of host deaths per unit time
$a_1 N_1 N_2$ = number of hosts attacked by parasites per unit time
dN_1/dt = rate of change in N_1 with time

If it is true that numbers of hosts born per unit time is given by $b_1 N_1$, a graph of b_1 (number of hosts born per unit time per host) against N_1 for any time interval should produce a straight line parallel to the N_1 axis. Patently, from Figs. 11.8, 11.9, and 11.10, empirical data yield nothing of the sort. Volterra (1928) makes the same false assumption as Lotka, when he asserts, "If there is only one species, or if the others have no influence on it, so that the circumstances of birth and death do not vary, we shall have, if N denotes the number of individuals,

$$\frac{dN}{dt} = nN - mN = (n - m)N$$

where t denotes time and n and m are constants, respectively the coefficients of birth and mortality." Obviously, n is not constant, and it is clear that the logistic (which uses the assumption that n is constant) is only a crude generalization, of no use in critical population analysis.

Pearl and Parker (1922) fitted their data to a curve of form

$$\log y = a - bx - c \log x \qquad (11.39)$$

where y denotes imagoes per mated female per day, and x denotes mean density of the mated population (measured as flies per bottle) over the whole 16-day period of the experiment. This equation gave a remarkably good fit to their data, and Pearl was moved to remark, "Plainly the curve is the expression of the law relating these two phenomena—rate of reproduction and density of population." However, (11.39) clearly is not the law, since it does not meet two of the conditions itemized in the introduction of this chapter. First, it yields no insight into the phenomenon, and second, it lacks generality. Equation (11.39) could not account for Allee type of curves because it has no component producing an increase in $\log y$ with increasing x. A really meaningful and general equation has to be capable of yielding Allee or *Drosophila* type of curves, since either type can be generated by the same population, merely by changing, for instance, the time the population has been cohabiting, the density (Kirchner, 1939), the degree of copulation (Crombie, 1943), the nutritional level (Robertson and Sang, 1944), or any of several other factors, singly or in combination.

Pearl (1932) proposed a new equation to describe *Drosophila* type of curves. Where y denotes eggs per 1,000 female-hours exposure; x represents mean density of population; and c, d, and K are constants, he suggested that

$$y - d = \frac{K}{x + c}$$

This equation suffers from the same limitations as the Pearl and Parker (1922) formula.

Andersen's (1957) fundamental assumption is that, above a certain limit of density, the fecundity is a linear function of the reciprocal of the density. The best blocks of data available for testing this assumption are those obtained from experiments covering a wide range of densities. We have plotted fecundity per female per unit time against the reciprocal of density for two such blocks, using data from Pearl's (1932) table 5, representing *Drosophila melanogaster* Meig. in quarter-pint bottles, and Utida's (1941c) table III, *Callosobruchus chinensis* (L.) under 24.8°C and 74 percent relative humidity. Andersen's assumption at best only approximates the facts at high densities. However, it yields no insight into the phenomenon we are exploring, and it tells us nothing about the part of the curve for which shape varies most from one case to another. Furthermore, it is this part of the curve that should describe the events most critical from the standpoint of population survival and growth (that is, those at low densities).

Fujita and Utida (1953) proposed a model to describe the effect of density on reproductive rate, reasoning from the equation for logistic growth. Their reasoning is based on a model that does not incorporate a sufficient number of terms to be realistic. The differential equation for logistic growth is

$$\frac{dN}{dt} = rN\,(K - N) \tag{11.40}$$

where N = population density at any point in time, during initial population growth

$\quad\;\; K$ = average maximum population density that can be supported by the environment

$\quad\;\; r$ = intrinsic rate of natural population increase, equal to instantaneous birth rate minus instantaneous death rate

To illustrate the meaning of "instantaneous" in this context, note that for exponential growth

$$\frac{dN}{dt} = rN \qquad \text{whence} \qquad N_t = N_o e^{rt}$$

or, for two consecutive points in time,

$$N_{t+1} = N_t e^r = N_t e^{(b-d)} \tag{11.41}$$

However, where B is the net birth rate measured from t to $t + 1$, and D is the net death rate measured over the same interval, we also have

$$N_{t+1} = N_t\,(B - D)$$

Therefore, $\qquad\qquad\qquad e^{b-d} = B - D$

Note, now, that all that (11.40) states is that rate of change in numbers with time will depend on the product of N, and *density-independent* instantaneous birth less death rates, all multiplied by a factor, $K - N$, to express the increasing effect of competition in decreasing birth rates and increasing death rates as density increases. There is no expression of the fact that *very low densities* can have a deleterious effect on birth rates, as well as very high densities. Thus, as might be expected, the Fujita-Utida model only holds for *Drosophila* type of curves.

Fujita (1954) subsequently made a more penetrating analysis of the effect of density on fecundity. He recognized the existence of three types of curves (we have been following his classification throughout this chapter) and made the following important observation: "The type of density effect of a given insect population is not maintained invariable during the course of a particular experiment, but changes systematically as the period of incubation is prolonged." Chapman (1928), Park (1932), and Rich (1956) all obtained data supporting this statement. Fujita thus concluded correctly that the density effect was a time-dependent property, and time had to be included in any equation for the effect of density on fecundity.

In the rest of this chapter, I shall present the rationale and derivation of the Fujita model, then attempt to fit it to available data. Then, after re-examining the logic of his model as critically as possible with the empirical data now available, and showing where the model needs modification, I shall develop and test a new model.

First, certain symbols must be defined:

N = population density (or population size, if we refer to an enclosed population)

E = total number of eggs laid from time 0 to time t

a = area occupied by single deposited egg; actually fraction of total space available for oviposition taken up by one egg, so that when $E = 1/a$,

$$1 - aE = 1 - \frac{a}{a} = (0)$$

ε = intrinsic rate of oviposition for given environmental condition

P = frequency of copulation per female (number of copulations per female per unit time)

k = proportionality constant

Fujita then writes

$$\frac{dE}{dt} = k\varepsilon NP (1 - aE) \tag{11.42}$$

Equation (11.42) is eminently reasonable. However, many writers, including Crombie (1943), MacLagan and Dunn (1935), Park (1936), Richards (1947), Robertson and Sang (1944), and Utida (1941b), have shown that environmental factors govern ε. Park (1936) showed experimentally that one could obtain curves relating ε to an index of an environmental parameter. Stanley (1934) worked out a mathematical model by which the decline in fecundity as a function of environmental conditions could be predicted, if one had all necessary measurements of the environment. There are not enough data available now to work out a Stanley-type equation for a natural population, so Fujita contented himself with detailed study only of P, hoping that by so doing he could produce a version of (11.42) that would be useful as a first approximation.

Fujita makes several assumptions. First, he assumes that copulation frequency is increased by increasing adult density. Furthermore, he states that, "if the collisions are made entirely in a random fashion, it may be assumed that P is represented by a Gaussian distribution function with respect to the average mutual distance between neighboring adults, r. Namely, we may put

$$P = \exp\left(-\sigma r^2\right)$$

where σ is a constant dependent primarily on the copulation characteristics of the given insect species."

He also assumes that in a two-dimensional universe,

$$N\sigma r^2 = \alpha$$

or

$$\sigma r^2 = \alpha/N$$

Hence

$$P = \exp(-\alpha/N) \tag{11.43}$$

A hidden assumption is contained in equation (11.42). We know that oviposition rate is some function of copulation frequency, which in turn is some function of density. In symbolic terms, we could write

$$\frac{dE}{dt} = f_1(P) = f_2(N)$$

Fujita devotes considerable attention to the form of function f_2, but immediately assumes that $dE/dt = f_1(P)$ is a simple rectilinear relationship. I know of no data that will support this assumption.

If we insert (11.43) into (11.42), we get

$$\frac{dE}{dt} = k\varepsilon \exp\left(-\alpha/N\right) N(1 - aE)$$

or

$$\frac{-\ln\left(1 - aE\right)}{a} = k\varepsilon \exp\left(-\alpha/N\right) Nt$$

or
$$1 - aE = \exp\left(-ak\varepsilon\, e^{-\alpha/N} Nt\right)$$

or
$$E = \frac{1}{a}\left[1 - \exp\left(-ak\varepsilon\, e^{-\alpha/N} Nt\right)\right]$$

But since, assuming a sex ratio of 50:50, the mean fecundity, F, per female per unit time is

$$F = \frac{E}{Nt/2} = \frac{2E}{Nt}$$

therefore
$$F = \frac{2}{Nta}\left[1 - \exp\left(-ak\varepsilon\, e^{-\alpha/N} Nt\right)\right] \tag{11.44}$$

Fujita's first assumption is suspect in view of findings by Utida (1941b) and MacLagan and Dunn (1935) that copulation frequency can decline at very high population densities, presumably because of interference. However, for the moment let us confine our attention to his second assumption.

If oviposition can occur in a volume rather than a plane, then

$$N\sigma r^3 = \alpha \tag{11.45}$$

This merely states that number of adults times volume/adult = total volume. It is true for a volume, as for a plane, that the shape of the probability distribution function for chances of contact is given by

$$P = \exp\left(-\sigma r^2\right) \tag{11.46}$$

But, from (11.45),

$$\sigma r^2 = \sigma^{1/3}\left(\frac{\alpha}{N}\right)^{2/3}$$

or, introducing a new constant,

$$\sigma r^2 = \left(\frac{\beta}{N}\right)^{2/3} \tag{11.47}$$

If we insert (11.47) in (11.46),

$$P = \exp\left[-\left(\frac{\beta}{N}\right)^{2/3}\right] \tag{11.48}$$

If we substitute for P in (11.42),

$$\frac{dE}{dt} = k\varepsilon N \exp\left[-\left(\frac{\beta}{N}\right)^{2/3}\right](1 - aE) \tag{11.49}$$

and proceed as in the derivation of (11.44), we obtain

$$F = \frac{2}{Nta} [1 - \exp(-ak\varepsilon e^{-(\beta/N)^{2/3}} Nt)] \qquad (11.50)$$

A simple procedure for testing the validity of (11.44) and (11.50) may be worked out as follows.

From (11.44),

$$1 - \frac{FNta}{2} = \exp(-ak\varepsilon e^{-(\alpha/N)} Nt)$$

$$\ln \left[\frac{2}{2 - FNta} \right] = ak\varepsilon e^{-(\alpha/N)} Nt$$

$$\ln \left[\frac{\ln\left(\dfrac{2}{2 - FNta}\right)}{Nt} \right] = \gamma - \frac{\alpha}{N} \qquad (11.51)$$

And similarly, where oviposition sites are in a volume,

$$\ln \left[\frac{\ln\left(\dfrac{2}{2 - FNta}\right)}{Nt} \right] = \gamma - (\beta/N)^{2/3} \qquad (11.52)$$

Since in some cases it may not be clear exactly how adult females search for oviposition sites, in order to make a really thorough test of Fujita's model, we should attempt to fit both (11.51) and (11.52) in these instances.

The best fit we can obtain for (11.51) or (11.52) is found by selecting (by trial and error) that value of a which minimizes the residual sum of squares from the regression of the transformation

$$\ln \left[\frac{\ln\left(\dfrac{2}{2 - FNta}\right)}{Nt} \right]$$

on $1/N$ or $1/N^{2/3}$. We can find out quickly which fine range of values for a we should be exploring by plotting

$$\frac{\ln\left(\dfrac{2}{2 - FNta}\right)}{Nt}$$

against $1/N$ or $1/N^{2/3}$ on semilog graph paper. When this is done, using Rich's (1956) 8-hour data, it is clear that Fujita's equation cannot be used to produce a straight-line plot; hence we must reconsider Fujita's assumptions.

Reexamining Fujita's basic equation

$$\frac{dE}{dt} = k\varepsilon NP(1 - aE)$$

we find that this allows for one factor (copulation and recopulation) that stimulates fecundity with increasing density, and one (competition for oviposition sites) that decreases fecundity with increasing density. There is sound empirical support for the assumption that both copulation (but not recopulation) and competition for oviposition sites operate. Park (1933) and Crombie (1942, 1943) clearly demonstrated the fecundity-elevating effect of copulation; Utida (1941a), Crombie (1942), and Ishida (1952) proved the existence of competition for oviposition sites.

There is another group of density-dependent factors, apart from competition for oviposition sites, which can depress fecundity and which Fujita has not taken into account. This is the group of factors we can refer to collectively as density-dependent interference; that is, sterility, fighting, interference with the mating process, and general disturbance of normal physiological functions.

A new model is clearly needed to incorporate these factors. The model will be expressed, first, as three partial differential equations describing the operation of each of the three components of the density-fecundity mechanism. Then the three will be combined to yield one integral equation.

First, we can write

$$\frac{\partial E}{\partial t} = k\varepsilon NPI(1 - aE) \tag{11.53}$$

This states that, all other factors being held constant, the rate of change of density of eggs with time, $\partial E/\partial t$, is equal to the product of the following six components:

k = proportionality constant
ε = intrinsic rate of oviposition for given environmental condition
N = population density
P = effect of probability of contact on fecundity [P is here defined differently than it was by Fujita; also, following Crombie (1942), we concede that no contact, after first, need be important]
I = effect of interference on fecundity
$1 - aE$ = effect of competition for oviposition sites on fecundity (a is defined as by Fujita)

Second, we need to determine the form of the relation between P and N. The best data for this purpose are those for rather low population densities. If we examine the data of Utida (1941b) and Ullyett (1945) in Fig. 11.10, for example, we are led to the conclusion that P and N are related by

$$\frac{\partial P}{\partial N} = bP(1 - P) \tag{11.54}$$

where b is a constant. The ascending limb of Ullyett's curve was well described by the logistic equation.

Third, it is necessary to know how N governs I. We can obtain a general expression for I by analyzing data from studies where the effect of N on I was measured separately. Smirnov and Polejaeff (1934) and Smirnov and Wiolovitsh (1934) measured the effect of population density on female sterility in two species of scale insects on lime and ash trees in Moscow parks and squares. Where I_{min} represents the minimum level to which I can be driven by increasing density, analysis of the data by Smirnov and his associates shows the relation, where f is a constant,

$$\frac{\partial I}{\partial N} = -f(I - I_{min})$$

or
$$\ln (I - I_{min}) = d - fN \tag{11.55}$$

This relation is illustrated in Fig. 11.11.

It is reasonable to assume that population density can decrease fecundity via other interference phenomena in the same way that it affects fecundity via sterility. The assumption that (11.55) is a general equation for the effect of population density on fecundity, operating through interference, will explain various observations discussed earlier in this chapter.

If we integrate (11.53), we get

$$E = \frac{1}{a}\left[1 - \exp\left(-ak\varepsilon NPIt\right)\right] \tag{11.56}$$

With a 50:50 sex ratio, the mean fecundity, as before, will be

$$F = \frac{2E}{Nt} = \frac{2}{aNt}\left[1 - \exp\left(-ak\varepsilon NPIt\right)\right] \tag{11.57}$$

From (11.54) and (11.55), respectively, we find that P and I are given by

$$P = \frac{1}{1 + e^{C - bn}}$$

$$I = I_{min} + e^{d - fN}$$

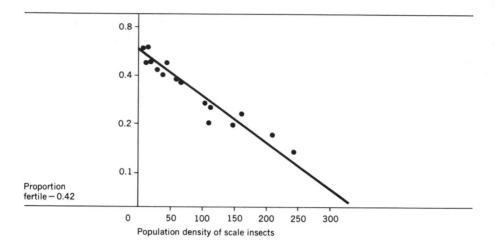

FIGURE 11.11

Test of equation (11.17) using data of Smirnov and Wiolovitsh (1934) on natural populations of Chionaspis salicis L.

and by substituting these into (11.57), we obtain

$$F = \frac{2}{aNt}\left\{1 - \exp\left[-ak\varepsilon N\left(\frac{I_{min} + e^{d-fN}}{1 + e^{C-bN}}\right)t\right]\right\} \tag{11.58}$$

The effects of changes in the various parameters on the shape of (11.58) are illustrated by Figs. 11.12, 11.13, 11.14, and 11.15. Additional insight into behavior of functions such as (11.58) may be gained from study of Fujita's (1954) figs. 4 and 5.

One more constant may be required to fit data with (11.58). Sometimes, as in Ullyett's (1945) data, the locus of F intercepts the F axis not at zero, but at some value A.

There is no point in testing (11.58) against Rich's (1956) data, since (11.58) has more parameters than the number of pairs of values Rich collected at any one time. However, it has been fitted to an observed Allee type of curve (Table 11.3) with 10 points, and an observed *Drosophila* type of curve (Table 11.4) with 14 points.

It will be noticed that $ak\varepsilon$ is equal to 0.0020 in both tables. In fitting the curves, $ak\varepsilon$ was deliberately held constant to make the point that these symbols in (11.58) are redundant. That is, the shape of the curve is completely defined by the other parameters. There is no such thing as "an intrinsic rate of oviposition for a given environmental condition," as this quantity is density-dependent. Hence the correct form of (11.58) is

$$F = A + \frac{2}{aNt}\left\{1 - \exp\left[-aNt\left(\frac{I_{\min} + e^{d-fN}}{1 + e^{C-bN}}\right)\right]\right\}$$

TABLE 11.3
*Simulation of data on oviposition by Anagasta kuhniella
(Zell.), using Equation (11.58) (Ullyett, 1945)*

Moths per cage	Observed number of eggs per female	Calculated number of eggs
6	86	87
10	86	88
20	99	97
24	108	108
28	126	126
30	132	134
38	136	135
44	125	128
52	116	118
56	96	96

Parameter values (days): $t = 2$, $a = 0.00054$, $ak\varepsilon = 0.0020$,
$c = 31.8$, $b = -0.70$, $d = 28.1$, $f = 0.493$, $I_{\min} = 0.0$,
$A = 86.0$.

Discussion

It will be clear to most readers that the model developed here merely represents an attempt to construct a reasonably satisfactory conceptual framework for the data now in the literature. However, a great deal more data would be required before a really satisfactory model for the effect of density on reproductive rate could be constructed. In all areas of resource management research, data-collection programs should be carefully conceived in terms of the mathematical models for which the data are to serve as input; otherwise the programs may turn out to be defective.

In the present instance, it is clear that several specific gaps in our knowledge have developed, and future experimental and field work on density and reproductive rate should be designed with these gaps in mind. Experiments are needed to cover a wide range of population levels, including very low densities, to aid in the interpretation of events in the field at endemic levels. Also, experiments covering a large number of different densities are necessary

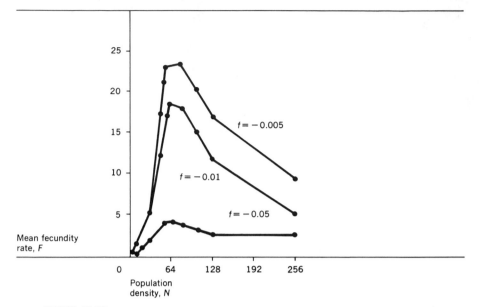

FIGURE 11.12

Behavior of function (11.58). Smaller negative f values reflect decreased effect of population density on interference. Other values used were $t = 1$, $a = 0.006$, $ak_\varepsilon = 0.002$, $c = 5.0$, $b = -0.10$, $d = 2.0$, and $I_{min} = 0.4$.

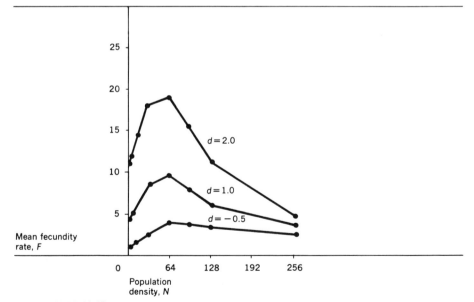

FIGURE 11.13

Behavior of function (11.58). Decreasing d values show effect of decreasing minimum level of interference (that is, $I_{min} + e^d$). Other values used were $t = 1$, $a = 0.006$, $ak_\varepsilon = 0.002$, $c = 1.5$, $b = -0.05$, $I_{min} = 0.4$, and $f = -0.01$.

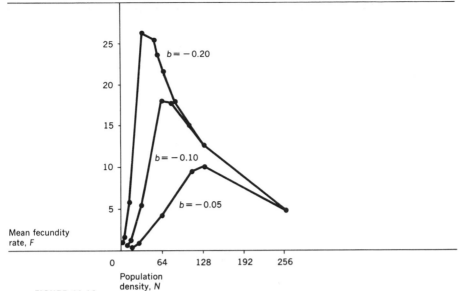

FIGURE 11.14

Behavior of function (11.58). Decreasing b values indicate increasing effect of probability of contact on fecundity. (The greater b is, the less fecundity is stimulated by increasing population densities.) Other values used were t = 1, a = 0.006, ak_ε = 0.002, c = 5.0, d = 2.0, f = −0.01, and I_{min} = 0.4.

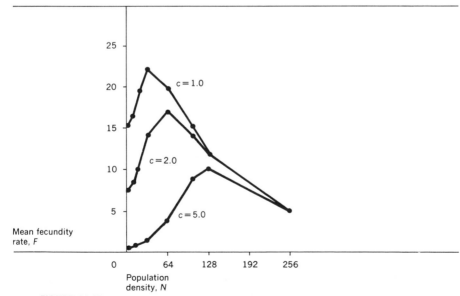

FIGURE 11.15

Behavior of function (11.58). Increasing c values reflect decreasing minimum fecundity rate. That is, the lower c is, the higher the fecundity rate will be at N = 2. Other values used were t = 1, a = 0.006, ak_ε = 0.002, d = 2.0, f = −0.1, b = −0.05, and I_{min} = 0.4.

since one degree of freedom is lost in testing an equation for each extra parameter used in the equation. Most realistic mathematical models of population phenomena require six or more parameters. It would therefore be desirable to have data on 20 or more different densities, with, say, 10 replicates or more at each density.

TABLE 11.4

Simulation of data on oviposition by Callosobruchus chinensis (L.), using Equation (11.58) (Utida, 1941c)

Population density	Mean oviposition rate per female at 30.4°C and 76% R.H.	
	Observed	Calculated
2	88.7	86.0
4	81.0	83.3
8	70.7	78.6
16	57.4	69.3
32	50.1	56.1
48	51.0	47.4
64	47.8	42.1
96	46.9	36.1
128	38.4	33.5
192	38.7	31.9
256	31.9	31.4
384	28.7	30.5
512	29.4	29.6
768	16.8	27.4

Parameter values: $t = 1.0$ (assumed, since t not given by Utida), $a = 0.000035$, $ak\varepsilon = 0.0020$, $c = 1.8$, $b = -0.01$, $d = -0.5$, $f = -0.027$, $I_{min} = 0.30$, $A = 0.0$.

More experiments are needed to check the form of equations (11.54) and (11.55). Analytical experiments such as those of Ishida (1952), Park (1933), and Utida (1941b) are required to separate the effects of interference and competition for oviposition sites. It is becoming apparent that the most immediate consequence of mathematical population analysis is to show that more analytical experiments on behavior are required to produce data of the type collected by Smirnov and his associates. It would be useful to know the effect of density on each component of interference separately. Not nearly enough data are available that reveal the effect of density on natural populations.

An extremely important gap in our knowledge that should be filled quickly concerns the effect of very low densities on mating behavior. Figure 11.10 suggests that fecundity rate may not rise much above its minimum value until a certain threshold population density has been surpassed.

Andersen (1965) has noted the failure of (11.58) to match the conditions of Ullyett's experiment. He has suggested another factor that may operate in reproduction. Suppose that undisturbed females are quiescent and have reduced metabolism. High population densities will lead to frequent disturbance, and to the squandering of an increased proportion of the female's metabolic activity on movement, as opposed to reproduction. Hence metabolism and rate of reproduction depend on density. Andersen proposes the sub-submodel

$$\tau = te^{K(N-N_0)}$$

to account for this, where t is chronological time, and τ is biological time, defined as the time needed at the lowest density (N_0) to accomplish a biological process taking time t at density N. The point is that only a components analysis of the type advocated by Holling (1965, 1966), in which the experiments are specifically designed to yield a systems model, will tell us which type of model is most satisfactory. Models of the type indicated here, where we attempt description of whole systems from experiments on fragments reported in literature, from experiments at different times and places and on many systems, are a poor stopgap at best. (The importance of the experimental program designed at the outset to yield a systems model is the motive for our concern with the research of Ivlev and Holling, discussed in the next two sections.)

In spite of these deficiencies in our knowledge, enough is known to indicate some conclusions that have implications for insect-pest control in particular, and resource management in general.

Equation (11.58) implies that a reduction in population density N may not in fact reduce the reproductive rate in a population. As indicated by Figs. 11.12 to 11.15, populations may produce an actual increase in the size of the progeny generation as a result of a considerable decrease in N. This is, of course, one of the explanations for the homeostatic responses to harvesting and pest control discussed in Chaps. 4 and 5.

If short-term screening of control methods and agents is employed, we may get very misleading comparisons. The poorest control measure from a long-term standpoint (for example, certain insecticides) may have a catastrophic short-term effect on a pest population, but may elicit a large-scale homeostatic reaction from the population. A control method that is by no means spectacular in its initial effort may be best from a long-term point of view because it gradually erodes the pest population's homeostatic capability. A systems model that incorporates a historical element is necessary to demonstrate this phenomenon convincingly. A truly objective comparison of control methods can only be made over more than one generation.

In general, all other things being equal, an Allee type of fecundity curve renders a population more vulnerable to extinction at low densities than does a *Drosophila* type of curve, and thus indicates a population more subject to control by any means. Therefore any genetic mechanism that could convert a pest population from *Drosophila* to Allee type could be of economic importance. Furthermore, a laboratory-produced *Drosophila* type of strain could be used to quickly spread deleterious alleles through a natural Allee-type pest population at endemic periods of a long-term sequence.

Rather more complex implications of (11.58) appear when we consider the interaction of chemical and biological control agents. Consider the following hypothetical situation: A *Drosophila*-type pest population is controlled in nature by two entomophagous insects, A, an Allee type of insect, and D, a *Drosophila* type. It is a basic tenet of modern ecological thinking that there will be lower masses, and hence densities, for large species (A and D) than for pests in any natural community (see Chap. 3). This is because of the low thermodynamic conversion efficiency from one level of a food pyramid to the next higher level and, in the case of a predator, may also be because of the greater linear dimensions of the predator (Hutchinson, 1959). Now suppose the population of the pest, of A, and of D are all reduced x percent by a non-species-selective insecticide. Clearly, A's position will be relatively worse than that of the pest, because of the great difficulty individuals of A will have in finding mates at very low densities. If A were the only entomophagous insect present, an insecticide could clearly help the pest in two ways. The pest's fecundity rate would shoot up, and the density level and improbability of extinction relative to A would be raised. Such ecological mechanisms, together with physiological and genetic mechanisms already well known, help explain why insecticides may not annihilate a pest after repeated application.

On the other hand, D does not have its long-term position relative to the pest made as tenuous as that of A. An implication is that if we want to use a combination of chemical and biological control agents, D type of entomophagous insects will fare better than the A type.

When we consider harvested animals in the light of equation (11.58), it is clear that increasing the intensity of exploitation, up to the optimal level, will increase the reproductive rate and the efficiency with which the population converts incident solar energy into biomass. This is because the harvesting process itself lowers N, and hence interference, and competition for sites where reproduction can occur (assuming that the harvested population, under low rates of exploitation, is dense enough for such effects to occur; this is typically the case).

11.5 FEEDING AND COMPETITION

The most comprehensive and penetrating quantitative research on the ecology of feeding and competition has been conducted by the fisheries biologist Ivlev (1961); this section is based largely on his findings. A distinguishing feature of Ivlev's mathematical models is that they have all been validated in experimental studies with a variety of fish species: goldfish, catfish, sunfish, carp, roach, perch, bleak, bream, and tench. In order to save space, we will confine our attention to the mathematical descriptions of his results; readers wishing more detail on the experimental program can find this in Ivlev's book and the extensive list of references therein.

Ivlev considers, first, the relation between the density of a prey population and the food ration available to each member of the predator population. Where

p = density of prey population
r = average size of ration consumed by each predator
R = maximal ration (amount each predator would eat per unit time in the absence of competitors)
ξ = constant

Ivlev found his data were described by the relation

$$\frac{dr}{dp} = \xi(R - r) \tag{11.59}$$

or

$$\int \frac{dr}{R - r} = \xi \int dp$$

or

$$- \ln (R - r) = \xi p + C \tag{11.60}$$

when $p = 0$, $r = 0$, and the constant of integration, $C = - \ln R$. Therefore (11.60) becomes

$$- \ln (R - r) = \xi p - \ln R \tag{11.61}$$

or

$$\ln\left(\frac{R}{R - r}\right) = \xi p \tag{11.62}$$

or
$$R - r = Re^{-\xi p} \tag{11.63}$$

or
$$r = R(1 - e^{-P}) \tag{11.64}$$

This happens to be the same equation Watt (1959b) arrived at independently to describe the effect of host population density on the number of eggs laid by insect parasites. Watt's equation, however, was expressed in the form

$$N_A = PK(1 - e^{-aN_o P^{1-b}}) \tag{11.65}$$

where P = density of parasites (or predators)

N_o = density of population vulnerable to attack by parasites or predators

N_A = number of prey or hosts attacked

K = maximum possible attack rate per unit time (that is, during period covered by equation) per attacking parasite or predator

a = searching efficiency

b = intraspecific competition pressure amongst the parasites or predators

Watt derived (11.65) from two differential equations: where A is attack rate per capita of attackers,

$$\frac{dN_A}{dN_o} = PA(PK - N_A) \tag{11.66}$$

which is the analog of (11.59), and

$$\frac{dA}{dP} = -b\frac{A}{P} \tag{11.67}$$

which states that attack rate per attacker is a diminishing function of P, the attacker population density. Equations (11.66) and (11.67), like Ivlev's (11.64), are based on analysis of many sets of data. The fact expressed in (11.59) and (11.66) is that any attacking organism has an upper limit to its attack rate, determined by environmental conditions, and as density of host or prey increase, attack rate approaches this fixed upper limit at a progressively decreasing rate. A typical situation is illustrated in Fig. 11.16. Unless enough types of data are available to obtain parameter values for the Holling model (described in Sec. 11.6), the best basic descriptor of feeding (or attack) rate, in terms of predator and prey, or host and parasite densities, is (11.65). The remainder of this chapter is devoted to extensions of this equation derived from Ivlev's experiments with fish.

Ivlev describes the distribution of food in space by ζ, the index of ag-

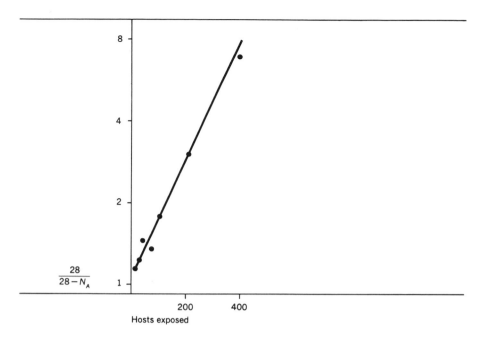

FIGURE 11.16

Test of equation $N_A = PK(1 - e^{-AN_oP})$ by plotting data in the form $ln(PK/PK - N_A) = AN_oP$. Here $PK = 28$, so $28/(28 - N_A)$ is plotted against N_o, hosts exposed, on semilog graph paper. The data describe searching by Dahlbominus fuscipennis (Zett.) for Neodiprion lecontei (Fitch) on a lawn, and are obtained from Burnett (1954). Fit is from Watt (1959b).

gregation, where

$$\zeta = \sqrt{(\Sigma \alpha^2/n)} \qquad (11.68)$$

and α and n are measured as follows. Imagine that we superimpose, on the environment within which food is distributed, a grid of squares. This grid contains large or small squares, whichever best describes the pattern of distribution of the food. The density of the food in each square is measured, and $\Sigma \alpha^2$ is obtained as the sum of squares of deviations of food densities in the individual squares from the mean food density in all squares. The denominator, n, is the number of squares on which the mean is based. Thus ζ is the standard deviation of the food densities in the squares about the mean density.

Ivlev points out that, as ζ increases, the concentration of food in particular places increases, and an actively moving animal can locate these concentrations and use them to increase his ration; thus, increasing the concentration of food, for an active searcher, has the same effect as increasing the mean food density in all squares. Hence Ivlev writes, as an analog of (11.64),

$$r = R(1 - e^{-\chi\zeta}) \tag{11.69}$$

where χ represents the coefficient of concentration of food.

Ivlev notes a difficulty with (11.69). Where food is absolutely evenly distributed, $\zeta = 0$, and (11.69) reduced to

$$r = R(1 - 1) = 0$$

which states that no food at all is available where food is evenly distributed in the environment. To correct for this, (11.69) is modified to

$$r = (R - p)(1 - e^{-\chi\zeta}) + \rho \tag{11.70}$$

When $\zeta = 0$, (11.70) reduces to $r = \rho$, where ρ is the ration consumed when there is perfectly even food distribution.

The problem now is to combine (11.64) and (11.70). That is,

$$r = f(p,\zeta) \tag{11.71}$$

$$dr = \frac{\partial f}{\partial p}\, dp + \frac{\partial f}{\partial \zeta}\, d\zeta \tag{11.72}$$

From (11.59) and (11.69), we have

$$\frac{dr}{dp} = \xi(R - r) \qquad \text{and} \qquad \frac{dr}{d\zeta} = \chi(R - r) \tag{11.73}$$

or
$$dr = \xi(R - r)\, dp + \chi(R - r)\, d\zeta \tag{11.74}$$

Equation (11.74) becomes, on integration,

$$r = R(1 - e^{-(\xi p + \chi\zeta)}) \tag{11.75}$$

This equation expresses the average food ration per individual in a feeding population as a function of the mean density and degree of aggregation (concentration) of food. Ivlev has shown that (11.75) describes data on both laboratory and natural populations of carp.

Ivlev measures the preference by a feeding species for particular types of food by using an electivity index, E.

$$E = \frac{r_i - P_i}{r_i + P_i} \tag{11.76}$$

where r_i = relative content of ith species of food in ration (as proportion of whole ration)

P_i = relative value of same ingredient in environment

Thus, if $r_i = P_i$, $E = 0$. As r_i becomes very large relative to P_i, E approaches (but never equals) 1. As r_i becomes very small relative to P_i, E approaches (but never equals) -1. Ivlev recognizes that electivity may have two components: E_p, or preference, and A, or accessibility. Thus we may write

$$E = E_p + A$$

or
$$A = E - E_p \tag{11.77}$$

When an experiment is designed so that all food is absolutely accessible ($A = 0$), $E = E_p$. Where $_pr_i$ represents the relative value of a particular species of food with absolute accessibility, we have, from (11.76) and (11.77),

$$A = \frac{r_i - P_i}{r_i + P_i} - \frac{_pr_i - P_i}{_pr_i + P_i} \tag{11.78}$$

Ivlev points out and demonstrates experimentally that electivity is affected by a host of factors, such as degree of satiation of predators, population density of the food animals, predation population density, duration of feeding on a specific food, size of predators and prey, mobility of prey, presence of shelter, protective mechanisms, food distribution and selection. Any experimental program on a resource management problem should be designed to expose any possible influence of such factors, by using an experimental design analogous to that required to produce input for (11.78).

Competition

Ivlev develops his own theory of competition inductively, choosing not to build on the mathematical formulations of earlier authors such as Lotka (1925) and Volterra (1926). This makes sense because, as I have pointed out earlier (1962), these models had an entirely a priori origin and there is consequently no guarantee that they bear any relation to the real world. Ivlev measures the force of competition as follows. Let

r^1 = ration of animal feeding in presence of other animals
r = ration of animal feeding in isolation
v = effect of competition

Then
$$v = \frac{r - r^1}{r} \tag{11.79}$$

from which
$$r^1 = r(1 - v) \tag{11.80}$$

Combining (11.75) and (11.80) gives

$$r^1 = R(1 - e^{-(\xi\rho + \chi\zeta)})(1 - v) \tag{11.81}$$

Ivlev has considered the question of how v varies with N. The basic equation he assumes is

$$\frac{dv}{dN} = bN^a$$

$$v = b \int N^a dN = \frac{bN^{a+1}}{a+1} + C \tag{11.82}$$

He postulates that $a = -(b+1)$, whereupon (11.82) becomes

$$v = \frac{bN^{-b}}{-b} + C = C - N^{-b} \tag{11.83}$$

Ivlev then reasons that since v should be 1.0 at $N = \infty$, and $v = 0$ at $N = 1$, C must be equal to 1.0. Therefore (11.83) becomes

$$v = 1 - N^{-b} \tag{11.84}$$

Ivlev has demonstrated (11.84) to be correct in many experiments. However, in some other cases, very much more complicated equations, such as

$$v = 1 - N^{-b} + \mu(N-1)^w e^{v^1(N-1)} \tag{11.85}$$

are required to describe the relation between v and N. A great deal more research on a variety of types of animals will be required to allow us insight into the physiological and ethological mechanisms producing a form such as (11.85).

Where pairs of species were competing, Ivlev found that the species-specific effects of competition could be expressed, for the first and second species, respectively, by pairs of equations such as

$$v_1^1 = 1 - N_1^{-b_1} + \lambda_2 N_2^{-b_2} \tag{11.86}$$

$$v_2^1 = 1 - N_2^{-b_2} + \lambda_1 N_1^{-b_1}$$

or
$$v_1^1 = 1 - N_1^{-b_1} + \lambda_2 [N_2^{-b_2} + \mu(N_2-1)^w e^{v(N_2-1)}]$$

$$v_2^1 = 1 - N_2^{-b_2} + \mu(N_2-1)^w e^{v(N_2-1)} + \lambda_1 N_1^{-b_1} \tag{11.87}$$

Nevertheless, a great deal of research on a wide variety of types of organisms will still be required to discern if (11.87) really is a basic model, universally applicable, or if it is simply an empirical interpolatory formula that happens to fit the data well in certain circumstances. Our motive for conducting the long, arduous search for an underlying, basic, and universally applicable model, which covers all cases merely by varying parameter values, is that such a model can become the key to manipulation of animal populations in nature. That is, if a general model is structured so that terms and variables in the model correspond to readily measurable attributes of an animal, we know, after making such measurements, how the animal will perform in a given ecological situation in advance of its being placed in that situation. Development of such models provides a tool for rapid prescreening of species to select the optimum species to fill a role in a specific ecological situation.

11.6 ATTACK

Our presentation of the mathematical theory of attack is based entirely on the two monographs by Holling (1965, 1966). The depth and comprehensive nature of his work clearly obviate the need for reference to earlier studies. The model is conceived primarily in terms of predators, though one for parasitism also must include the same basic components (Holling, 1961). In addition to being useful in computer programs for resource management strategy evaluation (Watt, 1964a,b), a realistic attack model can serve as the basis for a new type of prescreening program for biological control agents. Very careful prescreening is necessary because the net attack effect of parasites and predators of a pest is not increased by adding a new species of parasite or predator unless that parasite or predator meets certain desiderata (Watt, 1965). Computer simulation has already been used to explore the effect of varying the values of various factors on attack effectiveness (Holling, Brown, and Watt).

The attack model is treated in detail because, more than any other study yet published, it illustrates:

1. The tremendous power in analysis of complex systems when there is an intimate wedding of an experimental components analysis approach with computer simulation studies. It was only through this partnership that a realistic model of attack was constructed.

2. The great realism inherent in the Fortran language for describing biological processes of great complexity. This realism occurs because Fortran is, by design, a descriptive mathematical language of great power and flexibility. It incorporates within itself the features of many other branches of mathematics, including matrix algebra and logic.

3. The problems in numerical analysis and curve fitting created by the nonlinearities in biological models, which deal with realism on its own complex terms.

4. The difference equation approach to describing successive events in the systems model, even where a differential equation approach has been

used in deriving the submodels. That is, the systems model describes a sequential process, in which the output from each step becomes the input for the next step.

5. The lack of need to produce a grossly oversimplified model of reality when the computer has 32,000 ten-digit words of memory, does additions in 16 microseconds (millionths of a second), and reads and writes characters on magnetic tape at 75,000 characters per second. Since a computer with these characteristics rents for $125 per hour, we arrive at the amazing cost per unit operation of

$$\frac{1,000,000 \times 60 \times 60}{16 \times 12,500} = 18,000 \text{ additions per cent!}$$

Attack thresholds

An attack threshold is reached when the predator's hunger level has risen to the level at which it initiates search for prey. A basic feature of Holling's attack model is that attack thresholds are treated as constantly changing in response to the palatability of the available prey, and the hunger level. Let

HTK = maximum level to which attack threshold can be raised by experience
HT = attack threshold at any point in time
H = hunger level
E = amount that attack threshold is changed on one exposure
c and d = constants expressing degree of attractiveness of prey

$|HTK - HT|$ is greater, the greater the unfamiliarity of the prey. The smaller the difference (that is, the more familiar the prey is to the predator), the smaller will be the effect of one exposure to the prey. Holling concludes that E may be expressed as

$$E = (c + dH)(HTK - HT) \tag{11.88}$$

We can eliminate the constant c by setting $E = 0$ and $H = HTK$. Then $c = -dHTK$, and (11.88) reduces to

$$E = d(H - HTK)(HTK - HT) \tag{11.89}$$

Note that E can be negative if H is less than HTK (that is, if the animal is completely satiated).

We now introduce the palatability constants $AYUM$ and $HTKYM$ for a palatable prey, and $AIKY$ and $HTKKY$ for an unpalatable prey. (Note that the parameters are given Fortran designations so we can use the same label for constants in our computer programs.) Thus (11.89) becomes either

$$E = AYUM(H - HTKYM)(HT - HTKYM)$$

or $$E = AIKY(H - HTKKY)(HTKKY - HT) \qquad (11.90)$$

Whenever the attack threshold departs up or down from the threshold for an unfamiliar stimulus (HTS), it thereafter returns to that threshold unless moved again by a new stimulus. We assume that

$$\frac{d(HT - HTS)}{dt} = -AF(HT - HTS)$$

where t is time and AF is the instantaneous rate of forgetting. If h represents $HT - HTS$, this integrates to

$$h_t = h_0 e^{-AF \cdot t,}$$

or $$(HT - HTS) = (HTO - HTS)e^{-AF \cdot t}$$

where HTO represents the attack threshold at the starting time, t_0. Hence

$$HT = HTS + (HTO - HTS)e^{-AF \cdot t \cdot} \qquad (11.91)$$

Time elapsed between captures

Holling's attack model gives functions for the values, in succession, of the time intervals:

TD = time for digestive pause
TS = time spent searching for one prey
TP = time spent pursuing one prey
TE = time spent eating one prey

These four values sum to produce the time interval, TI, from the completion of eating one prey to the completion of eating the next. From TI, we can calculate the other information needed about the attack phenomenon. Note, for example, that if TA is the total time available for feeding, NA, the number of prey actually attacked, is given by having the computer add 1 to NA every time a TI cycle is completed until TA is used up. However, NA cannot be computed from the expression $NA = TA/TI$, because TI changes constantly with changing attack thresholds and hunger levels. Note also that TD will only take a value other than zero when the hunger level is below the attack threshold level of hunger. If the predator completes eating a prey, and is still so hungry that the hunger level exceeds the attack threshold, it immediately begins searching for the next prey.

The digestive pause

The general expression for the increase of hunger with time, starting with a state of complete satiation, is

$$H = HK(1 - e^{-AD\ TF})$$ (11.92)

where H = hunger level, measured in grams of food required to satiate

HK = maximum capacity of gut, in grams of food

AD = instantaneous digestion

TF = time of food deprivation from condition of complete satiation

This equation, like the others used by Holling, is the product of an extensive experimental program. What we need is an expression for TD, the digestive pause, not in terms of time from complete satiation, but in terms of time from the completion of eating the last meal, and the hunger level at that time. This can be obtained as follows. From (11.92)

$$TF = \frac{1}{AD} \ln \frac{HK}{HK - H}$$ (11.93)

Now, if $TF = TF0$ when $H = H0$, we want to know the hunger, H, T units of time later when $TF = TF1$.

Thus, $$TF1 = TF0 + T$$ (11.94)

From (11.93), we may write

$$TF1 = \frac{1}{AD} \ln \frac{HK}{HK - H}$$ (11.95)

and $$TF0 = \frac{1}{AD} \ln \frac{HK}{HK - H0}$$ (11.96)

Substituting (11.95) and (11.96) in (11.94) gives

$$\frac{1}{AD} \ln \frac{HK}{HK - H} = \frac{1}{AD} \ln \frac{HK}{HK - H0} + T$$

Solving for T,

$$T = \frac{1}{AD} \left(\ln \frac{HK}{HK - H} - \ln \frac{HK}{HK - H0} \right) = \frac{1}{AD} \ln \frac{HK - H0}{HK - H}$$ (11.97)

Now TD is the time required for the hunger level to rise from $H0$ to the hunger threshold for attack, so that, by substituting $HT0$ for H in (11.97), we have

$$TD = \frac{1}{AD} \ln \frac{HK - H0}{HK - HT0} \qquad H0 < HT0$$ (11.98)

where $HT0$ = level of the attack threshold at $T = 0$; that is, at the moment when one prey has been completely eaten.

If the hunger is equal to or greater than the attack threshold, then there is no digestive pause and

$$TD = 0 \qquad H0 \geqslant HT0 \tag{11.99}$$

Searching time

Holling develops his searching-time expression on the assumption that predators sweep out a surface. Our derivation is parallel to his, but assumes that a volume is swept out. A predator moving between two points, A and B, sweeps out a volume equal to the distance moved, times the area of the path, plus two half-spheres representing the area perceived behind the predator at point A and in front of the predator at point B. If the distance away from itself at which a predator will notice prey is *not* a function of hunger, then volume swept, AS, is given by

$$AS = (VR \cdot T \cdot \pi R^2) + \frac{4}{3}\pi R^3$$

where VR = speed of movement of predator
R = ½ diameter of path
T = time spent moving from A to B

However, R, the maximum distance away from the predator at which the prey is just noticed, changes with time as the predator becomes hungrier. Therefore

$$AS = \pi VR \int_{T=0}^{T} R^2 \cdot dT + \frac{4\pi}{6}(R_1^3 + R_2^3) \tag{11.100}$$

where R_1 = R at $T = 0$
R_2 = R at $T = T$

On the average, one prey will enter this swept-out area when

$$AS = \frac{1}{N0}$$

However, not every prey encountered is successfully captured, so one prey will be captured when

$$AS = \frac{1}{N0 \cdot SP \cdot SS} \tag{11.101}$$

where SP = pursuit success: the proportion of times that a pursuit ends with the predator successfully reaching the prey

SS = strike success: the proportion of times that a predator successfully captures a prey once it is reached

TS, the time taken to search for a prey destined to be successfully captured, can be obtained by equating (11.100) and (11.101), and making $T = TS$. In order to do this, we need to know the shape of the function for R, which on the basis of experimental research by Holling (1965) is given by

$$R = AKRGM(H - HT) \qquad (11.102)$$

where $AKRGM$ = a constant determined by the shape of the reactive field and by the rate at which the field expands with increasing hunger. However, both hunger and the attack threshold change continuously, so we must substitute for H and HT in (11.103) as follows:

$$R = AKRGM[HK - HTS + (H0 - HK)e^{-AD \cdot T} - (HT0 - HTS)e^{-AF \cdot T}]$$

$$(11.103)$$

By integrating from the time the digestive pause is completed to the time searching is completed, we obtain

$$\int_0^{TS} R^2 dt = (AKRGM)^2 \left\{ (HK^2 + HTS^2 - 2HK \cdot HTS)TS \right.$$

$$+ (H0 - HK)^2 \left[-\frac{e^{-2AD \cdot T}}{2AD} \right]_0^{TS} + (HT0 - HTS)^2 \left[-\frac{e^{-2AF \cdot T}}{2AF} \right]_0^{TS}$$

$$+ 2HK(H0 - HK) \left[-\frac{e^{-AD \cdot T}}{AD} \right]_0^{TS} - 2HK(HT0 - HTS) \left[-\frac{e^{-AF \cdot T}}{AF} \right]_0^{TS}$$

$$- 2HTS(H0 - HK) \left[-\frac{e^{-AD \cdot T}}{AD} \right]_0^{TS} + 2HTS(HT0 - HTS) \left[-\frac{e^{-AF \cdot T}}{AF} \right]_0^{TS}$$

$$\left. - 2(H0 - HK)(HT0 - HTS) \left[-\frac{e^{-(AD + AF)T}}{AD + AF} \right]_0^{TS} \right\}$$

$$= (AKRGM)^2 \left[(HK^2 + HTS^2 - 2HK \cdot HTS)TS \right.$$

$$+ \frac{(H0 - HK)^2}{2AD}(1 - e^{-2AD \cdot TS}) + \frac{(HT0 - HTS)^2}{2AF}(1 - e^{-2AF \cdot TS})$$

$$+ \frac{2HK(H0 - HK)}{AD}(1 - e^{-AD \cdot TS}) - \frac{2HK(HT0 - HTS)}{AF}(1 - e^{-AF \cdot TS})$$

$$-\frac{2HTS(H0 - HK)}{AD}(1 - e^{-AD \cdot TS}) + \frac{2HTS(HT0 - HTS)}{AF}(1 - e^{-AF \cdot TS})$$

$$-\frac{2(H0 - HK)(HT0 - HTS)}{AD + AF}(1 - e^{-(AD + AF)TS})\Bigg]$$

$$(11.104)$$

From (11.102), R_1 may be expressed as

$$R_1 = AKRGM(H0 - HT0) \qquad (11.105)$$

We can now obtain TS by equating (11.100) and (11.101), and substituting from (11.104) for

$$\int_{T=0}^{T=TS} R^2 \, dT$$

and from (11.105) for R_1 and R_2, in (11.100). This can only be done where $H0 \geqslant HT0$. Where $H0 < HT0$, TD is first calculated (that is, the time taken for $H0$ to rise to $HT0$) and $H0$ is made equal to $HT0$ in (11.104).

Either of two approaches can be used to solve for TS. One is to use a rapidly converging iterative process, and the other is to simplify the equation for TS by expanding each exponential term into a Taylor series, and retaining only that part of the series that makes a significant contribution to TS. The first procedure is best where the product $AD \cdot TS$ is small, or specifically, AD is 0.05 or less, and TS is less than 2 hours.

Pursuit time

After TD and TS have been calculated, the level of hunger and of the attack threshold, $H1$ and $HT1$, respectively, at the time of pursuit can be expressed as

$$H1 = HK + (H0 - HK)e^{-AD(TS + TD)}$$

and $\qquad HT1 = HTS + (HT0 - HTS)e^{-AF(TS + TD)}$

TP can then be calculated from

$$TP = \frac{AKRGM}{VP}(H1 - HT1)$$

where VP is the pursuit velocity. For some predators, pursuit may be extremely fast and may occupy only seconds whereas TS may be measured in hours. In such cases, little error is introduced by treating TP as zero.

Eating time

The following relationships pertain:

$$TE = KE \cdot W \qquad H1 \geqslant W$$

or
$$TE = KE \cdot H1 \qquad H1 < W$$

where KE is the time to eat 1 gram of food, and W is the weight of each prey. The predator will not eat all of W if its hunger level is only $H1$, so the minimum value of W is likely to be correct. Note, however, that in the case of large fish predators, eating time for small prey and perhaps also large prey is independent of prey biomass, and is probably a very short constant time of a few seconds.

The daily cycle

Having calculated TD, TS, TP, and TE, TI is obtained as their sum. The prey caught during the cycle is eaten, if it is palatable, and the hunger lowered from $H1$ to $H1 - WE$, where WE is the amount of the prey that is eaten.

$$WE = W \qquad H1 \geqslant W$$

$$WE = H1 \qquad H1 < W$$

Similarly, the attack threshold is changed from $HT1$ to $HT1 - E$, where E is calculated from either version of (11.90):

$$E = AYUM(H - HTKYM)(HT - HTKYM) \qquad HT > HTKYM$$

$$E = AIKY(H - HTKKY)(HTKKY - HT) \qquad HT > HTKKY$$

These new values for hunger and the attack threshold represent the values for $H0$ and $HT0$ needed to generate the next cycle.

Calculations for each day proceed, cycle by cycle, until the total time elapsed equals the length of time allowed for the feeding period. During the following nonfeeding period, no attack occurs and the hunger and attack threshold gradually change. To start calculation at the beginning of the next feeding period, the values of the hunger and attack threshold at the beginning of the period are calculated as follows:

$$HBEG = HK + (HLAS - HK)e^{-AD(24 - TA)}$$

where $HBEG$ = hunger at beginning of feeding period
$HLAS$ = hunger at end of last feeding period
TA = length of feeding period

and $$HTBEG = HTS + (HTLAS - HTS)e^{-AF(24-TA)}$$

where $HTBEG$ = attack threshold at beginning of feeding period
$HTLAS$ = attack threshold at end of last feeding period

Day after day of attacking is simulated until equilibrium is reached. The technique of deciding when equilibrium has been reached is outlined below.

The computer simulation program

The model described has been used for extensive computer simulation studies by Holling, Brown, and Watt. The model is designed to be completely general, in that it can be used to simulate attack by a wide variety of different kinds of predators. The computer program was originally developed on an IBM 1620 and subsequently on an IBM 7044, but with slight modifications it can be run on any computer that accepts Fortran in one of its variants.

Computer simulation studies have two purposes. First, they can be used to test understanding of the experimental situation and the extent to which this understanding is built into the mathematical model. Clearly, unless the computer output can faithfully mimic events in the historical (populational) experiments, our understanding and/or our model is defective, or deficient, or both.

Also, once a computer simulation program has been found to be a faithful mimic of reality, we can use experimentation on the computer as a supplement to actual experiments. For example, we can ask: How sensitive is attack effectiveness to changes in the value of $AKRGM$ in (11.102)? We can then simulate attack effectiveness for each of a wide range of $AKRGM$ values by using the computer program. This may be done, in turn, for each of the parameter values in the model, so we can find out the relative effects on attack effectiveness of each of the parameters. We are testing the sensitivity of the end result to variations in each of the parameters, and thus have a tool for systematizing the search after optimal biological control agents. Suppose we find that pursuit velocity is the most important factor determining the efficiency with which the trophic pyramid is cropped. Then we know that the optimal predator is the one whose pursuit velocity is the most efficient. (Efficiency in this context must be evaluated as weight gain per unit biomass of predators as a function of biomass of prey consumed.) Since intraspecific competition pressure among the predators can be an important factor determining predator efficiency, we must compare various predator species with respect to their cropping efficiency at each of several different predator population densities. Additional difficulty in comparing efficiency of various predator species arises because of the differential effects on their searching and pursuit effectiveness of environmental factors such as density and distribution of vegetation, turbidity, water temperature, oxygen content, etc. Experiments must be performed and components built into the systems model to deal with all such factors where systems analysis shows they have a significant effect. Such laborious comparisons can be speeded up enormously if computer experimentation is used as a supplement to laboratory experimentation (Bellman, 1962).

One of the most difficult problems in developing simulation programs of this type is in determining how to terminate loops, or cycles of operations. The difficulty arises because, for example, the predator may be in any of four different states at the end of a time period. Also, if an inordinate length of time is being taken in arriving at equilibrium, the cycle of operations involved in computing a particular case may be arbitrarily terminated at the end of a predetermined number of days, such as 10 days. In the present program, exiting from loops has been handled as follows. If $HT0$, the hunger threshold, has not risen to a preset level, the predator will take a long time to become hungry enough to attack, and the computer branches back to the first instruction to begin work on a new case. If this branching has not occurred by day 10, the computer loops back to instruction 1 in any case.

11.7   DISPERSAL

This section is of central importance in our discussion of mathematical systems description of resource management problems. First, as we have shown in Chap. 6, many large-scale biological phenomena such as furbearer cycles, insect-pest outbreaks, rodent eruptions, and epidemic and epizootic waves have a characteristic wavelike form, with effects radiating outward from an epicenter with the passage of time. This wavelike character has an economic implication, in that we can choose to control a phenomenon either early, when it is still at or near the epicenter, or late, when it has radiated out a considerable distance. Second, traditional mathematical methods are totally inadequate to describe the complex wave-phenomena in biology, where several waves originate at different times and places and radiate outward until they merge.

Skellam (1951) has presented the most comprehensive mathematical analysis and description of dispersal of which I am aware. His paper and the references there, and the papers by Dobzhansky and Wright (1943, 1947) should be consulted for techniques of analyzing dispersion by mathematical analysis, rather than by simulation.

Skellam's equation 26 most closely approximates the real situation in nature:

$$\frac{\partial \Psi}{\partial t} = \gamma^2 \left[\frac{\partial^2 \Psi}{\partial z^2} + \frac{1}{z} \frac{\partial \Psi}{\partial z} + \Psi(1 - \Psi) \right] \tag{11.106}$$

where Ψ = population density at particular time and place
$\quad t$ = time
$\quad z = 2r\gamma/a$
$\quad r$ = radial distance from origin
$\quad \gamma^2$ = intrinsic rate of natural increase
$\quad a^2$ = mean square dispersion rate per generation

Equation (11.106) describes the combined effect of diffusion and logistic growth in a radially symmetrical habitat. Each point in space is defined by a radial distance outward from the origin, or epicenter, of a wave of diffusion and population growth. All points lying a particular distance, r, out from the origin are treated as if they are identical. The equation expresses the rate of change of density as a function of (1) time, (2) radial distance from the origin, (3) the intrinsic rate of natural population increase, and (4) the dispersion rate per generation. The term $1 - \Psi$ indicates that the population-growth rate at any particular distance out from the epicenter approaches zero as the carrying capacity (maximum supportable population) for the habitat is reached. That is, since $\partial \Psi / \partial t$ reaches 0, Ψ reaches a fixed upper-maximum value for density, Ψ_{max}.

Let us now consider the mathematical difficulties of applying (11.106) to description and analysis of complex problems in resource management. The greatest difficulty has to do with the assignment of a particular density, Ψ, to a position in space. The only spatial coordinate in (11.106) is the polar coordinate r, for radius from the epicenter. As long as we only have *one* epicenter, this conceptual model works, but what happens when we have several epicenters? This is not an idle question, because in many natural situations there are many epicenters (Chap. 6), and even if there are not, we may deliberately create several epicenters for large-scale biological phenomena in the course of a control operation. Consider, for example, when we try to control an insect pest by ground release of parasites or predators, or a weed by release of phytophagous insects, or an invertebrate or vertebrate pest by release of a pathogen. In such cases, we have one or more epicenters from which the pests radiate out, and, at one or more subsequent times, we have one or more epicenters from which controlling agents radiate out. How do we position densities or other effects in space when there are several centers to the coordinate systems relative to which effects are being positioned? Also, how do we handle the effects when the waves radiating out from different epicenters merge?

It seems that the only reasonable answer is to describe the flow of events through time, and in space, in terms of a rectangular coordinate system that positions all causes and effects relative to real space, rather than a polar coordinate system that positions all effects relative to the place of origin of their ultimate causes.

A second difficulty in the description and analysis of dispersal phenomena is the sheer complexity of the bookkeeping because of the number of variables involved, the number of different points in space involved, and the number of different times at which we must record the variate values for the several variables at each point in space. We are led inexorably to computers used in conjunction with a pseudoalgebraic language such as Fortran, in which a succinct instruction program can specify all desired computations and output formats.

We will now consider the particular aspects of large-scale systems dispersal problems that arise in resource management strategy evaluation simulation, and how these are handled in Fortran. The main features of dispersal simulation using a rectangular coordinate system and a grid of squares can be visualized in terms of Fig. 11.17.

Suppose we have a grid with 15 rows and 15 columns of squares, as in Fig. 11.17. The computer will simulate dispersal by examining each of the 15×15, or 225, squares in turn, and will determine, for each of the eight surrounding squares, if there is a gradient of density from one square to the adjacent square. If there is, the computer calculates the density of animals subtracted from one square and added to the next. There are several tricks involved in this simple simulation routine, and we shall consider each of these in turn, referring to Fig. 11.17.

1. There will only be eight surrounding squares where the center square (of nine) is more than one row of squares removed from each edge of the 15×15 square. Otherwise, we have the situation as in B, a corner square, or C, an edge square. The computer must test to see if it is in an edge or corner of the large square before it attempts to simulate dispersal to or from the surrounding squares.

2. Dispersal from A to E, one of A's eight surrounding squares, could be handled at two different places in the computation: while E is the center square, movement could be treated as dispersal *to* E; while A is the center square, movement could be treated as dispersal *from* A. Thus, to avoid cal-

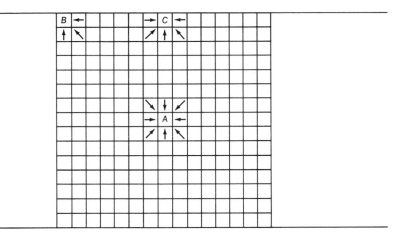

FIGURE 11.17
Simulation of dispersal in a grid.

culating each dispersal twice, by convention, while any given square is being considered as the center square (of nine squares), we only compute the dispersals *toward* that square.

3. Dispersal toward each square is computed on the basis of the density gradients from square to square *prior to the beginning of dispersal.* But, as the computer simulates dispersal in the 15 × 15 square, proceeding stepwise from first column and row to last column and row, it changes the density recorded for each square. Thus, we need two matrices, $P(I, J, 1)$ to record the density *after* dispersal, and $P(I, J, 2)$ to record the density *prior to* dispersal. Otherwise, the density in the last row and column would be different from that in the first row and column at the end of the calculation, even if we were only simulating symmetrical dispersion from the center square, just because of the order in which events had been simulated for the 15 rows and 15 columns of squares. $P(I, J, 1)$ is set equal to $P(I, J, 2)$, for all I and J values (15 of each in this case) at the start of the operation to simulate dispersal, but $P(I, J, 1)$ constantly changes through the calculation, whereas $P(I, J, 2)$ is unchanged.

The preceding discussion is illustrated by the following simple routine.

```
1   DO 12 I = 1,15,1
2   DO 12 J = 1,15,1
3   IL = MAX(1,I − 1)
4   IU = MIN(15,I + 1)
5   JL = MAX(1,J − 1)
```

```
 6   JU = MIN(15,J + 1)
 7   DO 12 K = IL,IU,1
 8   DO 12 L = JL,JU,1
 9   Z1 = MAX(0,PGRAD*(P(K,L,2) − P(I,J,2)))
10   P(I,J,1) = P(I,J,1) + Z1
11   P(K,L,1) = P(K,L,1) − Z1
12   CONTINUE
```

This routine tells the computer how to compute the effects of migration within a matrix 15 rows by 15 columns, for which the initial population densities are stored in the memory locations described as the matrix $P(I, J, 2)$, and for which the population densities produced by dispersal are to be stored in the matrix $P(I, J, 1)$.

Statement 1 informs the computer that it must perform the following steps, down to and including step 12, repeatedly for all rows of the 15×15 square. Statement 2 instructs the computer to perform the following steps, down to and including step 12, repeatedly for all 15 columns of the 15×15 square, *within each of the* 15 *rows.* The next four statements ensure that the computer will not deal with squares in the zero row or column, or the six-teenth row or column, and also tell the computer how to find the boundary rows and columns for the 3×3 matrix of squares surrounding whichever I,Jth square of the 15×15 matrix is being worked on by the computer in any particular cycle of operations.

Statement 9 computes the gradient in densities between the center square of the nine squares and one of its surrounding squares. If there is a positive gradient, step 9 computes migration from the surrounding square to the center square; this migration involves a number of individuals represented by the density excess in the surrounding square, multiplied by PGRAD, the probability of emigration. Step 10 adds the number immigrating to the memory element that is collecting density data on the center square, as affected by dispersal, and step 11 subtracts the corresponding number emigrating from the surrounding square. Step 12 is a dummy statement, which ends the cycle of operations for cases in which no operation is performed, as well as cases in which an operation is performed. CONTINUE is required because each cycle of operations within the loop, starting with step 1 and ending with step 12, must end at the *last* step in the loop. The rules of the Fortran language do not allow us to write a "DO loop" in which some operation cycles terminate at the *last step*, while incompleted cycles terminate at the *first step*. At a later stage in large programs containing this subroutine, $P(I, J, 2)$ is set equal to $P(I, J, 1)$ for all I, J values, to initiate the cycle of operations for the following year.

The procedure we have just described is in fact only the simplest of a large number of types of routines to simulate dispersal with the computer. Some of the extensions and modifications will indicate the great power, flexibility, and ease of working with Fortran programs.

1. The fundamental nature of the dispersal phenomenon can be changed in any of several different ways. For example, we could change statement 9 to Z1 = PGRAD * Y1 * Y1, so that the dispersal is proportional to the square of the density gradient. Second, we might assume, as in the case of small invertebrates or plant seeds, that thermal updrafts and winds lift the population high into the air, blow it for many miles, and then drop it. Thus, dispersal may not be to an adjacent square, but a square several squares removed. This can be handled in a number of different ways, depending on the nature of the process. One way is to increase the addends and subtractands in statements 3 to 6 to 6, say, instead of 1, and follow statement 8 with IF statements excluding squares close to the central square from the calculation of dispersal. (A grid 25 × 25 or larger may be required to simulate such a process.)

2. Fortran allows us to simulate dispersal that starts from several different epicenters, at several different times, and with different starting densities at each time and place, by the use of statements such as

$$P(14, 23, 2) = 473.16$$

which can be introduced at predetermined times. Alternatively, we can simulate release of standard numbers of organisms in a patterned grid of selected squares by the trio of instructions such as

```
       DO   1   I = 2, 20, 2
       DO   1   J = 2, 20, 2
   1       P(I, J, 2) = 40.0 + P(I,J,2)
```

This simulates the increase of density by 40.0 in every second row of squares from row 2 to row 20, and in every second column of squares from column 2 to column 20. We can select every *second* square by utilizing the feature of the instruction format

$$DO   N   I = I1, I2, I3$$

I3 gives the number of rows (or columns, as the case may be) that each I value is increased by at each iteration of the DO loop.

3. The most interesting extension we can deal with is where there is an asymmetrical environment, or some gradient-inducing factor such as a prevailing north wind. We would assume that we no longer have a single PGRAD

value, but rather a set of different probabilities of emigration depending on the direction in which emigration occurs. PGRAD is taken to be a 3×3 array indexed by K and L. Then the modification of the preceding program would consist of replacing Z1 by

$$Z1 = MAX(0,PGRAD(K + 1 - IL,L + 1 - JL)*(P(K,L,2) - P(I,J,2)))$$

Thus we can simulate any conceivable pattern of dispersal. The next step is to combine the simulation routine with routines to simulate population dynamics or the dynamics of epizootic or epidemic waves, as in the next section.

11.8 SIMULATION OF EPIDEMIC AND EPIZOOTIC WAVES

The reader should consult the comprehensive book by Bailey (1957) for a broad introduction to the literature and main developments of mathematical epidemiology. My approach is dictated by the arguments already presented in Chap. 6; namely, that large-scale biological phenomena may be triggered by a specific sequence of weather events. (Additional support for this view is found in Sargent, 1964, and the references there, for human populations.) I am not concerned with the solution of differential or difference equations produced when we assume *average* weather conditions, but rather with the development of systems of difference equations and strings of Fortran statements that can be used to mimic specific sequences of events, using tables with specific sequences of weather data as inputs. The desired Fortran programs can be developed easily by combining the methods of the previous section, on dispersal, with extensions of some of the ideas in the literature about mathematical models of epidemic and epizootic waves.

The particular technique I am using here·to develop models of waves of disease is based on extension of the difference equations model presented by Kermack and McKendrick (1927). This particular mathematical model lends itself to generalization in several different ways, and is known to be a realistic descriptor of existing bodies of data.

Kermack and McKendrick developed a model to describe the dynamics of the situation in which one or more infected persons enter a community of individuals more or less susceptible to the disease. The disease spreads from the affected to the unaffected by contact infection. Each infected person becomes sick, and ultimately recovers or dies. As the infection spreads, the number of susceptibles diminishes, and finally the epidemic terminates. An important feature of the Kermack and McKendrick model is that exhaustion of the supply of susceptibles is not required to terminate the epidemic; this can be brought about by the relationship between infectivity, recovery, and death rates. For each such set of rates, there is a critical threshold for population density. If the density is equal to or less than this threshold, the introduction of one or more infected persons does not give rise to an epidemic; if the population is only slightly denser than the threshold density, a small epidemic occurs. The size of the epidemic increases rapidly as the threshold density is exceeded, and in such a manner that the greater the population density at the beginning of the epidemic, the smaller it will be at the end. The epidemic increases as long as the density of the unaffected population is greater than the threshold density, but when this critical point is approximately reached, the epidemic begins to wane, and ultimately dies out. This point may be reached when only a small proportion of the susceptibles in an area have been affected. We will present the basic Kermack and McKendrick model, and then show how, by using Fortran on a large computer, we can deal with a generalized version of it that has (1) immigration and emigration, (2) variable infectivity rates, depending on weather, (3) three-factor and four-factor complexes [using May's (1961) terminology], and (4) stochastic, rather than deterministic, assumptions.

Let us assume we can divide the entire time period of an epidemic or epizootic wave into a series of separate, equal length intervals, and that infections occur only at the instant of passage from one time interval to the next. We need the following symbols to develop the Kermack and McKendrick model. All of the following symbols refer to the number per unit of area, and hence are measures of density.

$v_{t,\theta}$ = number of individuals at time t who have been infected for θ intervals ($v_{t,0}$ refers to number that has just become infected)

v_t = number of individuals infected at time t

y_t = total number of individuals who are ill at time t, or $\sum_{\theta=0}^{t} v_{t,\theta} = v_t$

x_t = number of individuals still unaffected at time t

N = total population density = $x_t + y_t + z_t$ where there is no immigration, emigration, birth, or death due to causes other than disease

ψ_θ = proportion of individuals who have been infected for θ intervals and who are removed by recovery and death

ϕ_θ = probability of unaffected individual being infected by individual who has been sick for θ intervals

At time $t = 0$, the number ill is $v_{0,0}$. We may schematize the epidemic process as follows.

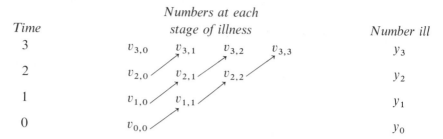

Time	Numbers at each stage of illness				Number ill
3	$v_{3,0}$	$v_{3,1}$	$v_{3,2}$	$v_{3,3}$	y_3
2	$v_{2,0}$	$v_{2,1}$	$v_{2,2}$		y_2
1	$v_{1,0}$	$v_{1,1}$			y_1
0	$v_{0,0}$				y_0

The number who are removed from each θ group at the end of the interval t is

$$\psi_\theta v_{t,\theta} = v_{t,\theta} - v_{t+1,\theta+1} \tag{11.107}$$

From (11.107), $\qquad v_{t,\theta} = v_{t-1,\theta-1}(1 - \psi_{\theta-1}) \tag{11.108}$

$$= v_{t-2,\theta-2}(1 - \psi_{\theta-1})(1 - \psi_{\theta-2})$$

$$v_{t,\theta} = v_{t-\theta,0}(B_\theta) \tag{11.109}$$

where $\qquad B_\theta = \prod_{t=1}^{\theta}(1 - \psi_{\theta-t}) \tag{11.110}$

If we assume that the chance of an infection is proportional to the product of the number infected and the number not yet infected, and that $x_t > 0$, we have

$$v_t = x_t \sum_{\theta=1}^{t} \phi_\theta v_{t,\theta} \tag{11.111}$$

also $\qquad x_t = N - \sum_{t=0}^{t} v_{t,\theta} - \sum_{t=0}^{t} z_t \tag{11.112}$

By substituting from (11.109) for $v_{t,\theta}$ in (11.111), we obtain

$$v_t = x_t \sum_{\theta=1}^{t} \phi_\theta B_\theta v_{t-\theta,0} \tag{11.113}$$

but since $\qquad v_{0,0} = v_0 + y_0 \tag{11.114}$

and $\qquad A_\theta = \phi_\theta B_\theta \tag{11.115}$

(11.113) becomes $\qquad v_t = x_t\left(\sum_{\theta=1}^{t} A_\theta v_{t-\theta} + A_t y_0\right) \tag{11.116}$

Substituting from (11.109) for $v_{t,\theta}$ in

$$y_t = \sum_{\theta=0}^{t} v_t$$

we get
$$y_t = \sum_{\theta=0}^{t} B_\theta v_{t-\theta,0} + B_t y_0 \qquad (11.117)$$

Now
$$x_{t+1} = x_t - v_t \qquad (11.118)$$

Therefore, from (11.116), we get the difference equation

$$x_{t+1} = x_t - x_t \left(\sum_{\theta=1}^{t} A_\theta v_{t-\theta} + A_t y_0 \right) \qquad (11.119)$$

Since $z_{t+1} - z_t$ is the number of persons removed at the end of interval t,

$$z_{t+1} = z_t + \sum_{\theta=1}^{t} \psi_\theta B_\theta v_{t-\theta} + \psi_t B_t y_0 \qquad (11.120)$$

From
$$x_t + y_t + z_t = N \qquad (11.121)$$

we get
$$y_{t+1} = y_t + (x_t - x_{t+1}) - (z_{t+1} - z_t) \qquad (11.122)$$

This merely states that the total, N, is unchanging, or that

$$x_{t+1} + y_{t+1} + z_{t+1} = y_t + x_t + z_t \qquad (11.123)$$

Equations (11.119), (11.120), and (11.122) give us a set of recurrence relations from which we can calculate the state of the population at any time, given a sequence of removal rates, ψ_1, ψ_2, ψ_3, etc., and a sequence of infectivity rates, ϕ_1, ϕ_2, ϕ_3, etc. for each time.

Kermack and McKendrick's classic threshold theorem follows readily if we consider the differential equation counterparts of (11.119), (11.120), and (11.122) when ψ and ϕ are constant with time, and equal to j and k, respectively. Then we get

$$\frac{dx}{dt} = -kxy \qquad (11.124)$$

$$\frac{dy}{dt} = kxy - jy \qquad (11.125)$$

$$\frac{dz}{dt} = jy \qquad (11.126)$$

The integral of dz with respect to time, over the entire period of the epidemic, is a measure of the magnitude of the epidemic. Since we wish to determine which factors determine this magnitude, and the nature of the functional relationship between the magnitude of the epidemic and the factors determining it, it is necessary to determine z_∞. This can be done as follows. From (11.121),

$$\frac{dz}{dt} = j(N - x - z) \tag{11.127}$$

From (11.124) and (11.126),

$$\frac{dx}{dz} = \frac{-kx}{j} \tag{11.128}$$

and since $z_0 = 0$, this integrates to yield

$$\ln x_t = \frac{-zk}{j} + \ln x_0$$

or
$$\ln \frac{x_0}{x_t} = \frac{zk}{j} \tag{11.129}$$

From (11.127) and (11.129) we get

$$\frac{dz}{dt} = j(N - x_0 e^{-kz/j} - z) \tag{11.130}$$

From this point on, the algebra can be condensed somewhat by setting $p = j/k$. Equation (11.130) cannot be integrated, but since we know that

$$e^x = 1 + x + \frac{x^2}{2!} + \frac{x^3}{3!}$$

etc., we can drop all but the first three terms in the series, and obtain

$$\frac{dz}{dt} = j(N - x_0) + z\left(\frac{x_0}{p} - 1\right) - \frac{x_0 z^2}{2p^2} \tag{11.131}$$

Several methods have been proposed for obtaining approximate values for z_∞, the number of individuals who have been removed from the ill group, by recovery or death, by the end of the epidemic. The simplest procedure, which gives the same result as considerably more complicated derivations, is to set $dz/dt = 0$ and $x_0 = N$ in (11.131). This approximation only holds if a relatively small proportion of the population is infected at $t = 0$, but it yields the same qualitative result as accurate values based on sophisticated mathematics.

From (11.131),

$$\frac{x_0}{p} - 1 = \frac{x_0 z_\infty}{2p^2} \tag{11.132}$$

$$z = \frac{2p^2}{x_0}\left(\frac{x_0}{p} - 1\right) \tag{11.133}$$

$$z = 2p\left(1 - \frac{p}{x_0}\right) \tag{11.134}$$

At the beginning of an epidemic, x_0, the number of individuals still unaffected at t_0 will be almost identical to N, the initial population size. Hence, if the ratio $j/k = p$ is equal to N, $z_\infty = 0$, which means that there can be no epidemic. If N slightly exceeds p, there will be an epidemic. Thus, $N_0 = p$ may be thought of as the threshold density of the population for an epidemic to occur. That is,

$$\text{Threshold density} = \frac{\text{instantaneous removal rate}}{\text{instantaneous infectivity rate}}$$

To put it another way, the greater the removal rate, the greater the threshold density required to produce an epidemic, whereas the greater the infectivity rate, the smaller the threshold density required to produce an epidemic. Suppose now, we consider the total population, N, as consisting of two components: N_0, and an "excess," n. Then we can write

$$N = N_0 + n = \frac{j}{k} + n \tag{11.135}$$

and, from (11.134),

$$z = \frac{2j}{kn}\left(N - \frac{j}{k}\right) = \frac{2jn}{kN} \tag{11.136}$$

or, since

$$\frac{j}{k} = N - n$$

$$z_\infty = \frac{2n}{N}(N - n) = 2n - \frac{2n^2}{N} \tag{11.137}$$

If n is small compared to N, then to a first approximation, if the balance between N, k, and j is such as to produce an epidemic at all, the deaths plus recoveries will equal $2n$. This is an odd result; it says that at the end of an epidemic, the deaths plus recoveries will be as far below N_0 as the excess was above it. The larger the initial density, the greater the drop below N_0 will be. The small population will get out of the epidemic larger than the very dense one. (There is little motive for enormous expansion of the world human population in these findings!)

An especially important feature of the Kermack and McKendrick model is that it provides a mechanism for terminating epidemics. The total number of sick individuals reaches a maximum when equation (11.125) is zero, that is, when

$$\frac{dy}{dt} = kxy - jy = 0$$

and at this point, $kxy = j$, or $x = j/k$. This means that the total number of sick individuals begins to decline when x, the number of individuals still unaffected, has dropped to N_0, the threshold density. From this point on, any infected individual is more likely to be removed by recovery or by death than to become a source of further infection, and so the epidemic terminates.

The difficulty with this model is that as it is made progessively more realistic, by adding in more complexity, the analytic difficulties become more and more formidable. We can circumvent all these difficulties, and in addition work with models of enormously greater complexity than anything previously considered, by the simple expedient of using Fortran programming and simulation studies on a large computer. In effect, we are dealing with precisely the same problem considered in the previous section, except that now our population is subdivided into three groups: sick, susceptible, and "removed" individuals. The last class can be subdivided into dead and recovered persons.

We now need to define a number of Fortran symbols: three vectors for rate of infectivity, rate of removal, and rate of death, RI (K), RR (K), and RD (K), respectively. Instead of the two arrays P (I, J, 1) and P (I, J, 2), we now need the following arrays:

V(I,J,K)	corresponding to $v_{t,\theta}$, with I and J defining position in large square, and K corresponding to θ
VSUM(I,J)	corresponding to $\Sigma v_{t,\theta}$
X(I,J,K)	corresponding to x_t
Y(I,J,K)	corresponding to y_t
Z(I,J)	corresponding to z_t, but cumulative
RECOV(I,J,K)	the component of Z that survives
DEAD(I,J)	the component of Z that dies
VISUM(I,J)	corresponding to $\Sigma \phi_\theta v_{t,\theta}$
R(I,J,K)	removal at t (z_t) for one θ
RSUM(I,J)	removal at t (z_t for all θ values)

To complete the process of adapting the program in the previous chapter, we need the following new symbols.

For the population-density gradient Y1, we now have Y1, a density gradient for ill individuals; X1, a density gradient for susceptibles; and RECOV1, a density gradient for recovered individuals. The emigrating density of ill, susceptible, and recovered individuals, respectively, is EMY, EMX, and EMREC.

The first part of the following program segment, down to step 20, is merely an elaboration of the programs in the previous section. The steps from 21 to 35, inclusive, summarize, in Fortran, the material in this section. For example, step 28 corresponds to equation (11.111), step 33 is a recurrence relation based on (11.117), and step 34 is (11.118). The reader should note, however, that this is only a segment of a much larger program. Other segments include instructions for computing the vectors for rate of infectivity, rate of removal, death rate, and gradient coefficients from tables of weather data; the laws governing population growth and nondisease mortality; initialization of "sum buckets" such as VSUM (I, J) (setting the bucket = 0.0); updating the V (I, J, K) array each time period, and so on. Note also that we could set up different weather conditions for each square of the grid, compute the vectors R1, RR, and RD separately for each square, and thus simulate spreading waves of upper-respiratory diseases, for example.

The aim of this section has been to show that phenomena of great complexity can be simulated on computers, and that, with the use of Fortran or related languages, writing the program is not too time consuming. The program can be extended in several directions, including those discussed in the previous section.

One last point should be noted. This type of research requires computers with the largest possible fast-access memory. With a 15 × 15 grid, the information being produced in this problem requires about 8,000 words. For a 25 × 25 grid, about 22,000 words are required. The storage for the actual machine-language program is of course additional to this.

```
 1   DO 20 I = J,15,1
 2   DO 20 J = 1,15,1
 3   IL = MAX(1,I-1)
 4   IU = MIN(15,I+1)
 5   JL = MAX(1,J-1)
 6   JU = MIN(15,J+1)
 7   DO 20 K = IL,IU,1
 8   DO 20 L = JL,JU,1
 9   Y1 = MAX(0,YGRAD*(Y(K,2,2) - Y(I,J,2)))
10   EMY = Y1*V(I,J,K1)/VSUM(I,J)
11   DO 20 K1 = 1,5
12   V(I,J,K1) = V(I,J,K1) + EMY
13   V(K,L,K1) = V(K,L,K1) - EMY
14   EMX = MAX(0,XGRAD*(X(K,2,2)-X(I,J,2)))
15   X(I,J,1) = X(I,J,1) + EMX
16   X(K,L,1) = X(K,L,1) - EMX
17   EMREC = MAX (0,REGRAD*(RECOV(K,L,2)-RECOV(I,J,2)))
```

```
18   RECOV(I,J,1) = RECOV(I,J,1) + EMREC
19   RECOV(K,L,1) = RECOV(K,L,1) - EMREC
20   CONTINUE
21   DO 35 I = 1,15,1
22   DO 35 J = 1,15,1
23   DO 30 K = 2,5,1
24   L = K - 1
25   R(I,J,K) = V(I,J,L)*RR(K)
26   RSUM(I,J) = RSUM(I,J) + R(I,J,K)
27   V(I,J,K) = V(I,J,L) - R(I,J,K)
28   VSUM(I,J) = VSUM(I,J) + V(I,J,K)
29   VISUM (I,J) = VISUM(I,J) + V(I,J,K)*RI(K)
30   DEAD(I,J) = RSUM(I,J)*RD(K)
31   RECOV(I,J,1) = RECOV(I,J,1) + RSUM(I,J) - DEAD(I,J)
32   V(I,J,1) = X(I,J,1)*VISUM(I,J)
33   Y(I,J,1) = V(I,J,1) + VSUM(I,J)
34   X(I,J,1) = X(I,J,1) - V(I,J,1)
35   Z(I,J) = Z(I,J) + RSUM
```

Three-factor and four-factor complexes can be handled by this system merely by increasing the number of arrays and scalers employed [see Bailey's (1957) equation set 4.31]. We can convert this program from a deterministic to a stochastic representation of an epidemic or epizootic wave by Monte Carlo techniques. That is, instead of working with a mean value for each of the various variables, the computer works with probability distributions. Bartlett (1957) has already done this for measles, using a computer program of considerable complexity.

11.9   COMMUNITY STRUCTURE AND STABILITY

In Chap. 3 we defined the information content of a genus, per specimen collected, as

$$I = \log \frac{N!}{N_1!N_2! \cdots N_i! \cdots N_n!} \tag{11.138}$$

where N is the number of individuals collected in the genus, and N_i is the number of individuals collected in the ith species in the genus (following Margalef, 1957).

We also noted that MacArthur (1955) had used the formula

$$S = - \sum_i p_i \log p_i \tag{11.139}$$

to describe the stability of a section of a trophic web. These two equations are different ways of stating the same thing (the degree of order found in a biological situation), and we shall now show how they are related. First, however, it is necessary to give an intuitive understanding of their meaning.

Shannon (in Shannon and Weaver, 1949) explained that (11.139) was derived to describe the following situation. Suppose we have a set of possible events with probabilities of occurrence p_1, p_2, p_3, . . . , p_n. Is it possible to discover a measure, H, of how much choice is involved in the selection of a particular event, or, stating the same thing in other terms, of how uncertain we are of the outcome of selection? Shannon derived the expression for H by working backward from three properties he proposed as desiderata for such a measure of choice (or uncertainty).

1. H should be a continuous function of p_i.
2. If all p_i are equal, then $p_i = 1/n$ and, clearly, H should be a monotonic, increasing function of n. This merely states that with equally likely events, the choice or uncertainty will be an increasing function of the number of possible events.
3. If a choice is decomposed into a sequence of successive choices, the original H should be the weighted sum of the individual values of H.

Thus, the equivalence of choices corresponding to Fig. 11.18 is

$$H\left(\frac{1}{2},\frac{1}{3},\frac{1}{6}\right) = H\left(\frac{1}{2},\frac{1}{2}\right) + \frac{1}{2}H\left(\frac{2}{3},\frac{1}{3}\right) \qquad (11.140)$$

Shannon's objective was to derive a measure, H, for which the above equation would be true. We shall refer to his desiderata as conditions 1, 2, and 3. Shannon reasoned as follows.

Let
$$H\left(\frac{1}{n},\frac{1}{n},\cdots,\frac{1}{n}\right) = A(n)$$

From condition 3, we can decompose a choice from s^m equally likely possibilities into a series of m choices, each from s equally likely possibilities. We can express this as

$$A(s^m) = mA(s)$$

and similarly
$$A(t^n) = nA(t) \qquad (11.141)$$

The object of the following derivation is to derive an expression for $A(t)$ that meets condition 3, as expressed in (11.141).

Now we can choose n, the number of choices in a series, to be arbitrarily

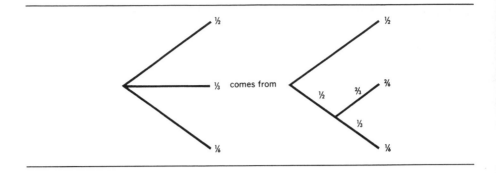

FIGURE 11.18

Decomposition of a choice into a sequence of successive choices.

large, and find a value of m that satisfies the relation

$$s^m \leqslant t^n \leqslant s^{(m+1)} \tag{11.142}$$

(We can be cavalier about the particular value of m chosen because all we seek in this derivation is the *form* of the expression for $A(t)$, not the *value*.) Taking logs across (11.142), and dividing by $n \log s$, we obtain the inequality

$$\frac{m}{n} \leqslant \frac{\log t}{\log s} \leqslant \left(\frac{m}{n} + \frac{1}{n}\right) \tag{11.143}$$

Since we have already decided to make n arbitrarily large, we can express (11.143) as

$$\left|\frac{m}{n} - \frac{\log t}{\log s}\right| < \epsilon \tag{11.144}$$

where t is arbitrarily small. Now, from condition 2, the monotonic property of $A(n)$, (11.141), is

$$A(s^m) \leqslant A(t^{\,n}) \leqslant A(s^{m+1})$$

so

$$mA(s) \leqslant nA(t) \leqslant (m+1)A(s)$$

Hence, dividing by $nA(s)$, we obtain

$$\frac{m}{n} \leqslant \frac{A(t)}{A(s)} \leqslant \frac{m+1}{n}$$

or

$$\left|\frac{m}{n} - \frac{A(t)}{A(s)}\right| < \epsilon$$

If we combine (11.144) and (11.145), we have

$$\left| \frac{A(t)}{A(s)} - \frac{\log t}{\log s} \right| \leqslant 2\epsilon$$

Now to state that the absolute value of a difference is vanishingly small is the same as to say that the expressions to the left and right of the minus sign are equal but of unequal sign. Thus we have

$$A(t) = -\frac{A(s)}{\log s} \log t \qquad (11.146)$$

But the series of s choices has an entirely arbitrary number of choices (s), and thus $A(s)/\log s$ may be treated as a constant, K. The choice of K is a matter of convenience, according to Shannon, and amounts to the choice of a unit of measure. Thus $A(t)$ is given by

$$A(t) = -K \log t \qquad (11.147)$$

and this is the appropriate expression of choice or uncertainty with equiprobable possibilities. Suppose, now, that the choices are not equiprobable; instead, we have a choice from n possibilities with measurable probabilities

$$p_i = \frac{n_i}{\sum_i n_i}$$

where the n_i are integers. A choice from Σn_i possibilities may be considered as a choice from n possibilities with probabilities $p_1, \ldots, p_n$ and then, if the ith probability is chosen, a choice from n_i with equal probabilities. Now using condition 3 again, we can write an expression analogous to (11.140), and equate the total choice from Σn_i as computed by two methods.

$$K \log \Sigma n_i = H(p_1, \ldots, p_n) + K\Sigma p_i \log n_i$$

Hence
$$H = K(\Sigma p_i \log \Sigma n_i - \Sigma p_i \log n_i) = -K\Sigma p_i \log \left(\frac{n_i}{\Sigma n_i} \right)$$

$$= -K\Sigma p_i \log p_i \qquad (11.148)$$

which is identical to (11.139) since K can take the value 1.0.

Now we can see that what MacArthur used as a measure of stability is in fact a measure of choice, or uncertainty in the selection of a particular event from an array of events with a known probability of occurrence. In effect, we measure the stability in a trophic web as the amount of choice available for a quantum of energy to move from one trophic level to a particular species that would consume it at another trophic level. The greater the number of kinds of eaters, the greater the choice, the higher the information content, and the greater the stability.

Margalef's (1957) idea of using (11.138) originated with Brillouin (1956), who reasoned as follows. Suppose we have a system with $N = N_1 + N_2 + \cdots + N_n$ states. Each state contains one and only one particle of a set of particles, where the N particles consist of N_1 particles of the first kind, N_2 of the second kind, N_n of the nth kind. The entropy (disorderliness) of such a system, according to Brillouin, can be defined as

$$B = \log \frac{N!}{N_1! N_2! \ldots N_n!} \qquad (11.149)$$

Baer (1953) showed that entropy, as measured by (11.149), is the negative of uncertainty, or choice, as measured by (11.148). If an observer, confronted with a system like that described by Brillouin, makes an observation at random of some state in the system, then the expected change in entropy of the system exactly equals the information entropy per state. To prove this, we note that if the particular state observed in a system happens to contain a particle of the ith kind, then the entropy of the system becomes

$$B^1 = \log \frac{(N-1)!}{N_1! \ldots (N_i - 1)! \ldots N_n!}$$

and the probability of the event is simply N_i/N if the observation is made at random. Thus, the expected change in entropy per system is

$$\Delta B = \sum_i \frac{N_i}{N} \log \frac{(N-1)!}{N_1! \ldots (N_i - 1)! \ldots N_n!} - \log \frac{N!}{N_1! \ldots N_n!} \qquad (11.150)$$

By adding and subtracting

$$\sum_i \frac{N_i}{N} \log \frac{N}{N_i}$$

in (11.150), we get

$$\Delta B = \sum_i \frac{N_i}{N} \log \frac{N!}{N_1! \ldots N_n!} - \log \frac{N!}{N_1! \ldots N_n!} + \sum_i \frac{N_i}{N} \log \frac{N_i}{N} \qquad (11.151)$$

$$= \sum_i p_i \log p_i \qquad (11.152)$$

Hence (11.138), the equation Margalef used to describe the information content of an array of individuals grouped into species, per individual, is an expression for entropy, or disorderliness, for which the counterpart expression of "negentropy," or orderliness, is (11.139), the expression used by MacArthur (1955). Use of this notion of entropy in ecology is unfortunate; for example, in a tropical rain forest, where orderliness is very great, the "entropy" measure would be maximal. To avoid confusion, it is better to regard (11.138) as a measure of the degree of organization, where high values indicate a high degree of organization.

The computing difficulty created by the factorials in (11.138) is nonexistent because in practice we use the second-order Stirling's formula (Feller, 1962, p. 64; Watt, 1964c), in which

$$\ln \frac{N!}{N_1! \ldots N_n!} = \ln N! - \sum_{i=1}^{n} (\ln N!)$$

$$= \ln \left[(2\pi)^{\frac{1}{2}} N^{+\frac{1}{2}} \exp \left(-N + \frac{1}{12N} \right) \right]$$

$$- \sum_{i=1}^{n} \left\{ \ln \left[(2\pi)^{\frac{1}{2}} N_i^{N_i+\frac{1}{2}} \exp \left(-N_i + \frac{1}{12N_i} \right) \right] \right\}$$

$$= 0.9190 + (N + 0.5) \ln N - N + 1/12N$$

$$- \sum_{i=1}^{n} [0.9190 + (N_i + 0.5) \ln N_i - N_i + 1/12N_i]$$

The Fortran programming system, particularly when used in a computer with a large memory (16,000 to 32,000 ten-digit "words"), is a tool of enormous potential for research on community ecology. The basic data of community ecology are often in the form of sets of large tables, and the Fortran language represents a very compact, powerful means of describing arithmetical operations on tables. The computations used to obtain the number of competitor species (Chap. 3) from Watt (1964c, 1965) were performed with an IBM-1620 computer of 4,000 words with instructions written in Fortran II. To illustrate the simplicity of the Fortran system, we will outline how the computations would have been performed with the use of a 32,000-word IBM-7044 computer with magnetic tape and instructions written in Fortran IV.

We will illustrate the technique for manipulating tables by showing how "mean number of insect species eating same host plants" is computed from a table, in which, for each serial number of an insect species (ISPEC), we are given the number of plant species that the insect species eats (IHOSTS), and the serial numbers of all the plant species eaten by the insect species in question, that is, IDENT (I), I = 1, IHOSTS.

Thus, the computer reads a table of the following form:

ISPEC	IHOSTS	IDENT (I), I = 1, IHOSTS
1	3	3, 14, 51
2	7	17, 18, 19, 37, 42, 49, 51
3	2	56, 57
4	5	46, 47, 53, 54, 55
etc.	etc.	etc.

We now wish to produce a new table, in which, for each ISPEC, we know the average number of competing insect species that eat the plant species eaten by the ISPEC in question. To illustrate, consider insect species 1, which eats plant species with serial numbers 3, 14, and 51. Suppose these plant species are eaten by 7, 22, and 41 species of insects. Then the average amount of competition on the plant species eaten by insect species 1 may be given by $(7 + 22 + 41)/3 = 23$.

We can perform such computations because any number, once read into the computer's memory, can be treated as an entry in a table, *or* as the number of the row or column in a table in which a particular item is to be entered or can be found. The following seven instruction steps illustrate the application of this principle (certain elements of the actual program have been removed from this sequence, in order not to detract attention from the main argument).

```
1   READ (5,50) ISPEC, IHOSTS, (IDENT (I), I = 1, IHOSTS)
2   DO 6 I = 1, IHOSTS
3   IIDENT = IDENT (I)
4   K = IB (IIDENT)
5   IA (IIDENT, K) = ISPEC
6   IB (IIDENT) = IB (IIDENT) + 1
7   GO TO 1
```

In effect, the computer converts a table, in which rows represent serial numbers of phytophagous insects, into a second table, IA (IIDENT, K), in which rows represent serial numbers of trees, and columns represent the serial numbers of insect species eating the trees.

In step 1, one line of the first table is read in. (Note that in some Fortran systems, IHOSTS would have to be read in one line prior to statement 1.) In step 2, the computer is told that for the line of information just read in, it is to perform the following steps for each tree serial number in turn, down to and including step 6. Step 3 is the key step. IDENT(I), the *i*th entry in a table of serial numbers, now becomes IIDENT, an index telling the computer the row of table IA into which the insect serial number, ISPEC, is to be inserted. However, the serial number ISPEC can only be entered in the correct row of table IA in the column, K, after the last entry in that row. Another table, IB, keeps track of the number of entries, K, in each row of IA. In step 4, the computer finds the updated value of K. In step 5, ISPEC is entered in the correct row and column of IA. Since we have now increased the number of entries in the IIDENT row of IA by 1, the same row of table IB must be increased by 1 in step 6. After steps 3 to 6 have been completed, the next card is read. (The computer is informed that *all* cards have been read by putting a blank card at the end of the input deck, and a statement between 1 and 2 that causes progression to step 8 when the blank card is encountered.)

From the new table of IA, now formed, the computer must calculate for each ISPEC, or insect-species serial number, the average number of insect species eating the plant-host species eaten by ISPEC. This is accomplished as follows.

Since the table, read in a line at a time in step 1, can be up to 52 columns wide and 1,200 rows long, making for a 62,400-word table, it is too large to store in the main memory of any of the current computer models. However, we need this table for the next step of the calculation, so it would either be read into a magnetic-tape reel, line by line on a big machine, after step 1, and read back in now, or, on a small machine, the input card deck would be reread. We would start with this rereading step, assuming we are reading from magnetic tape, and proceed as follows:

```
11   REWIND 1
12   READ (1) ISPEC, IHOSTS, (IDENT (I), I = 1,IHOSTS)
13   IF (ISPEC) 23, 23, 14
14   ITOTAL = 0
15   DO 17 I = 1, IHOSTS
16   IIDENT = IDENT (I)
17   ITOTAL = ITOTAL + IB (IIDENT)
18   ATOTAL = ITOTAL
19   AHOSTS = IHOSTS
20   AMEAN = ATOTAL/AHOSTS
21   WRITE (6, 51) ISPEC, ITOTAL, IHOSTS, AMEAN
22   GO TO 12
23   REWIND 1
24   END.
```

Step 11 rewinds the reel of magnetic tape on which the input data were stored. In step 12, the input data are reread, one line at a time. Step 13 instructs the computer to go to step 23 if all data have been read. Step 14 initializes the total number of insect species competitors. Step 15 instructs the computer to perform the following two steps for each tree species that ISPEC eats. In step 16, the tree serial number is used to identify a row in the vector IB. In step 17, the number of insect species eating tree species number IIDENT is added to ITOTAL, the total number of insect species eating the trees eaten by ISPEC. In steps 18 and 19, the two totals are converted from fixed point to floating point numbers, so significant digits following the decimal point will not be dropped in the quotient. In steps 20 and 21, the quotient is obtained, and read out. The process is repeated for the next insect species in step 22.

We have described a process of great potential, which can be extended to deal with ecological data analysis problems of great complexity. First, the method can be extended by making this routine a subroutine in a much larger program, which simultaneously deals with the effects of climatic stability, and density-dependent and site factors on stability of phytophagous species. Second, the method can be used to study the effects of not just two trophic levels and the links between them, but of several trophic levels and the various categories of trophic links relating all of them (as in Fig. 3.2).

11.10 THE NATURE AND ROLE OF MODELS IN BIOLOGY

Two very different types of models have been discussed, which represent two different philosophies about the use of models in science, require two different kinds of training, and have two different sets of implications for the future directions of science.

One of these types might be called analytical models. The most advanced efforts in this direction include the mathematical models of epidemics, reviewed in the book by Bailey (1957); the models of predator-prey, host-parasite, and interspecies competition systems, from the early work of Lotka and Volterra to the new work on stochastic models reviewed by Bartlett (1955); and the dispersal models of Skellam (1951). All these models have the following characteristics. They require a tremendous background in mathematical analysis, and in the newer techniques of stochastic models. The model builder soon discovers that if he builds models based on realistic assumptions, the resultant models are completely intractable mathematically. Therefore he quickly learns to develop models based on a small number of simple assumptions, and to accept that they may not correspond too closely to the corresponding phenomena operating in the real world. An entirely reasonable defense of this approach is that it does not drown or obscure the main elements of a process in a mass of detail. But the essence of biological reality is intimately bound up with detail, and with the nature of interactions among many variables. Statistical support for this statement lies in the discovery that, in certain situations, interaction effects are more important than main effects. As a consequence of the *modus operandi* in this approach, the model builder devotes the bulk of his ingenuity to discovering the implications of a small number of simple assumptions, and does not try to develop an elaborate and realistic set of assumptions. Thus the bulk of the effort in traditional biomathematics has been expended on mathematical, not biological, issues. As a result there has been surprisingly little feedback between biomathematics and biology: The vast majority of the classical biomathematical literature has been almost totally ignored by biologists working in the same problem areas and, similarly, the biomathematicians have largely ignored the findings of experimental and field biologists. Finally, a very important characteristic of all classical biomathematics is its assumption that temporal changes in the systems considered come from changing variate values of factors intrinsic to the system, not factors extrinsic to the system.

The other type of model includes Fortran programs and computer gaming, or simulation studies. This approach has been applied to extremely complex predator-prey problems by Holling (1965, 1966) and Holling, Brown, and Watt, and to tremendously complex problems of resource management strategy evaluation (Chap. 12). The emphasis in these models has been on incorporating large numbers of realistic assumptions into the model, for two reasons. The models were designed to provide specific answers to highly complex practical problems, and also, in view of the enormous capabilities of the computer, there was no need not to work with large numbers of realistic assumptions. A tremendous amount of effort goes into structuring the assumptions. A characteristic of such activity has been an enormous amount of feedback between the experimental work and the computer activity. This feedback has been particularly striking in the case of Holling's work. Finally, because of the ease with which computers handle tables of stored data, such as meterological records, it is easy, using simulation techniques, to consider the effects of changing variate values for factors extrinsic to a system. This is important because, as pointed out in Chap. 6, the particular sequence of weather variate values in time may be critical for many biological systems.

We now need to consider Monte Carlo methods. Monte Carlo refers to the technique of finding the probability array of output from a chance process by playing a game repeatedly on the computer, using as input, probability distributions of variate values derived from tables of random normal, chi-square, binomial, Poisson, or other deviates, or generated random deviates (Meyer, 1956). The difficulty with this procedure, in the case of resource management studies, has to do with what we mean by "chance." Chance is really a sequence of variate values of extrinsic origin imposed on a system, and thus is not at all under the control of the system. The most important component of chance in resource management studies is weather. As I have shown (Fig. 11.4), an important characteristic of sequences of weather variate values is that they may occur in very improbable sequences. Thus simulation studies can only be expected to mimic nature faithfully if variation about mean values from play to play in a game is generated, not by tables of random normal deviates, but rather by tables of actual historical weather records.

It is not at all clear at this writing what the ultimate role of stochastic models will be in biology. There are three possibilities.

1. Despite the formidable mathematical difficulties raised by these methods, their undoubtedly more realistic mimicking of events in nature may provide sufficient motivation to gain them wide acceptance and use.

2. The ultimate fate of stochastic models in biology may be foreshadowed by Beverton and Holt (1957). They say:

Considering population change as a stochastic process, in this way, often gives results appreciably different from those obtained with deterministic models (see Bailey, 1950; Bartlett, 1949; and D. G. Kendall, 1949), especially for the prediction of critical phenomena such as total extinction of the group. It is our belief, however, that, except in particular instances . . . , the multiplication of effort both in deriving the stochastic equations and in computing them would not have been justified when the standard of accuracy of our data, the complexity of the biotic system with which we are dealing, and the order of magnitude of the expected discrepancies, are all taken into account.

3. Another possibility is that, though stochastic models are difficult to manipulate, their usefulness justifies working with them in some manner, and Monte Carlo techniques will become very important in biological-mathematical model building.

I believe that possibility 2 will apply to analysis of phenomena involving large populations, and possibility 3, and sometimes 1, will apply to the analysis of small-population phenomena, in future biological-mathematical model building.

A question that often arises is: "What is the effect of population size, N, on the discrepancy between stochastically and deterministically derived predictions?" The simplest way to go about answering it is to use Monte

Carlo techniques (using a table of random normal deviates and a desk calculator) or else program an analogous routine for electronic calculator. A small table of random normal deviates is given by Deming (1938, pp. 252–254), and a very large table giving 100,000 normal deviates has been published by the Rand Corporation (1955).

The procedure is as follows (readers wishing to follow the development of equations in detail should consult Leslie (1958), whose equations are referred to here, as examples): Suppose we wish to predict population size N_{t+1} at time $t + 1$, as a function of population size N_t at time t.

1. Develop an equation relating N_{t+1} to N_t. Leslie (1958) uses

$$E(N_{t+1}) = \frac{\lambda Nt}{1 + \alpha N'} \qquad (11.153)$$

where E denotes expectation, or mean, and λ and α are related to constants in the logistic, from which (11.153) is derived.

2. Develop or assume a variance corresponding to the mean defined by (11.153). Leslie used

$$\text{var } (N_{t+1}) = 2E(N_{t+1}) \qquad (11.154)$$

[(11.154) follows from the theory of Markov chains; see Kendall (1949) or Bartlett (1956), argument concluding p. 70].

3. Calculate $E(N_{t+1})$, using a given N_t and the appropriate parameters [that is, calculate an equation corresponding to (11.153)].

4. Calculate the variance, σ^2, and from it σ [use an equation corresponding to (11.154)].

5. Obtain as many random normal deviates as are desired from the table. These deviates (Δ's) may be positive or negative; they represent fractions or multiples of the standard deviation distance from the mean. That is, the Gaussian deviate 0.215 corresponds to $106 + 0.215\,(20) = 1\dot{1}0.3$, for a normal distribution with mean 106 and standard deviation 20. The deviate -2.354 corresponds to $106 - 2.354\,(20) = 58.9$.

6. Generate the probability distribution for N_{t+1}, obtaining each value from the relation

$$N_{t+1} = E(N_{t+1}) \pm \sigma\Delta$$

By probability distribution, we mean a curve of frequencies of values or ranges of values for N_{t+1}, against N_{t+1}.

Leslie (1958) has already examined the effect of population size on the difference between stochastic and deterministic predictions, using Monte Carlo techniques. He predicted a mean value for ten replicates, where $N_0 = 15$, and the deterministic model yielded an answer 11 percent above the correct answer (derived from a stochastic model) for N_{t+10}. Where $N_0 = 300$, however, the deterministic prediction of N_{t+10} was only 2 percent over the stochastic prediction. This is an oversimplification of Leslie's results, since he considered several types of stochastic models; his paper should be consulted.

In general, where var $(N_{t+1}) = 2E(N_{t+1})$, the coefficient of variation at $t + 1$ will be

$$\frac{100\sigma}{\bar{x}} = \frac{100\sqrt{2x}}{\bar{x}} = \frac{141}{\sqrt{\bar{x}}}$$

If one population is R times the other, its coefficient of variation will be

$$\frac{141/\sqrt{R\bar{x}}}{141/\sqrt{\bar{x}}} = \frac{1}{\sqrt{R}}$$

as great as the other.

On the basis of Leslie's (1958) investigations, it does not seem worthwhile to consider the further analytic complication of introducing random variables into models unless a population has only a few hundred individuals.

One matter of the utmost importance for model building in biology has been inadequately considered in the literature; this is the technique for most realistically dealing with time. One of the most important classical techniques in biomathematics is exemplified by Ivlev (1961) (Sec. 11.5). His models assume that all experiments will be run for equal lengths of time, hence time is ignored in his equations, since it is considered to be a constant. The models are designed to predict the state of the system at the end of a time period as a function of the state of the system at the beginning of the time period. It is further assumed, implicitly, that all variables will change through time in a quantitative, but not a qualitative manner.

Holling, on the other hand, treats time as variable, and through his use of thresholds assumes that the very state of the system may undergo qualitative changes with the passage of time. This approach seems more reasonable, since biological systems are characterized by changes in qualitative states produced by thresholds (for example, searching behavior patterns are only initiated when the hunger threshold has been surpassed).

REFERENCES

Ahmad, T.: The Influence of Constant and Alternating Temperatures on the Development of Certain Stages of Insects, *Proc. Natl. Inst. Sci. India*, **2**:67–91 (1936).

Allen, W. W., and R. F. Smith: Some Factors Influencing the Efficiency of *Apanteles medicagines* (Muesebeck Hymenoptera: Braconidae) as a Parasite of the Alfalfa Caterpillar, *Colias philodice eurythere* Boisduval, *Hilgardia*, **28** (1):1–42 (1958).

Andersen, F. S.: The Effect of Density on the Birth and Death Rate, *Reprinted from Annual Rept. 1954–1955, Govt. Pest Infestation Lab.*, Springforbi, Denmark, pp. 1–27, 1957.

———: Simple Population Models and Their Application to the Formation of Complex Models, *Proc. XII Int. Congr. Entomol.*, London, pp. 620–622, 1965.

Andersen, K. T.: Die Lupinemblatterand äfer *Sitona griseus* F. und *Sitona gressonus* F., *Z. Angew. Entomol.*, **24**:325–356 (1938).

Andrewartha, H. G., and L. C. Birch: "The Distribution and Abundance of Animals," The University of Chicago Press, Chicago, 1954.

Baer, R. M.: Some General Remarks on Information Theory and Entropy, in R. Quastler (ed.), "Information Theory in Biology," The University of Illinois Press, Urbana, Ill., 1953.

Bailey, N. T. J.: A Simple Stochastic Epidemic, *Biometrika*, **37**:193–202 (1950).

———: "The Mathematical Theory of Epidemics," Hafner Publishing Company, Inc., New York, 1957.

Bartlett, M. S.: Some Evolutionary and Stochastic Processes, *J. Roy. Stat. Soc.*, ser. B, **11**:211–229 (1949).

———: "An Introduction to Stochastic Processes," Cambridge University Press, New York, 1966.

———: Measles Periodicity and Community Size, *J. Roy. Stat. Soc.*, ser. A., **120**:48–70 (1957).

Belehradek, J.: Temperature and Rate of Enzyme Action, *Nature*, **173**:170–171 (1954).

———: A Unified Theory of Cellular Rate Processes Based upon an Analysis of Temperatures Action, *Protoplasma*, **48**:53–71 (1957a).

———: Physiological Aspects of Heat and Cold, *Ann. Rev. Physiol.*, **19**:59–82 (1957b).

Bellman, R.: Mathematical Experimentation and Biological Research, *Fed. Proc.*, **21**:109–111 (1962).

——— and K. L. Cooke: "Differential-Difference Equations," Academic Press Inc., New York, 1963.

Beverton, R. J. H., and S. J. Holt: The Theory of Fishing, in M. Graham (ed.), "Sea Fisheries, Their Investigation in the United Kingdom," pp. 372–441, Edward Arnold (Publishers) Ltd., London, 1956.

——— and ———: On the Dynamics of Exploited Fish Populations, *Fish. Invest.*, ser. II, **19**:1–533 (1957).

Birch, L. C.: An Improved Method for Determining the Influence of Temperature on the Rate of Development of Insect Eggs, *Aust. J. Exptl. Biol. Med. Sci.*, **22**:277–283 (1944).

Bliss, C. I.: Temperature Characteristics for Prepupal Development in *Drosophila melanogaster*, *J. Gen. Physiol.*, **9**:467–495 (1926).

Bray, H. G., and K. White: "Kinetics and Thermodynamics in Biochemistry," J. and A. Churchill Ltd., London, 1957.

Brillouin, L.: "Science and Information Theory," Academic Press Inc., New York, 1956.

Browning, T. O.: The Influence of Temperature on the Rate of Development of Insects, with Special Reference to the Eggs of *Geryllubes commodas* Walker, *Austr. J. Sci. Res.*, ser. B, **5**:96–111 (1952).

Buchanan, R. E., and E. J. Fulmer: "Physiology and Biochemistry of Bacteria. vol. II. Effects of Environment upon Microorganisms," The William & Wilkins Company, Baltimore, 1930.

Burnett, T.: Influences of Natural Temperatures and Controlled Host Densities on Oviposition of an Insect Parasite, *Physiol. Zool.*, **27**:239–248 (1954).

Chadwick, L. E.: Temperature Dependence of Cholinesterase Activity, in "Influence of Temperature on Biological Systems," American Physiological Society, Washington, D.C., 1957.

Chapman, R. N.: The Quantitative Analysis of Environmental Factors, *Ecology*, **9**:111–122, (1928).

Chauvin, R.: "Physiologie de l'Insecte," 2d ed., L'Institute Nat'l. de la Recherche Agronomique, Paris, 1956.

Crombie, A. C.: The Effect of Crowding upon the Oviposition of Grain-Infesting Insects, *J. Exptl. Biol.*, **19**:311–340 (1942).

———: The Effect of Crowding upon the Natality of Grain-Infesting Insects, *Proc. Zool. Soc.* (A), **113**:77–98 (1943).

Crozier, W. J.: Biological Oxidation as a Function of Temperature, *J. Gen. Phys.*, **7**:189–216 (1924–1925).

Da Fonseca, J. P. Cancela: Influence de la Température Sensible sur le Développement des Insectes, *Brotéria*, **27**:145–152 (1958).

Davidson, J.: On the Speed of Development of Insect Eggs at Constant Tempeiatures, *Ausı. J. Exptl. Biol. Med. Sci.*, **20**:233–239 (1942).

———: On the Relationship between Temperature and Rate of Development of Insects at Constant Temperatures, *J. Animal Ecol.*, **13**:26–38 (1944).

Deming, W. E.: "Statistical Adjustment of Data," John Wiley & Sons, Inc., New York, 1938.

Dobzhansky, T., and S. Wright: Genetics of Natural Populations. X. Dispersion Rates in *Drosophila pseudoobscura*, *Genetics*, **28**:304–340 (1943).

——— and ———: Genetics of Natural Populations. XV. Rate of Diffusion of a Mutant Gene through a Population of *Drosophila pseudoobscura*, *Genetics*, **32**:303–324 (1947).

Feller, W.: "An Introduction to Probability Theory and Its Applications," vol. 1, 2d ed., John Wiley & Sons, Inc., New York, 1962.

Fry, F. E. J.: Effects of the Environment on Animal Activity, *U. Toronto Biol. Ser.*, no. 55, Publ. Ont. Fish Res. Lab. 68, 1947.

———: Temperature Compensation, *Ann. Rev. Physiol.*, **20**:207–224 (1958).

Fujita, H.: An Interpretation of the Changes in Type of the Population Density Effect upon the Oviposition Rate, *Ecology*, **35**:253–257 (1954).

———: and S. Utida: The Effect of Population Density on the Growth of an Animal Population, *Ecology*, **34**:488–498 (1953).

Goldberg, S.: "Introduction to Difference Equations," John Wiley & Sons, Inc., New York, 1958.

Grosenbaugh, L. R.: Generalization and Reparameterization of Some Sigmoid and Other Nonlinear Functions, *Biometrics*, **21**:708–714 (1965).

Hearon, J. Z.: Rate Behavior of Metabolic Systems, *Physiol. Rev.*, **32** (4):499–523 (1952).

Holling, C. S.: Principles of Insect Predation, *Ann. Rev. Entomol.*, **6**:163–182 (1961).

———: The Functional Response of Predators to Prey Density and Its Role in Mimicry and Population Regulation, *Mem. Entomol. Soc. Can.*, no. 45, 1965.

———: The Functional Response of Invertebrate Predators to Prey Density, *Mem. Entomol. Soc. Can.*, **1966** (48):1–85 (1966).

———, D. M. Brown, and K. E. F. Watt: "Simulation of Attack by Invertebrate Predators," manuscript in preparation.

Huffaker, C. B.: The Temperature Relations of the Immature Stages of the Malarial Mosquito, *Anopheles quadrimaculatus* Say, with a Comparison of the Developmental Power of a Constant and Variable Temperatures in Insect Metabolism, *Ann. Entomol. Soc. Am.*, **37**:1–27 (1944).

Hutchinson, G. E.: Homage to Santa Rosalia *or* Why are There so Many Kinds of Animals?, *Am. Nat.*, **93**:145–159 (1959).

Ishida, H.: Studies on the Density Effect and the Extent of Available Space in the Experimental Population of the Azuki Bean Weevil, *Res. Population Ecol.*, **1**:25–35 (1952).

Ito, M.: On the Meaning of the "Air Space" in the Population Growth of the Red-Rust Flour Beetle, *Tribolium castaneum* (Herbst.). Experimental Studies on a Role of Behavior in the Population of Injurious Insect. First Report, *Nippon Oyo Dobutsu Konchu Gaku Zasshi*, **11**:25–31 (1955).

Ivlev, V. S.: "Experimental Ecology of the Feeding of Fishes," Yale University Press, New Haven, Conn., 1961.

Janisch, E.: Uber die Temperaturabhangigkeit biologischer Vorgänge und ihre kurvenmassige Analyse, *Pflüger's Arch. Ges. Physiol.*, **209**:414–436 (1925).

———: "Das Exponentialgesetz als Grundlage einer vergleichenden Biologie," vol. 2, Abh. Theorie organ. Entwick, Berlin, 1927.

———: Die Lebens-und Entwicklungsdauer der Insekten als Temperaturfunktion, *Z. Wiss. Zool.*, **132**:176–186 (1928).

———: The Influence of Temperature on the Life-History of Insects, *Trans. Entomol. Soc. London*, **80**:137–168 (1932).

Johnson, F. H., H. Eyring, and M. J. Polissar: "The Kinetic Basis of Molecular Biology," John Wiley & Sons, Inc., New York, 1954.

Kavanau, J. L.: Enzyme Kinetics and the Rate of Biological Processes, *J. Gen. Physiol.*, **34**:193–209 (1950).

Kendall, D. G.: Stochastic Processes and Population Growth, *J. Roy. Stat. Soc.*, ser. B., **11**:230–265 (1949).

Kermack, W. O., and A. G. McKendrick: A Contribution to the Mathematical Theory of Epidemics, *Proc. Roy. Soc. London*, ser. A, **115**:700–721 (1927).

Kirchner, H. A.: Versuche über die Fruchtbarkeit von *Dixippus* (*Carausius*) *morosus* bei abgestofter Wohndichte und Raumgrösse, *Z. Angew. Entomol.*, **25**:151–160 (1939).

Kistiakowsky, G. B., and R. Lumry: Anomalous Temperature Effects in the Hydrolysis of Urea by Urease, *J. Am. Chem. Soc.*, **71**:2006–2013 (1949).

Leslie, P. H.: A Stochastic Model for Studying the Properties of Certain Biological Systems by Numerical Methods, *Biometrika*, **45**:16–31 (1958).

Lotka, A. J.: Contribution to Quantitative Parasitology, *J. Wash. Acad. Sci.*, **13**:152–158 (1923).

———: "Elements of Physical Biology," The Williams & Wilkins Company, Baltimore, 1925.

Luyet, B. J.: An Analysis of the Notions of Cooling and of Freezing Velocity, *Biodynamica*, **7**:293–335 (1957).

MacArthur, R.: Fluctuations of Animal Populations, and a Measure of Community Stability, *Ecology*, **36**:533, also, *Gen. Systems*, 3 (1955).

MacLagan, D. S., and E. Dunn: The Experimental Analysis of the Growth of an Insect Population, *Proc. Roy. Soc. Edinburgh*, **55**:126–139 (1935).

Margalef, D. R.: Information Theory in Ecology, (In Spanish) *Mem. Roy. Acad. Barcelona*, **23**:373–449 (1957) (Republished in English) *Gen. Systems*, 3:36–71 (1958).

Martini, E.: Uber die Kettenlinie und die Exponentiaikurve überhaupt als Bilder für die Abhangigkeit der Entwicklungsdaur von der Wärme, *Z. Angew. Entomol.*, **14**:273–284 (1928).

Matsazawa, H., H. Okamoto, and Y. Miyamoto: Application of Pradhan's Formula to the Data of Development of the Common Cabbage Butterfly, *Pieris rapae crucivora*, and Its Parasite, *Apanteles glomeratus*, *Kontyu*, **25**:89–93 (1957).

May, J. M.: "Studies in Disease Ecology," Hafner Publishing Company, Inc., New York, 1961.

Messenger, P. S., and N. E. Flitters: Effect of Constant Temperature Environments on the Egg Stage of Three Species of Hawaiian Fruit Flies, *Ann. Entomol. Soc. Am.*, **51**:109–119 (1958).

——— and ———: Effect of Variable Temperature Environments on Egg Development of Three Species of Fruit Flies, *Ann. Entomol. Soc. Am.*, **52** (2) (1959).

Meyer, H. A.: "Symposium on Monte Carlo Methods," John Wiley & Sons, Inc., New York, 1956.

Miyashita, K.: Outbreaks and Population Fluctuations of Insects, with Special Reference to Agricultural Insect Pests in Japan, *Bull. Natl. Inst. Agr. Sci.*, ser. C, no. 15, 1963.

Morris, R. F.: Single-Factor Analysis in Population Dynamics, *Ecology*, **40**:580–588 (1959).

——— (ed.): The Dynamics of Epidemic Spruce Budworm Populations, *Mem. Entomol. Soc. Canada*, **31**:1–332 (1963).

Murphy, G. M.: "Ordinary Differential Equations and Their Solutions," D. Van Nostrand Company, Inc., Princeton, N.J., 1960.

Neilson, M. M., and R. F. Morris: The Regulation of European Spruce Sawfly Numbers in the Maritime Provinces of Canada from 1937 to 1963, *Can. Entomol.*, **96**:773–784 (1964).

Noll, J.: Uber den Einfluss von konstanten und wechselnder Temperaturen auf die Entwicklungsgeschwindigkeit der Larven und Puppen von *Contarinia nasturtii*, *Anz. Schaedlingsbekaempf.*, **18**:73–78 (1942).

Park, T.: Studies in Population Physiology. I. The Relation of Numbers to Initial Population Growth in the Flower Beetle, *Tribolium confusum Duval*, *Ecology*, **13**:172–181 (1932).

———: Studies in Population Physiology. II. Factors Regulating Initial Growth of *Tribolium confusum* Populations, *J. Exptl. Zool.*, **65**:17–42 (1933).

———: Studies in Population Physiology. VI. The Effect of Differentially Conditioned Flour Upon the Fecundity and Fertility of *Tribolium confusum*, *J. Exptl. Zool.*, **73**:393–404 (1936).

Payne, N. M.: Freezing and Survival of Insects at Low Temperature, *J. Morphol.*, **18**:521–545 (1927).

Peairs, L. M.: Some Phases of the Relation of Temperature to the Development of Insects, *West Va. Univ. Agric. Exptl. Sta. Bull.*, **208**:1–62 (1927).

Pearl, R.: The Influence of Density of Population upon the Rate of Reproduction in *Drosophila*, *J. Exptl. Zool.*, **63**:57–85 (1932).

——— and S. L. Parker: On the Influence of Density of Population upon the Rate of Reproduction in *Drosophila*, *Proc. Natl. Acad. Sci.*, **8**:212–219 (1922).

Powsner, L.: The Effects of Temperature on the Duration of the Developmental Stages of *Drosophila melanogaster*, *Physiol. Zool.*, **8**:474–520 (1935).

Pradhan, S.: Insect Population Studies. II. Rate of Insect Development under Variable Temperature of the Field, *Proc. Natl. Inst. Sci. India*, **11**:74–80 (1945).

———: Insect Population Studies. III. Idea of Biograph and Biometer. (Instruments for Estimating Developmental Periods and Number of Generations of Insects in Nature.) *Proc. Natl. Inst. Sci. India*, **12** (6):287–331 (1946a).

———: Insect Population Studies. IV. Dynamics of Temperature Effect on Insect Development, *Proc. Natl. Inst. Sci. India*, **12** (7):333–412 (1946b).

Precht, H., J. Christophersen, and H. Hensel: "Temperatur und Laben," Springes, Berlin, 1955.

Rand Corporation: "A Million Random Digits with 100,000 Normal Deviates," The Free Press of Glencoe, New York, 1955.

Rich, E. R.: Egg Cannibalism and Fecundity in *Tribolium*, *Ecology*, **37**:109–120 (1956).

Richards, O. W.: Observations on Grain-Weevils, *Calandra* (Col., Curculionidae). I. General Biology and Oviposition, *Proc. Zool. Soc.*, **117**:1–43 (1947).

Ricker, W. E.: Stock and Recruitment, *J. Fish. Res. Bd. Can.*, **11**:559–623 (1954).

Robertson, F. W., and J. H. Sang: The Ecological Determinants of Population Growth in a *Drosophila* Culture. I. Fecundity of Adult Flies, *Proc. Roy. Soc.*, **132**:258–277 (1944).

Royce, W. F., and H. A. Schuck: Studies of Georges Bank Haddock. Part II: Prediction of the Catch, *U.S. Fish Wildlife Serv.*, *Fishery Bull.*, **56**:1–6 (1954).

Salt, R. W.: Time as a Factor in the Freezing of Undercooled Insects, *Can. J. Res. D.*, **28**:285–291 (1950).

Sargent, F.: The Environment and Human Health, *Arid Zone Res.*, **24**:19–32 (1964).

Satomura, H.: Relation between Temperature and Development and Death Rate in Larvae and Pupae of *Phyllotreta vittota*, *Oyo-Kontya*, **6**:1–9 (1950).

Schaefer, M. B.: Fisheries Dynamics and the Concept of Maximum Equilibrium Catch, *Proc. Gulf and Caribbean Fish. Inst,*, **6**:1–11 (1954).

Shannon, C., and W. Weaver: "The Mathematical Theory of Communication," The University of Illinois Press, Urbana, Ill., 1949.

Skellam, J. G.: Random Dispersal in Theoretical Populations, *Biometrika*, **38**:196–218 (1951).

Smirnov, E., and W. Polejaeff: Density of Population and Sterility of the Females in the Coccid *Lepidosaphes ulmi* L., *J. Anim. Ecol.*, **3**:29–40 (1934).

——— and N. Wiolovitsh: Uber dem Zusammenhang zwischen der populations-dichte und Eier-production der Weibchen bei der Schildlaus *Chionaspis salicis* L., *Z. Angew. Entomol.*, **20**:415–424 (1934).

Smith, Audrey U.: The Resistance of Animals to Cooling and Freezing, *Biol. Rev.*, **33**:197–253 (1958).

———: "Biological Effects of Freezing and Supercooling," Monographs of the Physiological Society, no. 9, Edward Arnold (Publishers) Ltd., London, 1961.

Stanley, J.: A Mathematical Theory of the Growth of Populations of the Flour Beetle, *Tribolium confusum Duval*. III. The Effect upon the Early Stages of Population Growth of Changes in the Nutritive Value, Palatability and Density of Packing of the Flour Medium, *Can. J. Res.*, **11**:728–732 (1934).

Turner, M. E., R. J. Monroe, and L. D. Homer: Generalized Kinetic Regression Analysis: Hypergeometric Kinetics, *Biometrics*, **19**:406–428 (1963).

———, ———, and H. J. Lucas, Jr.: Generalized Asymptotic Regression and Non-Linear Path Analysis, *Biometrics*, **17**:120–143 (1961).

Ullyett, G. C.: Oviposition by *Ephestia kuhniella* Zell, *J. Entomol. Soc. S. Africa*, **8**:53–59 (1945).

Utida, S.: Studies on Experimental Population of the Azuki Bean Weevil, *Callosobruchus chinensis* (L.). I. The Effect of Population Density on the Progeny Populations, *Mem. Coll. Agric. Kyoto Imp. Univ.*, **48**:1–30 (1941a).

————: Studies on Experimental Population of the Azuki Bean Weevil, *Callosobruchus chinensis* (L.). II. The Effect of Population Density on Progeny Populations under the Different Conditions of Atmospheric Moisture, *Mem. Coll. Agric. Kyoto Imp. Univ.*, **49**:1–20 (1941b).

————: Studies on Experimental Population of the Azuki Bean Weevil, *Callosobruchus chinensis* (L.). III. The Effect of Population Density upon the Mortalities of Different Stages of Life Cycle, *Mem. Coll. Agric. Kyoto Imp. Univ.*, **49**:21–42 (1941c).

Volterra, V.: Variazioni e fluttuazioni del numero d'individui in specie animali conviventi, *Mem. Acad. Lincei*, **2**:31–113 (1926).

————: Variations and Fluctuations of the Number of Individuals in Animal Species Living Together, *J. Conserv. Intern. Explor.*, **3**:3–51 (1928).

Von Bertalanffy, L.: A Quantitative Theory of Organic Growth, *Hum. Biol.*, **10**:181–213 (1938).

————: Problems of Organic Growth, *Nature*, **163**:156–158 (1949).

Watt, K. E. F.: Studies on Population Productivity. I. Three Approaches to the Optimum Yield Problem in Populations of *Tribolium confusum*, *Ecol. Monogr.*, **25**:269–290 (1955).

————: The Choice and Solution of Mathematical Models for Predicting and Maximizing the Yield of a Fishery, *J. Fish. Res. Bd. Can.*, **13**:613–645 (1956).

————: Studies on Population Productivity. II. Factors Governing Productivity in a Population of Smallmouth Bass, *Ecol. Monogr.*, **29**:367–392 (1959a).

————: A Mathematical Model for the Effect of Densities of Attacked and Attacking Species on the Number Attacked, *Can. Entomol.*, **91**:129–144 (1959b).

————: The Effect of Population Density on Fecundity in Insects, *Can. Entomol.*, **92**:674–695 (1960).

————: Mathematical Models for Use in Insect Pest Control, *Can. Entomol.*, suppl. 19, 62 pp., 1961.

————: Use of Mathematics in Population Ecology, *Ann. Rev. Ent.*, **7**:243–260 (1962a).

————: The Conceptual Formulation and Mathematical Solution of Practical Problems in Population Input-Output Dynamics, in E. D. LeCren and M. W. Holdgate (eds.), "The Exploitation of Natural Animal Populations," pp. 191–203, Blackwell Scientific Publications, Ltd., Oxford, 1962b.

————: The Use of Mathematics and Computers to Determine Optimal Strategy and Tactics for a Given Insect Pest Control Problem, *Can. Entomol.*, **96**:202–220 (1964a).

————: Computers and the Evaluation of Resource Management Strategies, *Am. Scientist*, **52**:408–418 (1964b).

————: Comments on Fluctuations of Animal Populations and Measures of Community Stability, *Can. Entomol.*, **96**:1434–1442 (1964c).

————: Community Stability and the Strategy of Biological Control, *Can. Entomol.*, **97**:887–895 (1965).

Wellington, W. G.: Qualitative Changes in Populations in Unstable Environments, *Can. Entomol.*, **96**:436–451 (1964).

Wigglesworth, V. B.: "The Principles of Insect Physiology," Methuen & Co., Ltd., London, 1950.

12 | SYSTEMS SIMULATION

This chapter introduces the reader to the techniques and concepts of systems simulation by using five examples of progressively greater complexity. The fundamental idea common to all five is that resource management strategies can be evaluated by reading a mathematical model of the resource into a computer, then exploring the consequences of various strategies on the computer to see which ones yield optimal results. This process is referred to as simulation, or gaming.

12.1 BUFFALO

The first example is very simple and artificial, and is introduced merely to illustrate the basic logical structure of a computer simulation program. Suppose there had been computers in 1800, and resource managers had been interested in discovering an optimum harvesting policy for buffalo. The question we ask is, "How many male and female buffalo older than 2 years can we remove each year without impairing the ability of the herd to replace the harvested animals?"

We start with a set of biological facts, all based on modern buffalo studies (Fuller, 1962).

1. For simplicity, and purposes of this illustration, we will ignore age-specific fecundity rates, and assume a probability of calving of 0.67 for all cows 2 years of age or older.

358

2. Fifty-three percent of these calves are males.

3. Forty percent of these calves survive the first summer, and 30 percent survive to breed.

4. After reaching breeding age, mortality is 0.5 percent per annum due to drowning; 5 percent per annum, tuberculosis; and 4.5 percent per annum, predation and other causes; giving a total of 10 percent mortality per annum.

The variables whose effects we wish to explore will be called QUOTAM and QUOTAF, for harvesting quota of males and females. IYEAR will stand for year, and other symbols are defined below. We assume a herd of 40 million (Roe, 1951), broken down as follows (from Fuller on modern herd structure):

ADM = adult males (2 years and older), 16,800,000
ADF = adult females, 16,800,000
YM = yearling males, 1,200,000
YF = yearling females, 1,200,000
CM = calf males, 2,000,000
CF = calf females, 2,000,000

The program consists of an outer loop in which the consequences of a trial pair of QUOTAM and QUOTAF values are explored for 100 years, and and inner loop (1 year of 100 tried). (Read-and-write formats are omitted.)

```
1   READ (5, 10) QUOTAM, QUOTAF
    ADM = 16800000.
    ADF = 16800000.
    YM = 1200000.
    YF = 1200000.
    CM = 2000000.
    CF = 2000000.
    IYEAR = 1800
    WRITE (6, 11) QUOTAM, QUOTAF
2   ADM = .9 * ADM + .75 * YM − QUOTAM
    ADF = .9 * ADF + .75 * YF − QUOTAF
    YM = .4 * CM
    YF = .4 * CF
    CM = .36 * ADF
    CF = .31 * ADF
    IYEAR = IYEAR + 1
    WRITE (6, 12) IYEAR, ADM, ADF, YM, YF, CM, CF
    IF (ADF) 1, 1, 3
3   IF (IYEAR − 1900) 2, 1, 1
    END.
```

We shall consider, first, the derivation of the various equations in the program, and second, the logical significance of the program structure and parts.

Each play of the game begins by reading in a new trial pair of QUOTAM and QUOTAF variate values, and setting all the other variables in the model back to the initial values. Then we write the QUOTAM and QUOTAF values on a line printer or magnetic tape, to identify this play of the game.

Statement 2 is derived as follows. Since adult mortality is 10 percent per year, survival is 90 percent per year, and the stock of adult males is only 0.9 times the stock the previous year. To this number we must add the number of yearling males that survive to the beginning of the second year of life. Since 40 percent of the calves survive the first summer and 30 percent survive to breed, 75 percent of those that survive the first summer survive to breed. From ADM and ADF we subtract the respective quotas. The equations for CM and CF follow from the fact that 53 percent of 0.67 is 0.36, and 47 percent of 0.67 is 0.31.

Now let us examine the logical structure of these equations a little more closely.

Note that statement 2 constitutes a difference equation, not a differential equation. (The difference between the two was explained in Sec. 11.2.) That is, we state that the difference between ADM this year and the previous year is a function of ADM and YM the previous year.

The particular way in which the program is written makes a very specific set of assumptions about the sequence of events through time. This is because a Fortran statement is an instruction to a computer to perform one or more operations in a specified order. Unless the instruction is modified by use of parentheses, the computer performs exponentiation first, then multiplication and division, then addition and subtraction. Also, the order in which the Fortran statements occur in the program implies a specific sequence of events through time. To illustrate, if we replaced statement 2 by ADM = .9* ADM + .75* .4* .36* ADF, the additions to the stock of adult males each year would be computed from the stock of adult females *now*, not the stock *2 years ago*, and would thus incorrectly ignore the 2 years required for a calf to mature. Since the Fortran statement implies a particular sequence of events in a dynamic process, we must take great care to ensure that the statement in fact simulates nature.

Some readers will be surprised at the amount of specific quantitative data used in such a short program. This reveals another characteristic of simulation studies: far from being a substitute for collection of data in the real world, they typically create a demand for more refined, comprehensive, and directed information-gathering programs than those that existed prior to the initiation of the computer experiments.

The basic structure of the program is noteworthy: a cycle of operations to be played repeatedly. The computer reads QUOTAM and QUOTAF, then simulates events for 100 years, or until ADF reaches zero or a negative value, then reads a new set of QUOTAM and QUOTAF values and repeats the cycle. In fact, Fortran programs for realistic simulation studies contain, not 20 statements, but up to several thousand. All simulation programs have this same basic structure of the repetitive game.

12.2 SALMON

The second simulation example is a computer program that has been used for the development of a salmon-gear-limitation policy by the government of the state of Washington (Royce et al., 1963).

This case differs from the preceding in several respects, all of which make it more realistic.

1. It is monumentally complex, requiring 1,000 equations to describe the fishery.

2. The buffalo example treated buffalo as if they behaved the same everywhere, all the time. But animals move through space, and populations behave differently at different places at the same time. The salmon study takes care of both these problems.

3. The buffalo example ignored the fact that human beings must catch or harvest a resource, get paid for so doing, and buy equipment. A realistic resource management simulation study will have technological and economic factors simulated, as well as biological factors. This implies that the research may be done by large groups of scientists that constitute interdisciplinary teams. In fact, the University of Washington salmon-gear-limitation study was conducted by a constitutional lawyer and a group of economists, biologists, and biostatisticians. The function of the lawyer was to determine if it was constitutional, in terms of United States and state of Washington law, for the state to limit the number of operators on a common property resource. He decided it probably was.

4. With a program of this complexity, the computer operations would not be conceived of at the outset as a string of Fortran instructions, because no mind could grasp all this complexity at once. Also, for the same reason, the program cannot be explained this way. Rather, in constructing and explaining the main features of the program, logical flow diagrams are used, as in Fig. 12.1.

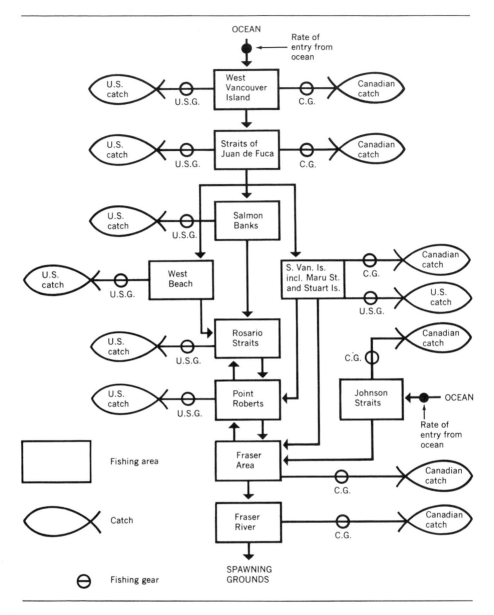

FIGURE 12.1
Flow chart for salmon simulation study (Royce et al., 1963).

Charts were made up for each of four species: sockeye, pink, silver, and chum, reproducing the habits of each species exactly (they all differ somewhat).

In addition to mimicking the details in space, the correct passage *times* for each step of the trip were simulated.

This was just the beginning in the steps of adding in complexity. Each type of gear has different efficiencies for different species in different areas. So we need tables stored in the computer (see Table 12.1).

TABLE 12.1
Relative gear efficiency, in numbers of gill-net units (Royce et al., 1963)

Area	Gear	Sockeye	Pink	Silver	Chum
1	Purse seines	8.0	13.0	4.0	4.0
2	Purse seines	8.0	9.5	2.0	3.8
	Reef nets	2.5	2.0	1.0	1.2
3	Purse seines	5.0	11.0	3.5	3.8
4	Purse seines		12.8	2.0	2.5

The program keeps separate track of the fixed cost, value of the catch, day's effort expended, and net earnings for each type of gear in each place. The major finding of this study was that the net return to the industry would be increased very considerably by restricting the number of operators. Broken down by type of gear, the findings can be charted as in Fig. 12.2. "Net earnings" is the gross value of the catch minus all fixed costs. With only a 50 percent increase in gear, the industry would actually *lose* money. With substandard runs, losses would be catastrophic unless gear was limited to half the present amount. Only a sequence of unusually good runs in recent years has concealed the fact that the industry is grossly inefficient. Even if a run was increased to four times the present level and the gear was dropped by half, the industry could harvest 77.9 percent of the run.

Note that the computer has made it practical to try out a large number of cases, which would be economically infeasible in practice, and all in a short time.

12.3 FOREST-INSECT-PEST CONTROL

The third example concerns the evaluation of strategies for control of an insect pest in a 10,000-square-mile tract of forest (Watt, 1964). Control strategies for many other types of animals could be simulated in analogous fashion.

This analysis has the following features not found in the salmon-gear-limitation study.

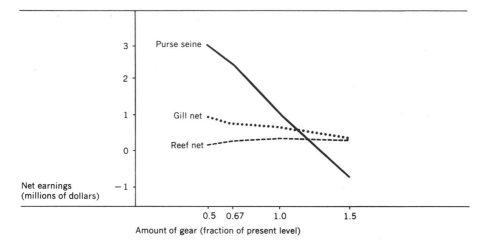

FIGURE 12.2

Effect of amount of gear on profitability of salmon fishery (Royce et al., 1963).

1. Dispersal of the pests themselves, and of some of the control agents (that is, parasites and disease) is not along lines, but outward in planes, wave-like, from epicenters or foci of biological activity. We are able to handle simulation of this wavelike activity by taking advantage of DO loops in Fortran language, and also by using magnetic tape to produce distribution maps very quickly (Sec. 11.7).

2. Since weather triggers many types of pest outbreaks, tables of weather data for 35 years were stored in the computer's memory. Weather, however, had to be handled in a special way, because weather at one location may trigger an outbreak, whereas weather at another location may not. This is an example of the type of phenomenon discussed in Sec. 6.2, in which an effect at one point on the earth's surface is a consequence of a cause at a prior time at some other point. That is, we may find an insect outbreak at a place where the weather patterns would never be such as to trigger an outbreak. This insect outbreak arose because at some other point there were weather patterns that triggered outbreaks and these outbreaks radiated and created an outbreak at this point. As we noted in Sec. 6.2, this situation, where an event at one place now may be the result of a prior event at some other place, is an important feature of many problems in resource management. Its implication for determining optimum management strategy is that space and time become important variables to be manipulated, as well as the type of control agent used. That is, we have the option of controlling a large-scale effect early, at the focus of the effect, or waiting until the effect has radiated out over a large

area before we do anything. If we are to make intelligent economic comparisons of the two broad classes of strategy, it is necessary to simulate not only the buildup of the effect in time at any point, but also its radial dispersal through space. Fortunately, the great memory (32,000 numbers or more) in the large new computers makes it feasible to simulate, in one calculation, events at a large number of map locations, and hence we can reproduce the spread of outbreaks from sequences of maps developed from actual outbreak histories (see, for example, the maps in Elliott, 1960).

We must program the computer to simulate waveform buildup and dispersal of populations, using weather data from those places where weather can trigger such widespread waves to simulate events at the foci of developing waves. The particular places used as the source of the critical sequences of weather data are selected by trial and error. That is, we build a mathematical model in which population size is the dependent variable, and there are a number of independent variables, including past weather. Then, in preliminary computer runs using this model, we test the sequences of weather data from various sources close to the known epidemic focus to determine which places give rise to data that can simulate the actually observed spatial and temporal pattern of outbreak. In my own statistical analyses of sequences of weather data (Sec. 11.3, and Watt, 1963, 1964), weather patterns have tended to be more uniform from place to place in relatively flat areas than in mountainous country. Nevertheless, the whole matter of the movement of storms, weather fronts, and various cloud types, and the consequent effect on biological activity on the ground is very complicated, and the object of some very interesting research (Wellington, 1964, and references therein).

But why is the sequence of weather effects unusual at certain points on the earth's surface? The answer lies in the statistical characteristics of the sequences of weather. For example, the mean monthly temperature for July might be very uniform at one place, and there might be a relative absence of clustering of years either well above or well below the mean, averaging over all years for which this datum is available. At another location a few miles away, the mean monthly temperature for July may vary from one year to the next and, in addition, several years in which the index is unusually high or unusually low may occur close together. Precisely this situation has been observed in sequences of weather data at different points in New Brunswick, Canada, for example. A major factor in the great population buildup of spruce budworm in the late years of the 1940 decade and early years of the 1950 decade was the most unusual sequence of weather conditions in the years 1944 to 1949, inclusive. At Green River, New Brunswick, 5 of those 6 years had far above average temperatures in the critical period for population development, June 1 to July 15 (Morris, 1963). At Edmundston, only 30 miles away, only 2 of the 6 years were unusually hot in that period, relative to the long-term Edmundston mean. Therefore, if tables of weather data from Edmundston are used in the computer, we cannot reproduce the outbreak history for spruce budworm, whereas if Green River data are used to simulate weather events at the outbreak focus, a history of population buildup and dispersal can be produced to closely mimic observed history.

3. In the salmon study, gear limitation was the only technique of population management being considered. But in many cases, a multiplicity of types of control are available to the resource manager. For example, in the case of insect-pest control, we can use insecticides, pathogens, parasites, predators, genetic manipulation, sex attractants, or manipulation of the environment to increase mortality of the pest or decrease mortality of its enemies. These different kinds of control agents can act very differently in space and time (Table 12.2). For example, insecticides have a big effect when applied, but may cause ecological, genetic, and physiological boomerang effects. Predators or parasites, on the other hand, may have a small effect just after being released, but this effect can build up for as long as 20 years (Smith, 1959), and the parasites and predators can spread, so that ultimately the pest is completely controlled (becomes much less abundant and much more stable from year to year). We need a computer program that simulates events over many years, so that economic comparisons can be made between slow acting, cumulative effects, and fast acting, "one shot" effects.

4. Insect pests can have a complex effect on the plants they eat, as can deer, birds, or any other type of herbivore. The response of perennials to being eaten is characterized by thresholds, lags, and cumulative effects (Belyea, 1952). Figures 12.3 and 12.4 show the threshold, since there is no tree mortality at all unless pest density rises above a certain minimum level; and also show the lag effect, since tree mortality (in percentage of all trees) continues after all the pests have died. Regression analysis also shows that there is a cumulative effect, because tree mortality is related to the *sum* of pest densities over a consecutive sequence of years, and not pest density in any particular year. The threshold reflects homeostasis, and interleaf competition in plants. The fact that defoliators have a lag effect and a cumulative effect on the plants they eat creates the need for a new type of complication in the computer simulation program: the computer must store data for several past years for each location being simulated. Each year in the computation for each place, the population density for the last year is set equal to the population density for the present year, all the past years of population data are updated, and the population density for the year furthest back in time is dropped.

TABLE 12.2

Comparison of mode of action of different kinds of insect-pest control techniques (Watt, 1964)

Control technique	Magnitude of initial effect, point of application	Subsequent reaction, point of application	Change in magnitude of effect with time	Spatial effect, time of application	Subsequent spatial effect
Insecticides	Very great	Physiological, populational, and genetic boomerang effects can occur	No effect after initial treatment	No spread apart from accidental spray drifts	No effect
Parasites and predators (assuming effective agent is used)	Very slight unless release is on vast scale	No boomerang effect except in rare cases (encapsulation of parasite by host larva)	Effect grows, sometimes slowly and for up to 20 years, to asymptotic level (Smith, 1959)	Spreads	Spreads
Pathogens (assuming effective agent is used)	Can be very great	Number of susceptibles decreases through death or development of immunity	Effect rises sharply, then declines slowly, sometimes over several years, but effect can rise sharply again when pest density increases	Spreads	Spreads
Sterilization by chemicals or X-rays	Very great	Slight, if any	No subsequent effect unless treatment is repeated	Spreads	No effect
Sex attractants	Can, in principle, be very great	None	No subsequent effects unless treatment is repeated	Spreads	No effect
Environmental modification	Can be very great	None	Effect persists	No spread	No effect

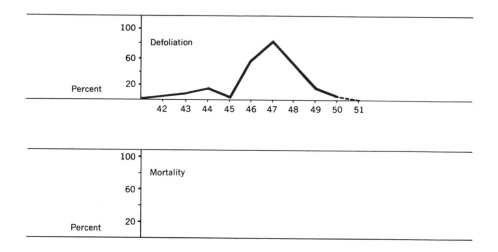

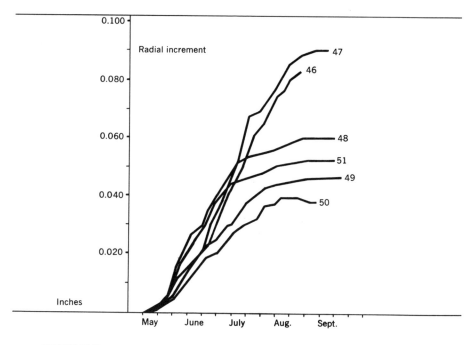

FIGURE 12.3

Southern border area.

Records of defoliation, mortality, and average radial increment of balsam fir for two conditions of spruce budworm infestation in the Lake Nipigon region (Belyea, 1952).

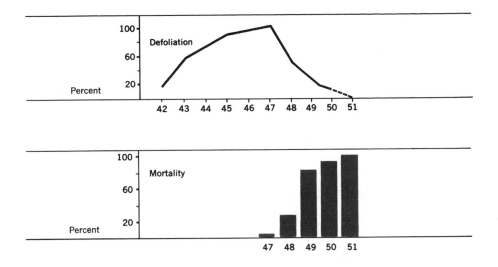

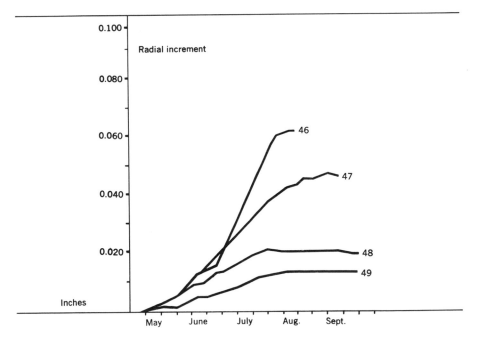

FIGURE 12.4

Black Sturgeon Lake area.

Records of defoliation, mortality, and average radial increment of balsam fir for two conditions of spruce budworm infestation in the Lake Nipigon region (Belyea, 1952).

5. Another feature introduced into this computer simulation study is that the density of insects this year is a function of insect density last year (following Morris, 1963). Thus the computer is programmed to simulate a density-dependent process.

6. The densities of insect parasites this year are treated as functions of last year's densities of insect parasites. Thus, as explained in Sec. 11.2, though our submodels and sub-submodels have been developed by using differential calculus, the equations obtained through integration of the differential equations are treated as difference equations.

The role of computer experimentation

Since this kind of work is new in biology, it is important that we ensure at the outset that there are no misunderstandings about its use. This type of research is not done merely to mimic a population system already existing in nature. Rather, we ask the following question: "How would a situation that presently exists be altered if we modified the mechanism of population regulation within ranges of values that we know from other studies are realistic?" For example, our introducing into the following model a parasite that spreads at a particular array of rate values does not mean that we already have a parasite that spreads this way in controlling spruce budworm (the insect from which most of the data for this model were taken). Rather, our aim is to find out what *would happen* if a parasite having physiological parameters similar to those of insects whose dispersion rates have been measured were used in the hypothetical situation under study.

In general, there are three aims of strategy evaluation simulation.

1. We wish to determine if certain strategies or combinations of strategies are consistently superior or inferior to others.

2. We wish to determine if there are any broad features of particular strategies or combinations of strategies that are relatively invariant, even when various independent variables such as weather, site factors, dispersion rates, or epidemiological parameters are run through a wide range of values on the computer. Putting it mathematically, we wish to gain insight into the geometry and topology of the "policy solution space."

3. We always hope, of course, that serendipity will work for us: that the computer output will suggest something that comes as a happy surprise, which otherwise might never have occurred to us. We would be particularly interested to find, for example, that economic damage by a pest is extraordinarily sensitive to changes in values of a parameter for a particular control strategy (for example, searching efficiency of parasites).

Of fundamental importance in computer experimentation is the self-teaching side effect. That is, we build a simple model of a phenomenon into the computer by means of a program, observe the discrepancy between the behavior of real and "model" systems, and improve the model on this basis. While we may start with inadequate and unrealistic assumptions, the pattern of the output will quickly suggest to us how assumptions must be modified to make the model more realistic. In other words, we create a feedback loop, involving the computer program, the data, and the mind of man, in which the pace of activity in the loop is largely stimulated by the computer, the fastest component. Such computer experimentation is an ideal means of suggesting how key experiments to obtain new data can clarify our understanding of a phenomenon with a minimum amount of actual experimentation. Computer experimentation has played an important role in such diverse areas as rocket design, urban and airport traffic control, and inventory management in large factory operations.

Realism in computer experimentation

A number of considerations determine the degree of realism we strive to incorporate into the computer program for strategy evaluation. We want the model to contain sufficiently realistic data, forms of functions, and parameter values so that the program has some relationship to reality and is of pedagogic and heuristic value. But we cannot make a model that approximates reality very closely at present, because we lack the data. Rather than laboring this point, we will give five examples.

1. There are no good studies on parasite dispersion in nature analogous to those of Dobzhansky and Wright (1943) on *Drosophila pseudoobscura*. Indeed, quantitative studies on insect dispersion are almost totally lacking.

2. There is an almost total lack of detailed quantitative information on parasite population buildup after parasites have been released. Therefore we do not know the form of the function describing such population growth. [The paper by Smith (1959) is the only available check on parasite population buildup over a long period.]

3. There is very little good information on the effects of insecticides following application. This may sound amazing in view of the vast number of papers on the number of insects killed within days, weeks, or months of insecticidal treatment. However, from a population standpoint, these data are of very little use. What we want to know is the number of insects remaining in the *generation* following spraying since this is the datum of long-term population significance. It does not matter if a spray killed 95 percent of the insects within an hour after application if the average density of the insects

fails to drop satisfactorily from generation to generation in the face of such treatment.

4. We need to know the effect on trees of varying densities of insects on a particular plot over 15- or 20-year periods. This information is needed because the effects of defoliation on tree mortality are complicated by thresholds, lag effects with a lag of several years, and cumulative effects that may cumulate for many years.

5. We have relatively little quantitative data for the effects of defoliators on trees of different ages and sizes over long periods of time.

One last word about realism is in order. Do not make the mistake of thinking that, since most of the data for this simulation study came from the spruce budworm, this program faithfully mimics the spruce budworm. The program only *approximately* mimics spruce budworm population dynamics or effects on trees because of the aforementioned shortage of data. This simulation study only involves a realistic array of parameter values and forms of functions and sets of input data describing *in a very general way* events pertaining to any hypothetical insect pest.

The motives for computer experimentation

1. The cost of large-scale insect-pest control programs is tremendous. For example, if 2 million acres are sprayed, the cost can be as much as $1.2 million. But if we do not attempt control, losses may be much greater. Even a small percentage saving on the sum of losses plus cost of control would justify the cost of considerable strategy evaluation research.

2. Strategy evaluation research in nature would be prohibitively expensive if we were actually to test many strategies, and combinations of strategies, and follow the effects of each treatment by adequate sampling for many years. It is precisely this comparative cheapness of computer experimentation relative to actual experimentation that has caused the widespread adoption of computer experimentation in science.

3. The large number of types of insect-pest control strategies becoming available is making the decision-making process more complex. Futhermore, the various control methods now available operate in different ways, with respect to change of effect through time and space (see Table 12.2). An additional level of complexity enters the problem if we consider preventive control as well as emergency "save the foliage" control. Then we have to consider the comparative effects of spraying many years in advance of an epidemic, at the incipient outbreak focus, versus spraying when the trees are in danger of being 50 percent defoliated, etc. Such comparisons can be

simulated for all types of control, singly and in combination, merely by altering the threshold pest density at which each type of control is applied. The lower the threshold, the earlier the control is applied.

The simulation study mimicked a 10,000-square-mile area, consisting of 625 four-mile × four-mile squares arranged in a 25 × 25 grid of small squares. All computations were performed separately for each of the 625 squares. It was assumed that the 6.4 million acres were full of evenly spaced balsam firs 8 inches in diameter at breast height, 262 per acre, each of which grew exactly 0.095 inch in radial increment at breast height each year under optimum conditions. We assume further that the marketable lumber in the trees had a stump value of $4.50 per cord (128 cubic feet, including air spaces, taken to be 85 cubic feet of solid wood in this study).

Fifteen alternative methods of control were considered:

1. Do nothing.
2. Spray 1 pound of DDT per acre (from a fleet of aircraft).
3. Spray ½ pound of DDT per acre (from a fleet of aircraft).
4. Spray ¼ pound of DDT per acre (from a fleet of aircraft).
5. Release parasites (from cages on the ground).
6. Spray viruses in emulsion (from a fleet of aircraft).
7–9. Use method 2 plus 5 or 6, or both.
10–12. Use method 3 plus 5 or 6, or both.
13–15. Use method 4 plus 5 or 6, or both.

We wish to find the strategy, and the timing of use, that minimizes the sum of control costs plus losses due to inadequate control. Control should cost $0.60 to $2.00 an acre per annum to be feasible.

Very complex computer programs like this are conceived of in terms of loops, or cycles of operations. The "outer loop" computes the output for one case, using one set of trial parameter values describing the characteristics of the pest, the control strategies, etc.

An "intermediate loop" simulates the events for 1 year within a case.

An "inner loop" simulates the events for 1 year in a 4-mile × 4-mile square. Included in this inner loop are calculations simulating the effect on each square of past events in that square and in eight surrounding squares (or less, if we are at a boundary or corner of the big square). The inner loop simulates the effect of weather, past pest densities, disease, parasites and insecticides on abundance of the pest, and in turn simulates the effects of the pest on tree growth and survival. Finally, the inner loop keeps cumulating sums of all control costs, and financial losses from tree damage, and the grand cumulative total of costs and losses.

At the outset of each computer run, a set of 35 years of weather data is read.

For each case, the following input parameters are read:

1. Two parameters expressing the effect of mean maximum temperatures over the period 1 June to 15 July, inclusive, on pest survival (TC and TK).

2. Three parameters expressing the amount of migration of pests, parasites, and incidence of disease as proportions of the gradient from one 4-mile × 4-mile square to the next one (PGRAD, PAGRAD, DIGRAD).

3. The control strategies used. This array is read as five fixed-point numbers, each of which has the value 1 if the strategy is to be followed, and 0 if the strategy is not to be followed. Application of the strategy in each case is further contingent on the possibility of the population of pests having risen above a control threshold density peculiar to each strategy. The thresholds are read in for each case as five floating-point numbers. The strategies and symbols are as follows:

Spray 1 pound DDT per acre from fleet of aircraft   ISW1, CT 1
Spray ½ pound DDT per acre from fleet of aircraft   ISW2, CT 2
Spray ¼ pound DDT per acre from fleet of aircraft   ISW3, CT 3
Release parasites from cages on ground   ISW4, CT 4
Spray viruses in emulsion from fleet of aircraft   ISW5, CT 5

4. Constants describing the density of parasites released and the arrangement in space of the grid of points at which parasites are released. Parasites are released on the ground in cages. This procedure is simulated by the following three instructions:

$$DO\ 4\ I = I1,\ I2,\ I3$$

$$DO\ 4\ J = J1,\ J2,\ J3$$

$$4\ PAR\ (I,\ J,\ 2) = APAR$$

This instruction means, "Set the number of parasites released last year in the square in row I, column J, equal to APAR, where the first and last I and J values are I1 and J1, and I2 and J2, respectively." I3 and J3 give the spacing between squares in which cages of parasites are set out. For example,

$$DO\ 4\ I = 4,\ 10,\ 6$$

$$DO\ 4\ J = 4,\ 10,\ 6$$

$$4\ PAR\ (I,\ J,\ 2) = APAR$$

means that we only released parasites in 4 of the 625 squares: 4, 4; 4, 10; 10, 4; and 10, 10.

All of I1, I2, I3, J1, J2, J3, and APAR are read as input parameters at the beginning of each case.

5. EGGS, which is the egg complement.

6. SE, which is the searching efficiency.

7. PARCOM, which is the effect of intraspecific competition among the parasites on parasite efficiency per parasite.

8. One parameter, DISB, is required to describe the change in incidence of disease through time at any point.

9. Three parameters are required to describe the genetic and ecological post-spray generation reaction to spraying. [The data in Morris (1963) make it quite clear that such a reaction does occur.] These parameters are R1, R2, and R3. To illustrate the use of these factors, which apply to the three different levels of DDT spray treatment, suppose R1 = 8. Whenever control treatment 1 is applied to the I, Jth square, RNEXT (I, J) is set equal to R1. This will multiply by 8 the generation survival for the offspring of the larvae that survive spraying. For the generation two generations after treatment, RNEXT (I, J) will once again have the value 1.0, unless of course the first spray treatment was applied 2 years in succession.

10. Two final input parameters are required to describe the cost of control operations. PCOST is the total cost of parasite release, and DISC is the amount per acre by which we multiply cost of insecticide spraying to obtain the cost of aerial spraying of virus.

Details of organization of the computer program

The insect from which we obtained most of the data and empirical relationships used to construct this program was the spruce budworm. This choice was made purely because more is known about the population dynamics of the budworm and its effects on balsam firs than is known about any other insect.

Pest and parasite populations are expressed as numbers per 10 square feet of branch surface because typical field studies express population density in these sampling units; hence reasonable ranges of values are known. To convert such units to numbers per acre, for forests of the type we are considering, we multiply by 30,000 (Morris, 1955). For example, if we add 1,000 mated female parasites to a 4-mile × 4-mile square, the parasite population in that square is increased by

$$\frac{1,000}{4 \times 4 \times 640 \times 30,000} = 0.00000325/10 \text{ sq. ft. branch surface}$$

The key to the operation of this program is what happens in the innermost loop; therefore we shall describe it first. In order that the reader can get a broad picture of the computations before becoming lost in detail, we will describe stepwise, at the outset, the general plan of the inner loop computations. Origin of all data and mathematical models used is explained subsequently.

1. Since migration of pests, parasites, and virus vectors can occur from one year to another, each inner loop must begin with computation of the effect of this migration in the preceding year on the pest, parasite, and disease states in the I, Jth square in the present year.

2. Given the output from 1, we compute the present pest density for the I, Jth square, using the appropriate weather datum.

3. Under switch control, we compute the effect of those strategies of control being tested in the case under consideration. For example, if switch 1 is in the "1," or "on," position, whenever P(I, J, 1) exceeds CT1, P(I, J, 1) is multiplied by 0.01, and 16 square miles are added to the area sprayed. [P(I, J, 1) is the pest population density in the I, Jth square.]

4. The effects of the insect pests on the growth and mortality of trees are computed, using the equations and procedures sketched on pages 382 and 383. The threshold for 50 percent defoliation is computed, and the computer determines if trees in the 16 square miles are 50 percent defoliated. If so, this area is added to the running sum of area in a hazardous condition.

5. The computer returns to the first step of the innermost loop, if this is not the last of the 625 squares.

The intermediate loop

At the beginning of each intermediate loop, the computer performs the following steps:

1. The year is set equal to 1924 + K. This is because we want to be able to compare the output with data from actual years, beginning with 1925.

2. The sum buckets for area defoliated, and proportion of the trees dead in 6.4 million acres, are set equal to zero.

3. The weather index for the year is removed from the table of weather indices.

Then the computer loops through the innermost loop 625 times. After the innermost loop has been repeated 625 times, the computer performs the following steps.

4. The proportion of the trees dead in the whole area is obtained by dividing the sum of the proportions dead in the 625 small areas by 625. This follows from the relation: Mean proportion of trees dead in 625 areas = (total of proportions dead in 625 areas)/625.

5. The cost of tree mortality is computed from the relation "cost = (proportion dead) $453.312 million," because the value of the entire 6.4 million acres is $453.312 million.

6. Parasite release is simulated. The computer is programmed so that this can only happen once, by setting a switch to the "1" position immediately after release is simulated. Release can only occur if this switch is in the "0" position. This is done because experience has shown that if a parasite is not successful after the first (correctly conducted) release, "booster" release will be of no avail. Cost of the parasite release is added to the cost of control operation, and research overhead is also included.

7. Costs of aerial spraying of disease and insecticides are computed, using a cost per acre figure that is a function of the acreage sprayed.

8. The grand total of losses and costs is summed.

9. The output tape is printed.

10. The year is increased by 1.

11. All state vectors are updated.

12. The computer loops back to the beginning of the intermediate loop, provided that some trees in the 6.4 million acres are still alive, and provided that less than 36 intermediate loops have been computed for this case.

The outer loop

Each outer loop has the following opening steps:

1. The parameters for the case are read and printed on the output tape.

2. All vectors, matrices, scalars, and sum buckets are set equal to zero. RNEXT is set equal to 1.0 for all 625 squares.

3. The year number is set equal to 1.

4. The population for the center square (of 625) is set equal to 0.01.

Then, after repeating the intermediate loop 35 or fewer times, we return to outer-loop step 1. Before any outer loops are processed, the following initial steps are performed by the computer:

1. Stored labels and table headings are printed on the output tape.

2. The 35-element vector of weather indices is read.

Memory requirements

For population densities of the pest itself, we require the three-dimensional array:

P(I, J, K) for I = 1, ... , 25
             J = 1, ... , 25
             K = 1, ... , 7

I and J = position in large square
    K = 1 signifies this year
    K = 2 signifies last year
    K = 3 signifies two years ago, etc.

We also require TMAX (L), for L = 1, ... , 35 where TMAX represents mean maximum temperature from 1 June to 15 July, inclusive, at Green River, New Brunswick, for a 35-year period.

PLAST (I,J) represents the proportion of the trees that died in each square last year.

PAR (I,J,M) where M = 1 or 2 represents the parasite population density in each square this year and last year.

DISIN (I,J,M) represents the incidence of dead and diseased for M = 1 or 2.

RNEXT (I,J) represents the boomerang effect produced in each square by spray treatment in that square the preceding year.

The total memory requirements of this problem are 8,160 words, not including the program and subroutines.

Sources used for the data and equations in the program

Most biologists will be prepared to concede that simulation studies are useful, but will argue that it is rarely possible to conduct simulation because of the paucity of data available for constructing the mathematical model. This is defeatism; often a reasonably satisfactory model (for preliminary simulation work) can be constructed by taking bits of information from various sources in the literature and fitting these together like pieces of a jigsaw puzzle. Admittedly, this process leaves much to be desired, but it does have the advantage of giving us a crude model that can be used to show where future data-collecting effort should be concentrated, or redirected. To illustrate how bits and pieces can be fitted together, we present the following explanation of how the forest-insect-pest strategy evaluation simulator was developed.

Data on structure of the forest From Vincent (1955, p. 10) we see that, for mature softwood stands in the Green River, New Brunswick, watershed, the marketable softwood volume per acre was 2,097 cubic feet in 1950; the corresponding figure for 1945 was 1,933 cubic feet. The mean is 2,015 cubic feet. Assume, to simplify subsequent calculations, that the stand consisted of only balsam firs 8 inches in diameter at breast height. We read in tables 13 and 14 in Jarvis (1960) that, for balsam firs in the Goulais River watershed in 1946, there were an average of 10.3 eight-inch trees per acre, representing 79.1 cubic feet. Therefore it is sufficiently accurate for our purposes to assume that one 8-inch diameter balsam fir represents 7.68 cubic feet of marketable timber, and 1 acre contains $2,015/7.68 = 262$ trees. We assume that 8-inch diameter trees grow an average of 0.095-inch radial increment each year (Belyea, 1952, figs. 2 and 3). While Belyea's data are for Ontario and our other data are for New Brunswick, 0.095 inch seems reasonable relative to the data of Mott, Nairn, and Cook (1957, figs. 1 and 2).

Population dynamics of the pest Morris (1963) has found that spruce budworm population fluctuations in New Brunswick may be simulated surprisingly well by the following simple empirical formula. (Note that this formula and all the others following are written in Fortran.)

$$BN = EXPF (.98 + .76 * LOGF(P) + C * (Z - K))$$

in which BN = population this year
       P = population last year
       Z = mean maximum daily temperature from 1 June to 15 July, inclusive, at Green River, New Brunswick, in degrees Fahrenheit
C and K = constants, taking the values 0.18 and 66.5, respectively, with our weather data

The insecticide program We assume that the insecticide used will be DDT, and that it will be sprayed by fleets of aircraft at dosages per acre of 1, ½, or ¼ pound. We further assume that these dosages will cause mortalities of 99, 97.5, and 95 percent.

Further, we assume that the cost per acre of spraying will be $3.00 if an acreage equal to or less than 200,000 acres is sprayed. If more than 200,000 acres are sprayed, the cost per acre drops gradually to a lower asymptote of $0.60 per acre at 2 million acres. We assume that the effect of acreage sprayed on cost per unit can be described by the following simple empirical relationship:

$$\text{COST} = \text{EXPF} (1.2774 - .000000894 * \text{ASPRAY}),$$

$$200,000 < \text{ASPRAY} < 2,000,000$$

where COST is the cost per acre sprayed, and ASPRAY is the acreage sprayed.

Parasites

Where N_A = number of pests attacked by parasites

N_O = number vulnerable to attack

P = parasite population density

$a, b,$ and K = constants equivalent to previously defined Fortran variables SE, PARCOM, and EGGS

$$N_A = PK (1 - e^{-aN_o P^{1-b}})$$

is known to describe parasite attack in nature (Watt, 1959; Miller, 1959, 1960). Therefore, S_A, the proportion of N_o surviving attack is given by

$$S_A = 1 - \frac{PK}{N_o} (1 - e^{-aN_o P^{1-b}})$$

We know from previous work that values for K, a, and b should be in the ranges 50 to 200, 0.0005 to 0.005, and 1.7 to 2.2, respectively (Miller 1959, 1960).

Cost of biological control has been calculated by determining an array of typical costs per parasite of released parasites, then computing the number of parasites required to produce an economically useful result, and multiplying the cost per parasite by the number required.

I have assumed costs per released parasite of $0.02 to $2.00 apiece.

Data on rates of spread of parasites per annum are rare. Nevertheless, there are enough clues in the literature to indicate the order of magnitude of such rates. Glendenning (1933) reported, for example, that *Blastothrix sericea* Dalm., a chalcid, was about 1 percent as dense 2 miles away from the liberation point as at the liberation point, 2 years after release. Since the squares in this simulation study were 4 miles wide, Glendenning's data correspond to a PAGRAD of under 0.01 in 2 years, or something less than this per year. How much less, we do not know, since spread rates of an insect after the first year depend on population buildup in the first year. Since many parasites are much larger organisms than *Blastothrix*, I have used PAGRAD values of 0.01, 0.04, and 0.08 in this simulation study.

Disease   The most useful set of data for constructing a model for disease-incidence change in time and space, in insects infected with virus, is that of Bird and Burk (1961). Change in disease incidence at a point is evidently reasonably well described by the relationship.

$$DISIN\ (I,J,1) = DISIN\ (I,J,2) * DISB$$

Values in the range 0.2 to 0.8 seem reasonable for DISB, in the light of Bird and Burk's data, and values in the range 0.001 to 0.04 seem reasonable for DIGRAD.

Costs for aerial spraying of virus on a volume basis are not yet known; however, a reasonable set of values could be obtained by multiplying the dependent variable in the function for insecticide costs per acre by 1, 2, 3, and 4. We assume that for aerial spraying of viruses, as in the case of aerial spraying of insecticide, cost per acre sprayed depends on the number of acres sprayed. The constant by which we multiply the cost per acre as compared with the insecticide cost will be labeled DISC.

Economic effects of pests on defoliation, mortality, and growth The loss within 6.4 million acres due to dead trees is a product of the following terms:

1. Proportion of dead trees in whole area.
2. The number of cords of marketable timber represented by the proportion dead.
3. The value per cord.

We begin by assuming that the value of a standing balsam fir is \$4.50 per cord (D. A. Wilson, personal communication). We also assume that there are 2,015 cubic feet of marketable timber per acre, or that the value of the entire 6.4 million acres is

$$\frac{2,015}{85} \times 4.50 \times 6.4 = \$682.729\ \text{million}$$

The proportion of dead trees in this area can be computed by using an expression based on analysis of the data of McGugan and Blais (1959) and Belyea (1952). If we examine fig. 3 in Belyea's paper, and fig. 2 in McGugan and Blais', we note the following:

1. Tree mortality lags behind defoliation. The proportion of trees dead (TMORT) may actually be increasing 4 years after budworm densities have begun to decline. At this time budworms may have all but vanished from the tree.
2. Defoliation has a cumulative effect on tree mortality. High defoliation levels in any given year do not cause mortality unless preceded or followed by high defoliation in adjacent years.
3. There is a threshold effect: there is no mortality at all unless the cumulative defoliation exceeds a certain level (G).

After analyzing the data in these two papers in conjunction with those used in Morris (1963), and after revising the model on the basis of comparisons of computer output with data in Elliott (1960), we arrive at the following simple procedure for computing the proportion of trees dead. Statement 5 indicates that even if defoliation is declining, tree mortality will rise from year to year if the cumulative defoliation over the past 3 years has been high enough.

```
      G = P(I,J,7) + P(I,J,6) + P(I,J,5)
      IF (G − 150.0) 1, 1, 3
1     IF (G − 50.0) 2, 4, 4
2     TMORT = PLAST(I,J)
      GO TO 8
3     TMORT = .0067 * G − .000006*G*G −.87
      IF(TMORT − PLAST(I,J)) 4, 4, 6
4     IF (PLAST(I,J)) 2, 2, 5
5     TMORT = PLAST(I,J) + .10
6     IF(TMORT − 1.0) 8, 8, 7
7     TMORT = 1.0
8     TPMORT = TPMORT + TMORT.
```

After the inner loop has been executed for all 625 squares, TPMORT is divided by 625, to yield the proportion of dead trees in the entire forest.

An expression for growth lost as a function of larval densities in previous years was developed as follows.

1. From Belyea (1952), it was possible to obtain a function,

$$Y = a - bX$$

where Y = radial growth increment at breast height in 8-inch diameter trees as proportion of 0.095 inch

X = cumulative percentage defoliation 4, 5, and 6 years previously

2. From the Green River data files, we can get the conversion from cumulative defoliation in a 3-year period to larval densities in a 3-year period (Morris, 1963).

3. From Jarvis (1960), we are able to compute a relationship between cubic feet per tree and diameter at breast height.

4. From 1, 2, and 3 we can obtain a relationship between cubic feet per tree lost as a function of larval densities, 4, 5, and 6 years previously.

5. From 4, and data given previously, we can compute financial loss for the 6.4 million acres as a result of larval densities 4, 5, and 6 years previously.

Where GL is the cost of lost growth per 16 square miles in millions of dollars, and G is defined as before,

```
GL = (.004631 * LOGF(G) − .001333) * (1.0 − PMORT)
IF (GL) 1, 1, 2
1 GL = 0.0
```

A final datum we need for the innermost loop concerns defoliation. One of the most useful indices of the state of the forest is the area 50 percent defoliated, because at this level of defoliation we have a situation hazardous to continued tree vigor.

The larval density at which the trees are 50 percent defoliated depends on the amount of defoliation the tree has been subjected to in previous years. Again, we encounter the principle that the amount of attack an entity can withstand depends on the amount of attack it has already been forced to withstand. The following procedure for computing whether the 16-square-mile area is 50 percent defoliated is based on analysis of Green River data (Morris, 1963). (TDEFOL is threshold larval density this year which causes 50 percent defoliation.)

```
DO 1 N = 2, 7
1 SUMLAR = SUMLAR + P(I,J,N)
  TDEFOL = 60.0 − .06 * SUMLAR
  IF(P(I,J,1) − TDEFOL) 2, 3, 3
3 ADEFOL = ADEFOL + 16.0
```

(The last step states that an area 50 percent or more defoliated is increased by 16 square miles.) The techniques of Sec. 11.7 were used to simulate dispersal of pests, parasites, and disease.

Output

For each case, the computer prints as output all input data, for ease in interpreting the results. Also, for each year in each case, the computer prints maps showing the population density of the pests, the parasites, and the incidence of diseased and dead pests in one corner quarter of the large square. Finally, for each year of each case, the computer prints an updated operating summary giving the area critically defoliated, the proportion of the trees dead in 10,000 square miles, and a financial statement. The financial statement gives separate cumulative dollar costs of all control treatments. Finally, the financial statement gives us the cumulative sum of all costs and losses. This is the item we seek to minimize in the computer experimentation.

Conclusions from the simulation study

The first conclusion from this study is that a computer program can be devised to yield output remarkably similar to events in nature, even when our information is as scanty as at present.

For example, in the control case where we assumed a PGRAD of 0.020, the following sequence of events was observed. In 1925, the population in the center square was 0.05 larvae per 10 square feet of branch surface, and the eight surrounding squares had populations below 0.01 larvae per 10 square feet. By 1942, the population had spread throughout the 10,000 square miles and 336 square miles were 50 percent or more defoliated, but there were no tree deaths and no lost tree growth. By 1943, there was $157,073 worth of trees dead and $224,288 worth of lost tree growth. By 1944, tree deaths increased to $271,497 worth, but lost tree growth had a cumulative value of $11,860,184. By 1948, 70 percent of the trees were dead. However, the pest population reached its peak in 1947 with 235 larvae per 10 square feet of branch surface at the outbreak epicenter, and from then on, larval densities dropped, the drop only occasionally being broken by slight rises. But trees continued to die after larval densities had sunk to about 20, and the entire forest was dead by 1959.

In the first group of experiments, we used no controls, but sought to determine the effect of varying PGRAD on the total damage done by the pest in a given time. The following tabulation illustrates the pattern of events.

PGRAD	*Cumulative total loss by 1947 in millions of dollars*
.02	319.28
.04	435.46
.08	597.04
.16	734.39

(The value of the stand in any year can be exceeded by cumulative dollar losses because lost tree growth can accumulate.)

Doubling the spread rate of a pest does not double the damage because it takes time for the pest to build up.

The next set of experiments was designed to get general impressions about the consequences of using various control strategies, singly and in combination. Results are summarized in the following array of figures:

$$\text{ISW1} = 1 \quad \text{CT1} = 15 \quad \text{R1} = 4.0 \quad \text{PGRAD} = .16$$

Cumulative costs and losses by 1957 in millions of dollars

Tree deaths	0.000
Lost tree growth	99.239
Cost of spray operation	0.268
Total	99.507

ISW2 = 1   CT2 = 25   R2 = 4.0   PGRAD = .16

Cumulative costs and losses by 1957 in millions of dollars

Tree deaths	0.000
Lost tree growth	178.991
Cost of spray operation	0.266
Total	179.257

ISW4 = 1   CT4 = 15   PGRAD = .16
PAGRAD = .01   SE = .0005   APAR = .00000325
PARCOM = 2.00   EGGS = 50.0
All trees in forest dead by 1950.

ISW5 = 1   CT5 = 15   PGRAD = .16

Cumulative costs and losses by 1957 in millions of dollars

Tree deaths	621.661
Lost tree growth	215.492
Cost of spraying virus	1.000
Total	838.153

In general, the qualitative patterns following use of the three types of control are quite different. If what we have programmed for the computer is at all realistic, the explanations for these differences are as follows.

In the case of insecticides, the pest is never allowed to become sufficiently dense to damage the tree permanently. In the case of disease and parasites, the pest never gets as dense as it would if no control were applied, but it is allowed to persist at a sufficiently high density to kill the trees after several years. The key point in this situation is that a submaximal pest density can kill trees if it lasts long enough. Therefore it is not enough to keep the pests below a density at which they can kill trees quickly; they must also be kept below a density at which they can kill trees slowly. The cumulative nature of tree damage makes long sequences of moderate-damage years just as lethal as short sequences of high-damage years, in situations where annual loss of foliage exceeds annual foliage production, or where annual foliage production is progressively reduced as a result of the killing of buds and twigs.

The major sets of experiments in this study were designed to explore the effect on parasite and virus efficiency of changing various parameters.

The principal points found in the parasite simulation studies are summarized in Table 12.3. First, timing of control is tremendously important. Even where a parasite has the biological attributes required to annihilate the pest, the parasite cannot do this in time to prevent tremendous tree mortality if it is released when CT4 = 50. This point strongly supports the philosophy of preventive control rather than "emergency save-the-foliage control." Even though we release parasites in the center 4-mile × 4-mile square of every 25 such squares, the parasite does not have enough years to spread out and build up in numbers after release, if we only release it when a pest is about to have explosive population growth. The parasites must be released at least 10 years prior to the peak year of a pest outbreak to achieve maximum effectiveness.

The next point made by Table 12.3 is that at this level of PAGRAD, PAGRAD has surprisingly little effect. However, we may be exploring too high values of PAGRAD; perhaps values in the range 0.001 to 0.005 would be more realistic for small parasites, such as chalcids.

A fourfold change in searching efficiency produces a much greater change in effect than a fourfold change in parasite egg complement, within the range of values we are exploring. Nevertheless, all such findings from simulation studies should be viewed with deep suspicion until we can test them with careful field and laboratory studies. Computer experimentation does suggest mechanisms to look for, and ranges of parameter values of particularly critical interest.

Simulation shows us the ranges of values within which biological parameters could reasonably lie. For example, nine combinations of SE and EGGS values could not occur in nature because such parasites would annihilate their hosts and soon become extinct themselves. Interestingly, the parasites now attacking the spruce budworm in nature could not possibly control it, according to Table 12.3 (in fact, they don't).

Interaction effects abound in our results. Note, for example, that the slope of the parasite effectiveness versus egg complement curve is enormously affected by the searching efficiency.

The results of simulation experiments on the effect of varying parameters in control by disease are summarized in Table 12.4. The most striking feature of Table 12.4 is that DIGRAD and DISB, in the range of values we are exploring, have very little effect on effectiveness of disease. When one examines the computer listings in detail, the reason for this becomes apparent. The dominant factor determining the effectiveness of aerial spray of virus is not

TABLE 12.3

Effect of changing parasite parameters on effectiveness of pest control by parasites (Watt, 1964)

CT4	PAGRAD	Cost of dead trees to and including 1947 in millions of dollars when PARCOM = 2.0								
		SE = .0005			SE = .0020			SE = .0050		
		50 eggs	100 eggs	200 eggs	50 eggs	100 eggs	200 eggs	50 eggs	100 eggs	200 eggs
15.0	0.01	580.1	534.5	424.4	453.4	213.7	0.0	154.7	0.0	0.0
	0.04	577.9	530.1	418.2	439.0	193.7	0.0	113.9	0.0	0.0
	0.08	576.8	528.7	415.2	433.8	182.7	0.0	100.2	0.0	0.0
25.0	0.08	584.4	544.8	451.2	477.5	264.3	25.2	219.9	66.0	0.7
50.0	0.08	614.8	613.6	606.0	614.8	613.6	603.4	614.8	613.6	602.9

TABLE 12.4

Factors affecting the effectiveness of disease as a control measure (*Watt, 1964*)

DIGRAD	DISB	Cost of dead trees up to and including 1951, in millions of dollars		
		CT5 = 25	CT5 = 50	CT5 = 100
.001	.500	181.5	325.8	536.6
.010	.500	181.4	325.8	538.2
.001	.750	180.3	308.3	537.5
.010	.750	180.1	308.2	539.0
.040	.750	179.5	307.8	539.4

the biological parameters of the virus, in these ranges of values, but the threshold pest level at which we decide to spray virus. The earlier we spray virus relative to the year of potential peak-pest density, the more likely we are to prevent the pest from attaining densities at which it can kill trees slowly. By releasing virus too late, the pests have their density diminished sixfold or so, but this is not enough to save the trees.

Admittedly, all such conclusions are suspect until a great deal more data are available to support the computer findings and help us improve the computer program. It does appear that preventive control is vastly better than save-the-foliage control, no matter what type of control we use. (In a real situation, of course, this conclusion could be tempered by weather conditions, or other considerations such as side effects of insecticides on wildlife, which might cause stalling of insecticide application until the last possible moment.)

The principal conclusion from this simulation study is that our knowledge of biological systems in the field of resource management is full of gaping holes. For example, we are very short of quantitative data on long-range dispersal, on invertebrate epidemiology, and on insect-plant relations over long periods. It is to be hoped that new field research will fill these gaps. Also, it is to be hoped that large computers will come to play an increasingly important role in biological systems management problems.

The information yielded by large-scale biological simulation studies

In view of the great effort and expense involved in simulation studies of the type reported here, it behooves us to determine precisely how much information we gain from such studies—which would otherwise be unavailable.

1. Clearly, the major immediate benefit is that we learn which of the

control strategies presently available to us is best. More particularly, we find out which strategies are best for specific situations, and conversely, the optimum timing of application and deployment in space for any particular control agent.

2. Broad features of the whole control situation are revealed: the critical importance of the threshold densities at which we decide to take action for ultimate economic losses, and the great value in controlling a pest when it is at low densities and confined to a focus of subsequent epidemic outburst; the importance of applying parasites and disease early; and that a low level of damage to trees for a long time can be as lethal as high damage levels over a much shorter period.

3. Simulation studies show us the relative sensitivity of control effectiveness, for any control agent, to various factors that regulate efficiency. For example, we see that the efficiency of parasites is far less sensitive to dispersal rate than to searching efficiency or to egg complement, if our model is sound. Also, we see that the effect of a fourfold increase in egg complement depends largely on the searching efficiency. Computer simulation helps in the search for optimum biological control agents, or optimum agents of any other kind, and also suggests critical experiments. Note, however, that in order to be of maximal use, computer models have to be very detailed and realistic, as in Sec. 11.6.

4. Simulation studies are important in focusing research attention on gaps in our knowledge or understanding of various systems.

Several difficulties have appeared in this strategy evaluation research. One of the most basic lies within the domain of forest economics. How much money has really been lost when a tree is killed by defoliators? This depends on several factors, all of which pose certain problems of measurement. When a tree is killed by excessive defoliation, it does not lose all its value immediately. Its value drops gradually as more and more of the tree rots. Therefore the dollar loss when a tree is killed by pests depends on when the tree is salvaged relative to the time of death. Also, what is the dollar loss: the stump value, or that plus the value added to the harvested tree as it is converted into newsprint, furniture, or construction material? This matter is further complicated because the tree may be cut in one community, then shipped by rail to another. Are we interested only in the dollar loss to the community that would cut the tree, or the sum of all losses to all communities that process the tree? A further economic factor arises because, in economic life, the number of units of goods available always affects the price per unit: Even though there has been a real biological loss to society, the economic loss could, in principle, be offset by increasing the value per unit of the surviving trees. Much more econometric analysis of such questions is required.

Another difficulty is that the relation between defoliator population density and tree growth and survival is complex. There are lag effects, changing physiological thresholds as the tree suffers more damage, and cumulative effects, all of which necessitate long runs (many years) of data from the same plots in order to work out the mathematical relation between density and growth and survival of trees. Our knowledge of these relationships for deer, insects, rabbits, and other defoliators is very scanty.

12.4 WATER RESOURCE SYSTEMS

Several new types of complexity enter simulation studies when the object of research interest is not a species, such as buffalo, salmon, or a forest-insect pest, but a unit, such as a lake, river, system of waterways, an ocean or section thereof, area of forest, or all the chaparral country in California. These resource units can be used in many ways; for example, water systems can be used for recreation, irrigation, power, urban use, commercial fishing, etc. Multiple use almost always implies conflicting interests (Chap. 7), and this in turn leads to a host of problems. How do we compare various management strategies when some strategies clearly favor one interest group, and hurt others? What ultimate yardstick for measuring the value of a management policy is fair to all interests? How do we assess the economic or other values of recreation, so that the recreational value of a policy is measured fairly relative to the irrigation or power productivity, which is much easier to evaluate? What method of economic evaluation will deal with profitability, yet leave something more than a wasteland to future generations?

The first type of problem shows the need to rank alternative management policies by some measurable, objective, quantitative yardstick that is economically sound, yet does not lead to deterioration of resources. We can illustrate the difficulty by discussion of the technique currently used by resources economists (Marglin, in Maass, 1962).

We require the following symbols:

$E_t(\mathbf{y}_t) =$ gross benefits in year t from particular management policy, expressed as function of vector of factors $\mathbf{y}_t$ determining gross benefits; variables $\mathbf{y}_t$ might be design parameters in system of dams, turbines, and irrigation canals, or variables defining forest management, insect-pest control, and fire-control policy, etc.

$M_t(\mathbf{x}) =$ operation, maintenance, and replacement costs in year t as function of vector of variables $\mathbf{x}$

$K(\mathbf{x})$ = initial capital costs [for example, for water resource system, $K(\mathbf{x})$ would relate to construction; for forests, to bush roads, buildings, and equipment]

Whenever money is invested, we must weigh the return against alternative investment at compound interest rate r. Marglin arrives at the following expression for the efficiency net benefit, or ranking function, for a multipurpose water resource system with an economic life of T years.

$$R = \sum_{t=1}^{T} \left(\frac{E_t(\mathbf{y_t}) - M_t(\mathbf{x})}{(1 + r)^t} \right) - K(\mathbf{x}) \qquad (12.1)$$

We could calculate the value of R for each of several different systems management policies, using computer simulation, then decide on the policy for managing the resource that maximized R. There is, however, a basic fallacy in this approach. It will be noted that because of the denominator $(1 + r)^t$, a given value of $E_t(\mathbf{y_t}) - M_t(\mathbf{x})$ is worth less, the further into the future it is obtained. Therefore there is no premium on saving any resource; the optimum policy, according to (12.1), is always that which uses up the resource at the fastest rate without inordinately increasing $M_t(\mathbf{x})$. In other words, (12.1) justifies wiping out all our natural resources at the earliest economically and technologically feasible date.

There are two possible approaches to correcting for this fallacy. One is to eliminate the denominator term $(1 + r)^t$, but this is unsatisfactory because profitability is then given short shrift in computing R. A better method is to use (12.1) to obtain, for any broad resource strategy, an array of R values for different sets of $\mathbf{y_t}$ and $\mathbf{x}$ values. Then we use what the economists and mathematicians call "marginal conditions," "side conditions," or "constraints" to eliminate those cases that annually remove more than the annual maximum sustainable yield for the resource (as explained in Secs. 4.1, 5.2, and 11.1; note that the constraints are the source of conflict in conflict of interest situations). We can combine economic and biological considerations in the structure of our ranking function, R, by incorporating economic considerations into the function itself, and biological considerations into the constraints. (Chapter 13 considers the significance of constraints in more detail.)

We now need to consider in somewhat more detail the method of evaluating $E_t(\mathbf{y_t})$. For many of the gross benefits, evaluation of the dollar worth is entirely straightforward. In the case of annual benefits from irrigation, energy, and flood control, we are dealing with such $\mathbf{y_t}$ elements as

y_1 = millions of bushels harvested

y_2 = kilowatt hours

y_3 = flood damage avoided

However, how do we handle such an item as

y_4 = man-hours of recreation?

One way is to add up the total amount of money that fishermen, hikers, campers, tourists, and hunters spent to get to and return from a place, and sleep, eat, hunt, hike, camp, or fish there. For example, Mahoney (1960) has found that in 1955, resident California freshwater anglers spent about $226,884,935, with a mean annual expenditure of $217.89, representing a mean daily expenditure of $14.27 over a mean of about 15 days. The derivation of this figure and corresponding figures for saltwater fishing is given in Table 12.5 (Mahoney, table 7). Table 12.6 (Benson, 1961, table 13) compares corresponding United States and Canadian national estimates.

To get a rough idea of the relative importance of the power output from water systems compared to the recreational output of water systems, we can use the following figures. The current retail price of electricity is roughly 2 cents per kilowatt hour in central California, averaging over times of the year and type of use (urban, rural, industrial, etc.). From Landsberg, Fischman, and Fisher (1963, table A10-21), the per customer consumption of electricity (residential plus commercial) in the United States is very roughly 21,000 kilowatt hours per year. This represents a per customer per year delivered retail value of very roughly $420. However, only 500,000 California citizens fish in fresh water (Mahoney, 1960), whereas almost the entire population uses electricity. Therefore the relative importance of electricity to recreational fishing in the Californian economy is very roughly

$$\frac{\$420 \times 20{,}000{,}000}{\$218 \times 500{,}000} = 77:1$$

Clearly, from these crude calculations, the recreational use of freshwater resource systems is not important relative to power use, if only profitability is the criterion of importance. Most citizens, however, would presumably argue that in comparing alternative strategies for power, irrigation, and urban use of water resources, the optimum strategy would be that with the most positive effect, or the least deleterious effect on the recreational use of the water. The most convenient way to handle this mathematically is as a constraint.

In summary, various types of design for a water resource system can be evaluated by means of computer simulation which evaluates the profitability of each design throughout T years of operating by using a ranking function. The design then selected has the highest R, *and* meets various side conditions imposed by the need not to destroy any natural resources and the desire to use the system for recreation as well as power and irrigation.

392   *The methods of resource management*

TABLE 12.5
Distribution of expenditures, resident-citizen anglers, 1955 (Mahoney, 1960)

Type of expenditure	Freshwater fishing			Saltwater fishing		
	Total statewide expenditure, in thousands	*Expenditure per angler*	*Percentage of total*	*Total statewide expenditure, in thousands*	*Expenditure per angler*	*Percentage of total*
Transportation	$65,797	$63.19	29.0	$23,838	$37.23	26.3
Food	53,772	51.64	23.7	18,399	28.73	20.3
General purpose equipment	36,755	35.30	16.2	10,151	15.85	11.2
Fishing equipment	24,050	23.10	10.6	10,423	16.28	11.5
Lodging	11,344	10.89	5.0	3,444	5.38	3.8
Rentals	10,664	10.24	4.7	3,807	5.94	4.2
Bait	9,529	9.15	4.2	3,988	6.23	4.4
Gas and oil (for boats and outboard motors)	4,538	4.36	2.0	2,900	4.53	3.2
Angling license	2,495	2.40	1.1	997	1.56	1.1
Repair and maintenance	2,042	1.96	0.9	1,268	1.98	1.4
Extra vehicles	2,042	1.96	0.9	363	0.57	0.4
Party and charter boat fees	1,588	1.53	0.7	10,243	15.99	11.3
Publications	1,361	1.30	0.6	544	0.85	0.6
Club dues	907	0.87	0.4	272	0.42	0.3
Totals	$226,884	$217.89	100.0	$90,637	$141.54	100.0

TABLE 12.6

Comparison of seasonal and daily expenditures of sport fishermen and hunters in Canada and the United States* (Benson, 1961)

Category	Canada† (Persons 14 years of age and older) 1961		United States‡ (Persons 12 years of age and older) 1960		1965	
	Season	Day	Season	Day	Season	Day
All sport fishing and hunting	$180.25	$9.03	$126.57	$5.85	$114.42	$5.03
All sport fishing	143.13	9.50	106.26	5.78	91.98	4.82
Freshwater fishing	138.30	9.53	95.25	5.36	77.38	4.21
All hunting	110.44	8.16	79.34	6.03	79.49	5.53
Big-game hunting	85.42	10.62	55.07	8.82	73.38	10.50
Small-game hunting	49.62	5.19	59.98	5.25	50.30	4.16
Waterfowl hunting	79.03	9.07	45.74	5.90	59.79	5.95

*A major source of difference between the Canadian and United States surveys is the approach to vehicle mileage. The United States surveys used an estimate of out-of-pocket expenses, while the Canadian survey allowed a fixed amount of $0.075 per car mile. Automobile or private vehicle expenses amount to 12.8 percent of the total expenditures of the United States 1950 survey; 13.8 percent of the United States 1960 survey; and 21.7 percent of the Canadian 1961 survey.
†This survey.
‡U.S. Department of the Interior (1956 and 1961).

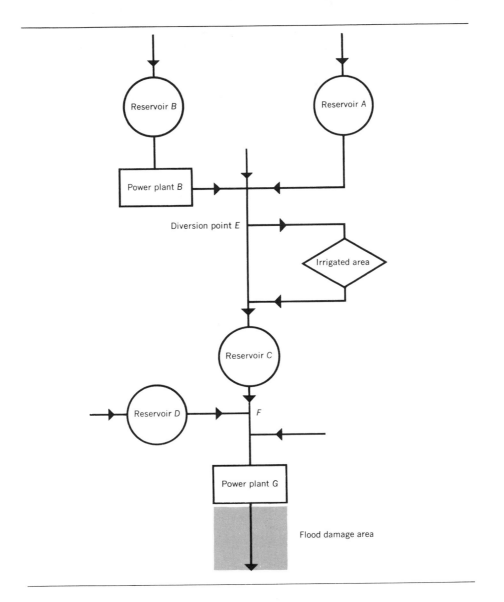

FIGURE 12.5
Schematic outline of the simplified river-basin system (Maass, 1962).

Maass (1962) described various approaches to the mathematical and computer simulation analysis of a hypothetical water resource system; this study is by far the most comprehensive operational-research study of natural resource systems management of which I am aware. There were 12 variables in the study: the storage capacities of three reservoirs, the capacities of two power plants, the allocations of active and dead storage to a reservoir feeding one of the power plants, the allocation of flood-control storage to the other three reservoirs, and the annual target outputs for irrigation and energy.

The entire problem considered in the simulation study was posed as follows. "Given:

1. a certain combination of reservoirs, power plants, and irrigation diversion and distribution facilities;

2. target levels of irrigation and energy outputs;

3. specified allocations of reservoir capacity for active, dead, and flood storage;

4. a representative series of monthly runoff values, and 6-hour flows for flood months;

5. a specific operating procedure;

then by routing the available flows through the reservoirs, power plants, and irrigation system for an extended period of years, such as 50, determine the physical outputs and the magnitude of the net benefits created."

The game is played repeatedly, while we hunt for the combination of operating procedures that produces a maximum value for the ranking function. This hunting, because of the large number of combinations that must be tried, uses systematic search techniques (Sec. 13.4).

In order to construct the simulation program, it was necessary for Maass to take the following steps, which illustrate the procedure required for any large resource system of this type.

1. The structure of the entire system had to be outlined schematically (see Fig. 12.5), and mathematical equations had to be developed describing the internal functional relationships of all parts of the system to each other. These relationships were derived from analysis of data on various water resource systems already operating in the United States. The following array indicates the types of relationships involved:

Dependent variable	*Independent variable*
Irrigation benefits, in dollars per acre foot	Annual target output for irrigation, in acre feet
Capital costs of irrigation diversion, distribution, and pumping works	Annual target output for irrigation
Capital cost of power plant	Installed capacity of power plant (megawatts)

Net effective power load (feet)

Reservoir capacity (acre feet)

Flood damage (dollars)

Stream flows in cubic feet per second

Capital cost

Reservoir capacity

2. The runoff data were prepared in suitable form for input to the simulation program; 60 years of observed or synthetic data were used in each simulation study.

3. An operating procedure for systems management had to be formulated. For example, the following February through August policy was followed:

a. Release water from reservoir C until target output is met, the water capacity of power plant G is reached, or reservoir C is emptied.

b. Draw water from reservoir D until the target output is attained, the water capacity of power plant G is reached, or reservoir D is emptied.

c. Make additional releases from reservoir A, if this is still possible, until the target output is realized, the water capacity of power plant G is reached, or reservoir A is emptied.

d. Draw additional water from reservoir B, if this is still possible, until the target output is met, the water capacities of power plants B and G are both reached, or reservoir B is drawn down to dead-storage level.

e. Make specific provision for flood-storage capacity during April, May, and June in any combination of reservoirs B, C, and D. A set of subroutines was developed for this operation. An example of a sample subroutine is: if a flood occurs in April from melting snow, make no releases during the month. If it does not, lower each reservoir during April until the total specified flood-storage capacity has been provided by May 1.

f. During March, April, and May route flood-storage releases through the turbines at power plants B and G up to their turbine capacities, and use the waters issuing from reservoir B to satisfy the target output for irrigation.

This is only a partial list of the detailed operating procedures built into the computer program. It does show that a tremendous amount of information and insight into the details of actual management decision-making processes is required to formulate computer simulation programs for resource management strategy studies.

4. The operating procedure was coded into sets of Fortran instructions.

5. The rules for computing the dollar value of system outputs were coded into Fortran.

6. Output formats for the computer were arranged so that all physical and economic results of a computer simulation using particular operating strategies would be readily available.

It is not necessary to give further details about this study here, since they can be found in the original publication. The general outline of the program given here makes the important point that this type of study has profound implication for the organization and philosophy of research programs on resource management.

The object of the research program is to study a very complicated system in its entirety, not an isolated fragment thereof. Therefore it follows that a principal characteristic of the research will be an attempt to mimic reality in detail, even where this involves massive complexity. In this case, the computer simulated the behavior of about 250 variables. The information demands of such a program are so great that the research depends on a team effort involving a great variety of talents, and a tightly organized inter-disciplinary team. The Harvard group numbered about 50, drawn from such diverse fields as public administration, economics, engineering, and agriculture.

As with the salmon and the spruce budworm study, the model employed considers movement of entities through space (water, in this case) as well as changes in system states through time. Finally, as noted in Sec. 11.10, the spruce budworm study and the Harvard water resources study, like other large-scale biological systems management studies, represent a clear break with traditional types of mathematical models in one very important respect. Instead of modeling the behavior of a system through time under *average variate values* for variables extrinsic to the system (that is, a "steady-state system"), we mimic the responses of the system to a *specific temporal sequence of variate values* for variables extrinsic to the system. Once again, we wish to emphasize the importance of discovering the consequences of the specific effects traceable to the statistical properties of such sequences of values, rather than the average effect over time of such a sequence. Some of the most important events that can occur in a system may never be demonstrated if we assume average extrinsic conditions, or some artificially constructed time series not based on data from actual sequences in nature.

12.5 MANAGEMENT OF CHAPARRAL

The ultimate challenge in computer simulation studies on resource manage-ment arises when we attempt to simulate events in a heterogeneous environment, with diverse species of plants and animals that can affect each other via a multitude of causal pathways, and with all events under the influence of man and a variety of extrinsic factors such as weather and fire, and where plant and animal succession can occur. Succession will occur when there is a sustained shift in environmental conditions, so that an area that was formerly optimal for one assemblage of plants and animals becomes optimal for another assemblage, and the second gradually replaces the first. An example of such a heterogeneous habitat is the chaparral in California.

The particular association of woody shrubs referred to as chaparral occurs in California to the west of the Sierra–Cascade Divide, at altitudes above those of the great Central Valley (that is, 500 feet in the north to 1,250 feet in the south) but below those of the yellow pine belt (3,000 feet in the north; 5,000 feet in the south). The most characteristic plant of the chaparral is chamise, a low shrub that grows 2 to 12 feet high, often in dense thickets virtually impenetrable except by small animals. Many other types of plants grow in chaparral, including several species of dwarf oaks; two genera of shrubs, *Caenothus* and manzanita (*Arctostaphylos*); and many other less important shrub genera. A large number of these shrubs are of great importance for deer browse and are used to a lesser extent by livestock.

Perhaps the most tricky aspect of simulating chaparral on a computer is that a variety of complex and subtle mechanisms are at work in chaparral country to promote rapid plant succession:

Fire Summers can be long, very hot, and very dry in chaparral country, and large fires are a common fact of life. Because fires can change the composition of the plant community, they can also change the animal community, only the lag is longer. In general, fires increase the animal biomass in chaparral, largely as a result of improved food supply. In many areas, however, fires have remarkably little effect on the species composition of shrub communities because of the shrubs' adaptations to fire, such as crown sprouting.

Grazing Grazing can shift the balance of competition between plants in both the herb layer and the shrub layer. If two species of plants, A and B, coexist and compete, then a buildup in the grazing pressure on A relative to that on B will produce an increase in incidence of B relative to that of A (succession will have occurred). Since different kinds of animals (for example, cattle, deer, and sheep) have different preferences for plants, then in effect we can regulate the direction and velocity of plant succession by regulating the "mix" of animal species in the grazing population. Grazing can affect succession not only because different species of grazers have different plant preferences, but also because different plant species have different tolerances to high levels of grazing.

Direct control of plant succession by man This can be achieved by reseeding, by mechanical or chemical control (herbicides, silvicides), or by fertilization. Erosion control can also regulate plant succession.

Plant diseases and insect pests These, too, can regulate plant succession.

Weather Weather will, of course, regulate plant succession.

Simulation of chaparral is complicated, not only because of the variety of influences operating on the plants, but also because of the variety of influences operating on the animals. Each species will be affected, and affected differently than the other species, by weather, fire, the type and amount of plant species available, the nutrient content of the plants, hunting, and intra- and interspecific competition from other browsers and grazers. Furthermore, within each species of animals, different age groups will react differently to various factors. Development of a computer simulation program for the chaparral proceeds as follows.

Step 1   The computer's memory must be set up to simulate the physical arrangement of the chaparral at the starting point in the simulation. That is, we simulate the way in which chamise thickets, dwarf oaks, and other shrub species are aggregated, distributed, and interspersed by allocating these to squares within a grid stored in the memory. Data describing the actual spatial arrangement of the various species of plants in chaparral must be obtained from quadrat sampling in nature. A convenient unit to represent chaparral in the computer is a 220-yard × 220-yard square ($\frac{1}{8}$ mile × $\frac{1}{8}$ mile, or an areal unit $\frac{1}{64}$ of a square mile). Suppose we simulate events in an area 2 miles × 10 miles; that is, 16 squares wide and 80 squares long, requiring 1,280 memory locations for each type of information stored for our 20-square-mile unit. The status of each of the plant species in each of the 1,280 squares will be stored as the most probable biomass of foliage production of that plant per annum in the given square. We will also allocate a predetermined number of sheep, cattle, deer, and other important animals to each of the quadrats at the beginning of the period, the appropriate number to be determined by sampling studies.

Step 2   It is then necessary to develop, either from existing data or from new field work, a set of difference equations expressing the biomass of each type of plant foliage production, and the numbers of each type of animal in each $\frac{1}{64}$-square-mile quadrat as a function of the system state at the prior point in time. Time intervals for such computations can be as long or as short as we choose to make them: the time unit might be an entire year, or less, depending on what insight we gain into the dynamics of chaparral production by analyzing actual data. The difference equations must be designed to allow for dispersal, local extinction of a species, or appearance of a species where it had not hitherto occurred. All possible relevant causal relationships must be incorporated into the system of difference equations.

Step 3   We need to construct, from actual field data, typical long-term sequences of variate values for relevant variables extrinsic to the chaparral ecosystem, such as weather and fire.

Step 4   Operating policies for managing the chaparral must be formulated. These will describe hunting legislation, stocking densities of cattle and sheep, and possible operating procedures with respect to controlled burning, reseeding, brush control, watershed management, and other desirable procedures. As with the two previous studies, these operating procedures will follow IF statements. That is, we will only use a particular policy if the conditions are right for it to be used.

Step 5   Procedures for computing the cost of all operations, gross return from all system outputs, and net profit must be incorporated into the program.

Step 6   All the data and procedures in steps 1 to 5 must be expressed in the form of data input and sequences of Fortran instructions for the computer.

Step 7   Appropriate input-output formats must be devised: financial data, operating summaries, etc.

There is a simulator already in existence which has most features in common with that just proposed; this is the Harvard Forest Management Simulator of Gould and O'Regan (1965). This program is primarily for use in studying the economics of forest management, but in addition, it contains subroutines to simulate the consequences of fires and storms. The program includes a great deal of detail on the economic aspects of natural resource management. Yearly harvest is determined by price and by predetermined policies. Stocking densities for cattle and sheep should presumably be handled the same way. The CASH1 subroutine of Gould and O'Regan computes income, current and cumulative interest and other costs, taxes, and net income. Liabilities, assets, and net worth are determined. To make a natural resource simulator of maximum usefulness, these items should always be incorporated. Indeed, the usefulness of any simulator is only limited by the imagination of the research team that produces it.

REFERENCES

Belyea, R. M.: Death and Deterioration of Balsam Fir Weakened by Spruce Budworm Defoliation in Ontario, *J. For.*, **50**:729–738 (1952).

Benson, D. A.: "Fishing and Hunting in Canada," Canadian Wildlife Service, National Parks Branch, Department of Northern Affairs and National Resources, Ottawa, 1961.

Bird, F. T., and J. M. Burk: Artificially Disseminated Virus as a Factor Controlling the European Spruce Sawfly, *Diprion hercyniae* (Htg.), in the Absence of Introduced Parasites, *Can. Entomol.*, **93**:228–238 (1961).

Dobzhansky, T., and S. Wright: Genetics of Natural Populations X. Dispersion Rates in *Drosophila pseudoobscura*, *Genetics*, **28**:304–340 (1943).

Elliott, K. R.: A History of Recent Infestations of the Spruce Budworm in North-Western Ontario, and an Estimate of Resultant Timber Losses, *For. Chron.*, **36**:61–82 (1960).

Fuller, W. A.: The Biology and Management of the Bison of Wood Buffalo National Park, *Can. Dept. Northern Affairs Natl. Resources, Wildl. Mgt. Bull.*, ser. 1, no. 16, 1962.

Glendenning, R.: A Successful Parasite Introduction into British Columbia, *Can. Entomol.*, **65**:169–171 (1933).

Gould, E. M., and W. G. O'Regan: Harvard Forest Papers, Simulation, a Step toward Better Forest Planning, *Harvard Forest*, no. 13, 1965.

Jarvis, J. M.: Forty-five Years Growth on the Goulais River Watershed, *Can. Dept. Northern Affairs Natl. Resources, For. Res. Div. Tech. Note*, no. 84, 1960.

Landsberg, H. H., L. L. Fischman, and J. L. Fisher: "Resources in America's Future," The Johns Hopkins Press, Baltimore, 1963.

Maass, A. (ed.): "Design of Water-Resource Systems; New Techniques for Relating Economic Objectives, Engineering Analysis, and Governmental Planning," Harvard University Press, Cambridge, Mass., 1962.

McGugan, B. M., and J. R. Blais: Spruce Budworm Parasite Studies in Northwestern Ontario, *Can. Entomol.*, **91**:758–783 (1959).

Mahoney, J.: An Economic Evaluation of California's Sport Fisheries, *Calif. Fish and Game*, **46** (2):199–209 (1960).

Miller, C. A.: The Interaction of the Spruce Budworm, *Choristoneura fumiferana* (Clem.), and the Parasite *Apanteles fumiferanae* (Vier.), *Can. Entomol.*, **91**:457–477 (1959).

————:The Interaction of the Spruce Budworm, *Choristoneura fumiferana* (Clem.), and the Parasite *Glypta fumiferanae* (Vier.), *Can. Entomol.*, **92**:839–850 (1960).

Morris, R. F.: The Development of Sampling Techniques for Forest Insect Defoliators, with Particular Reference to the Spruce Budworm, *Can. J. Zool.*, **33**:225–294 (1955).

———— (ed.): The Dynamics of Epidemic Spruce Budworm Populations, *Mem. Entomol. Soc. Can.*, no. 31, 1963.

Mott, D. G., L. D. Nairn, and J. A. Cook: Radial Growth in Forest Trees and Effects of Insect Defoliation, *For. Sci.*, **3**:286–304 (1957).

Roe, F. G.: "The North American Buffalo," University of Toronto Press, Toronto, Canada, 1951.

Royce, W. F., D. E. Bevan, J. A. Crutchfield, G. J. Paulik, and R. F. Fletcher: Salmon Gear Limitation in Northern Washington Waters, *Univ. Wash. Publ. in Fish.*, new ser., **2** (1):1–123 (1963).

Smith, R. W.: Status in Ontario of *Collyria calcitrator* (Grav.) (Hymenoptera:Ichneumonidae) and of *Pediobius beneficus* (Gahan.) (Hymenoptera:Eulophidae) as Parasites of the European Wheat Stem Sawfly, *Cephus pygmaeus* (L.) (Hymenoptera:Cephidae), *Can. Entomol.*, **91**:697–700 (1959).

Vincent, A. B.: Development of a Balsam Fir and White Spruce Forest in North-Western New Brunswick, *Can. Dept. Northern Affairs Natl. Resources, For. Res. Div. Tech. Note*, no. 6, 1955.

Watt, K. E. F.: A Mathematical Model for the Effect of Densities of Attacked and Attacking Species on the Number Attacked, *Can. Entomol.*, **91**:129–144 (1959).

————: Dynamic Programming, "Look Ahead Programming," and the Strategy of Insect Pest Control, *Can. Entomol.*, **95**:525–536 (1963).

————: The Use of Mathematics and Computers to Determine Optimal Strategy and Tactics for a Given Insect Pest Control Problem, *Can. Entomol.*, **96**:202–220 (1964).

Wellington, W. G.: Qualitative Changes in Populations in Unstable Environments, *Can. Entomol.*, **96**:436–451 (1964).

$\int_{\int}^{3}$ | SYSTEMS OPTIMIZATION

13.1 EVALUATION OF STRATEGIES

Many problems in the management of natural resources reduce to the basic problem of evaluating strategies: confronted by a problem, for which many alternative solutions and combinations of solutions are available, how do we determine which strategies of solution are best? This problem of deciding on an optimal strategy or combination thereof is far more complicated than it appears. We shall develop our explanation of the matter in three steps. First, we will consider a number of typical, practical problems in resource management, with a view to discovering if these problems have any features in common. Second, we will analyze the elements the various problems have in common, in order to clarify the precise nature of the desiderata that must be found in any proposed line of mathematical solution if it is to be completely satisfactory. Finally, we shall survey the various forms of mathematical solution available for problems of this general type in order to determine which classes of techniques best match the logical structure of problems in resource management.

1. The first problem relates to forest cutting policy. Growth of a forest stand depends on a great many factors, including the structure of the management policy. If annual cut is too low, it results in delaying the harvest of young stands until they have passed the established rotation age. This leads to

numerous well-known difficulties caused by insect and pathogen damage directed primarily against old trees. At the other extreme, if the annual cut is too high, it leads to understocked stands and another set of difficulties. Clearly, annual allowable cut may be thought of as a dependent variable governed by a number of factors, including the magnitude of annual cut in the same stand at a previous point in time. The function has some maximum point and the object of any mathematical analysis is to determine that management policy which leads to a maximum annual sustained yield. This problem of determining how to maximize or minimize some function is referred to by mathematicians as the *extremum problem*.

2. The second problem is the classical optimum-yield problem of fisheries. Productivity of fish stocks varies constantly through time under the influence of many factors: weather, such as changing temperature, direction, velocity or constancy of water currents; disease; population density; competition from other species; predation by other animals; and predation by man (that is, fishing). If we fish too little, the fish population left in the water after fishing will build up to densities at which intraspecific competition for food stunts fish growth and diminishes the probability of survival. Where density-independent factors are limiting, underfishing causes excessive wastage of the resource to natural mortality. If we fish too hard, too few adults are left behind to spawn future generations of fish, and the stock goes into a decline. Many of the factors in this situation are beyond influence by man (weather, food for the fish). Nevertheless, we can have considerable influence over the situation by means of legislation affecting not only the numbers caught, but also mesh size and other factors determining the size and hence age of the fish caught. In some cases, as in the "put and take" trout-stream-angling fisheries in many states of the United States, it may be economically feasible to supplement natural reproduction through a stocking program. In other cases, such stocking will not be worthwhile, as noted in Sec. 5.4. Note, however, that whatever action we take against the fish stock at any point in time, provided we have removed enough fish to have a real effect on the stock, the situation we create in one fishing season determines the situation we will find at the beginning of the next. The mechanism of this effect has already been explained in Sec. 4.1. No matter whether we are harvesting whales, deer, pheasants, mink, cattle, or zebras, the problem is fundamentally the same optimum-yield problem in all cases.

3. This is the same problem as the first two, only turned upside down, as it were, since it involves minimizing the survival of pests, whether insects, rodents, lampreys, disease-carrying snails, or weeds. This is also an extremum problem, only instead of wishing to find the highest point in an n-dimensional

hypersurface, and the values of the independent variables that produce this maximum, we wish to find a minimum value.

The survival of a destructive insect pest may be thought of as a dependent variable under the influence of a large number of independent variables, some of which are subject to control by man. As in the preceding cases, the solution to the problem is bound by a variety of constraints. For example, there is a limit to the amount of money that can be spent on control for the given tract of forest if operation of the tract is to be profitable. Also, no control measure can be used if it has a deleterious effect on the structure of the forest. Again, no matter what we do, it may affect the situation we will have to deal with in the future.

Let us now consider the elements that the three problems have in common:

1. In each case, we have a dependent variable that is a function of a considerable number of independent variables, only some of which are subject to manipulation by man.

2. In each case, there are constraints on the solution to the problem because of economic feasibility, equipment characteristics, side effects of the process, or some other factor.

3. These constraints are not equational: they are in the form of inequalities. That is, instead of having statements of the form, "management of this forest is only profitable if overhead is exactly equal to X," we have inequational constraints of the form, "management of this forest is only profitable if all operating costs add up to an amount equal to or less than X."

4. In all three problems, not only is the equation relating the dependent variable to the independent variables a function of many factors, but the inequations expressing the constraints may also be a function of many factors. For example, when we say that all costs must be equal to or less than X, this is expressed mathematically by a sum of terms in which each term expresses one particular kind of cost.

5. In all these problems, one of the dimensions of the problem must be time, if we are to consider the problems from the most completely realistic point of view. This is because every time we make a set of decisions, we alter all the variables in the system being managed. The next time we have to make a set of decisions, we have to deal with variate values for all the variables produced by the previous set of decisions. In such cases, in which for each of n times there were m variables about which we could make decisions, the final product of the whole sequence of decisions is a function of mn variables. This way of looking at decision processes recognizes that we are dealing with a multistage decision process.

Another aspect of all resource management problems also forces us to consider them as multistage decision processes. This aspect follows from the nature of resource management strategies. Many of these may constitute an optimal policy if used only once, or a small number of times, but may be grossly suboptimal if used repeatedly. That is, optimal policies may not necessarily be able to follow themselves. A few examples may make this point clear. Consider the strategy of controlled burning to improve range quality in chaparral country. Burning may be highly effective at intervals, but no one would suggest burning repeatedly. Consider insect-pest control. If an insect pest is present at very high densities, and emergency save-the-foliage control is necessary, then aerial spraying of vast tracts of forest may be in order. When densities have fallen to much lower levels, such control may not be necessary or desirable for economic or biological reasons. At such times, ground release of biological-control agents in cages or air drops of virus in emulsion over known loci of the epidemic outbreak might be optimal instead of insecticide. Optimal policy depends on the situation encountered, but the situation encountered depends on our previous policies. It is necessary to think in terms, not of "one-shot" solutions, but of sequences of control actions, the *net* effect of which is optimal.

We shall now consider, beginning with the simplest approach, eight mathematical techniques exhibiting increasing degrees of realism, which can be used when dealing with complex problems in the area of management decisions. Which of these we select in any instance depends on either, or both of, the fundamental nature of the problem and the way in which we choose to formulate it. We shall discuss the eight approaches very briefly in turn, pointing out in each case what characteristics our formulation of the problem must have in order for the method to be applicable. Some of the methods are patently inapplicable to solution of extremum problems in most practical situations. They are included to add to the intelligibility of the discussion, since all eight methods are to some degree conceptually or methodologically related.

1. First, consider the situation in which there are no constraints, and a dependent variable, Y, is a function of one other variable, X, and the form of the relationship is given by

$$Y = a + bX - cX^2 \tag{13.1}$$

Equation (13.1) describes a parabola. We could plot a graph of Y versus X, and read off the value of X that produced this maximum. Where there are two independent variables, X_1 and X_2, we could plot a family of graphs against X_1, one for each value of X_2. This family of graphs could then be used to find the values $X_{1\,max}$ and $X_{2\,max}$ that produce the maximum value of Y. However, once a problem has more than two independent variables, this method of finding maximum-producing values of the independent variables by using graphs clearly gets out of hand very quickly as the number of variables increases. Hence, we turn to an algebraic analog of the graphical process.

By reconsidering (13.1) and remembering from calculus that at $X_1 = X_{max}$, $dY/dX = 0$, we find that the value of X_1 may be determined as follows:

$$\frac{dY}{dX} = b - 2cX$$

and when $dY/dX = 0$, $X = b/2c$. Similarly, for two independent variables, when

$$Y = b_0 + b_1X_1 - b_2X_1^2 + b_3X_2 - b_4X_2^2$$

we can extend this procedure from a plane to a space, and find

$$\frac{\partial Y}{\partial X_1} = b_1 - 2b_2X_1$$

and

$$\frac{\partial Y}{\partial X_2} = b_3 - 2b_4X_2$$

so that at $Y = Y_{max}$,

$$X_1 = \frac{b_1}{2b_2}$$

and

$$X_2 = \frac{b_3}{2b_4}$$

This approach can be generalized to any number of variables, but two objections are immediately apparent. First, this procedure does not recognize the existence of constraints, and second, the simultaneous solution of a very large number of nonlinear equations could raise formidable mathematical difficulties.

Nevertheless, a principle underlying all eight extremum methods is exposed. These methods can all tell us, for the appropriate problem formulation, which value each independent variable should have if the dependent variable is to be maximized or minimized. The value zero, which can arise, indicates that a certain procedure should not be followed at all.

2. Let us now make the problem statement somewhat more complex,

and more realistic. Let us assume that we have no control over certain of the independent variables. This can be expressed mathematically as follows:

$$Y = f(X_1, X_2, X_3, X_4, \ldots, X_n)$$

If we suppose we only have control over X_1, X_2, X_3, we can treat the remaining X_i's as constants, and take partial derivatives only with respect to X_1, X_2, and X_3.

3. We have now introduced the notion of a constraint into our problem formulation, but the approach still isn't very realistic. This is because various types of restrictions, or constraints, are often related. A simple example occurs when the total contribution of all components of overhead must not exceed a certain dollar value. We might say that the basic equation to be maximized or minimized is

$$Y = f(X_1, X_2, X_3, X_4, X_5)$$

subject to the constraint that

$$c_1 X_1 + c_2 X_2 + c_3 X_3 \leqslant C$$

in which C is the total cost of overhead that must not be exceeded, and the c_i's are costs per unit. Considerably more complex versions of this basic class of problems can be handled, in principle, by using Lagrange multipliers. That is, we can maximize

$$Y = f(X_1, X_2, X_3)$$

given the constraints, or "side conditions" that

$$s_1(X_1, X_2, X_3) = 0$$

and

$$s_2(X_1, X_2, X_3) = 0$$

4. We still do not have a very realistic expression of the problem. Constraints expressed in equations say that something is equal to something else. But in the foregoing analysis of the common elements of resource management problems, we found that a typical constraint takes the following form: "The total cost of all overhead must not exceed a certain value." This means that the total cost must be, not equal to, but equal to or less than, a certain value. In other words, our problem can be formulated more realistically as follows. Maximize the profit function

$$c_1 X_1 + c_2 X_2 + \cdots + c_n X_n$$

subject to the conditions:

$$X_1 \geqslant 0, \, X_2 \geqslant 0, \, X_3 \geqslant 0, \, \ldots, \, X_n \geqslant 0$$

$$a_{11}X_1 + A_{12}X_2 + \cdots + a_{1n}X_n \leqslant b_1$$

$$a_{21}X_1 + a_{22}X_2 + \cdots + a_{2n}X_n \leqslant b_2$$

$$\cdots \cdots \cdots \cdots \cdots \cdots \cdots \cdots$$

$$a_{m1}X_1 + a_{m2}X_2 + \cdots + a_{mn}X_n \leqslant b_m$$

where m = number of resources

n = number of commodities produced

a_{ij} = number of units of resource i required to produce one unit of commodity j

b_i = maximum number of units of resource i available

c_j = profit per unit of commodity j produced

X_j = level of activity (amount produced) of jth commodity

In other words, inequality constraints are recognized as such in this formulation of the problem. The field of applied mathematics that deals with such problems is called *linear programming*, and was invented by George Dantzig in 1947 while working for the United States Air Force. It is related to mathematical game theory.

Linear programming and the next two techniques, which are called nonlinear programming and dynamic programming, respectively, are generally referred to collectively as *mathematical programming*.

5. A formulation of programming problems that is based on more realistic assumptions than linear programming is the nonlinear counterpart, nonlinear programming. Unfortunately, this field is only about 16 years old, and is difficult mathematically.

6. To this point, we have ignored the role of time in our problem formulations. However, there is a quite different way of looking at management problems, which seems more natural for the world of plants and animals, all of which are a product of their history. What we need, ideally, is a method of formulating problems that recognizes:

   a. That the states of complex systems change through time,

   b. That the state of a complex system at any point in time is in part determined by its state at a previous time, and

   c. That we may wish to make decisions about the system, and alter the system, at each of a sequence of N times, when the system is in different states at consecutive times in the sequence of N.

Suppose, first, we consider a brute-force approach to such problems, to indicate why a more sophisticated approach is necessary. Suppose we have an M-variable system, in which we could vary each of the M variables through L levels at each of a sequence of N times. Suppose we wish to maximize some output, Y, from this N-stage decision process. Since there are M variables that can take L levels, each regulating the system at each of N times, Y may be thought of as a function of MN variables. Since the classical approach to this situation, which takes partial derivatives with respect to each of the MN variables, is hopelessly difficult, search techniques on a computer are usually employed. Suppose we wish to try 10 different values for each of the MN variables, to see which combination maximizes Y. For $M = 1$, and a 10-stage sequence of decisions, there are 10^{10} different combinations of values for which we must compute Y.

Suppose $M = 10$, and $N = 100$, which is getting considerably more realistic, then there are $10^{1,000}$ different Y values to be computed. Clearly, the computational requirements for a brute-force approach to such problems soon get out of hand. What is needed is some means of reducing the dimensions of the problem.

About 14 years ago, Bellman (1957, 1961) perceived the existence of this problem, and the means of solving it. Suppose, to illustrate the means of solution, we consider the simplest possible multistage decision process, in which our system may be described as being in the state p_1. We made a decision, denoted by q_1, which transforms the state to p_2. That is,

$$p_2 = T(p_1, q_1)$$

Then we make a second decision, q_2, producing the state p_3, so that

$$p_3 = T(p_2, q_2)$$

This process continues until we produce the final state, p_n, by making decision q_{n-1} on state p_{n-1}. Our object is to determine, for a given p_1, that sequence of decisions $q_1, q_2, \ldots, q_n$ that maximizes Y, where

$$Y(p_1, p_2, \ldots, p_n; q_1, q_2, \ldots, q_n)$$

is called a *criterion function*, or return function.

The technique Bellman uses for reducing the dimensionality of this problem is to invoke two powerful notions of modern mathematics: the *Markov property* and the *optimality principle*. As applied in this context, the Markov property states that after decision q_i, we want the outcome of the remaining decisions (that is, Y, the total return) to depend on only p_{i+1} and the remaining q's, from q_{i+1} to q_n. The optimality principle, which is the other basic notion underlying dynamic programming, states that a policy is only optimal if, for any initial state and initial decision, the remaining decisions are optimal, given the state resulting from the first decision. Let us use the notation $f_n(p_1)$ to represent the maximum return from an n-stage process with initial state p_1. It follows immediately from the Markov property and the optimality principle that

$$f_n(p_1) = \operatorname*{Max}_{q_1} \{g(p_1, q_1) + f_{n-1}[T(p_1, q_1)]\}$$

in which $g(p_1, q_1)$ represents the return from the first decision q_1, applied to the first state, p_1; and $p_2 = T(p_1, q_1)$.

In other words, the computational approach we follow in determining how to obtain $f_n(p_1)$ is to determine the q_1 that maximizes $g(p_1, q_1)$. Then, for the $T(p_1, q_1)$ produced by the value of q_1 that maximized $g(p_1, q_1)$, we seek the q_2 that maximizes $g(p_2, q_2)$, and so on.

Let us see now what has been accomplished by translating management problems into these terms. The advantages of the dynamic programming approach are as follows:

a. Dynamic programming is realistic with regard to the role of time, because systems change their states through time in part because of how we managed the systems at previous points in time.

b. The dynamic programming approach allows us to make an enormous reduction in the volume of computation required to handle a problem that considers the effects of time realistically. It decomposes a single MN-dimensional problem into a sequence of MN-dimensional problems. To illustrate, suppose M is 2 and N is 5. Using a brute-force search technique, and 10 trial values for each of the MN variables, the number of sets of calculations of the return or criterion function is

$$10^{MN} = 10^{2 \cdot 5} = 10^{10}$$

By using dynamic programming, and rejecting all the q_i values but the one that produces maximum $g(p_i, q_i)$ at each step, the number of calculations of $g(p_i, q_i)$ is only

$$5(10^2) = 500$$

c. Dynamic programming uses mathematics to determine optimum policy, rather than optimum parameter values for a given policy.

d. Dynamic programming will often be useful in showing what not to do, in addition to what to do. This follows since at each step in the calculation of $f_n(p_i)$, we drop those q_i values that produce submaximum values of $g(p_i, q_i)$, so those q_i values do not figure in calculation of $p_{i+1} = T(p_i, q_i)$ for the next step.

e. Dynamic programming is sufficiently flexible to allow us to consider all the economic problems of resource management simultaneously, not just a particular problem, such as the pest-control problem, lifted out of context. This is important because the best policy for one aspect of forest management may not be best when all aspects are considered jointly. To illustrate, the best control policy from the standpoint of minimizing pest survival may not be best from the standpoint of maximizing profit from a forest resource.

The reason that the dynamic programming approach is so flexible is that p_i and q_i need not be numbers, but may be vectors. That is, $\mathbf{p_i}$ may be a "state" vector representing a number of variables describing the state of the forest, and $\mathbf{q_i}$ may be a decision vector describing a number of control measures about which we can make decisions: the cutting rate, the expenditure for fire control, the characteristics of the fire-control operation, the expenditure for pest control, the characteristics of the pest-control operation, and so on. We will consider this aspect of dynamic programming in more detail in the next section.

f. The concept of dynamic programming may be generalized considerably to cope with adaptive control processes, precisely because dynamic programming is designed to deal with the change of states through time. The fundamental notion of the adaptive control process is that of feedback control, or a closed loop of operations in which there is a self-correcting feature. Research and management can be part of a closed loop as explained in Sec. 13.5.

7. Another new class of methods for seeking optimum points in hypersurfaces is based on the idea of speeding up the search for optimum points by following the steepest gradients up or down, depending on whether we seek a maximum or minimum. These methods have become very important in chemical engineering, in large part because of the pioneering efforts of G. E. P. Box. The methods are applicable to problems of finding optimum strategy in many resource management problems, and will be discussed in detail in Sec. 13.4. Again, this is one of the new classes of techniques that would be totally infeasible without the large-memory, fast new computers.

8. Fan (1966) has developed the maximum principle idea of the Russian mathematician, Pontryagin, to produce a new set of powerful procedures for optimizing continuous processes. This technique can be combined with dynamic programming to yield methods of great power indeed. Unfortunately, the mathematical level of these methods precludes their being understood by almost all biologists.

13.2 DYNAMIC PROGRAMMING

The central idea of dynamic programming is the *principle of optimality*. Bellman (1961, p. 57, and elsewhere) defines this principle as follows. An optimal policy has the property that, whatever the initial state and the initial decision are, the remaining decisions must constitute an optimal policy with regard to the state resulting from the first decision. Wouk (1962) has pointed out that the principle is a special case of recursive definition or recursive formulation, and he restates the principle as follows. A policy is an optimum policy if and only if it is optimum over every subperiod of the *n*-period problem. To discover the policy that indeed is optimum, we construct a recursive model that, as output from each stage of an *n*-stage process, produces the input for the next stage. The model may be quite elaborate, and may at each stage use data stored in tables, such as sequences of weather data. When the model is constructed, we use it to play games that repeatedly simulate the system in nature. By trial and error, the optimum strategy is discovered, though as explained in the previous section, use of the optimality principle enormously reduces the amount of trial and error required. The issue raised when we apply dynamic programming in biology concerns the meaning of "optimum." As explained in Chaps. 3, 4, and 5, one of the most important problems in any resource management program is learning how best to cope with the homeostatic capability of populations. That is, if we kill as pests, or remove for food, most of the members of a population, the intensity of the struggle for existence among the remainder is reduced. Various compensating mechanisms will come into play, such as increased survival and reproductive rates, and reduced dispersal to unfavorable habitats or niches. The compensating mechanisms take time to operate, and there may be overcompensation or undercompensation, depending on the state of the population, the precise nature of our action against the population, and the state of the environment at any stage in the sequence.

We thus may not be able to evaluate various actions on the basis of their results one stage in the sequence after they were applied. It may take two or more stages before even the initial consequences of the compensatory mechanisms reveal themselves. Considerable computer experimentation and simulation may be required to define precisely the criterion function used for evaluating the relative success of different strategies. For example, what is the best measure of optimal strategy: that strategy which has the best result this year, that which has the best result this year followed by the best result next year, or that which has an optimal total result for the two consecutive years? All three of these criterion functions give different answers as to which strategy is best.

Much work in resource management to date has been woefully short-sighted in that strategies are typically evaluated on the basis of their impact as measured shortly after the strategy is applied. This method of evaluation completely ignores the fact that populations bounce back from catastrophes, and the most important element in any strategy evaluation is in determining how far and how fast this bouncing back occurs. It is particularly important that strategies be evaluated on this basis when we consider insecticides, rodenticides, fungicides, herbicides, and silvicides, whose effects do not persist after one application. The boomerang effect that shows up in the sprayed or dusted population one, two, and three generations after treatment is the key item that must be considered in comparing long-term efficiencies of various spray treatments. Dynamic programming is particularly important in evaluating management policies for biological systems, since it is conceived of as a method for evaluating policies over an n-stage sequence. Slight modification of the basic dynamic programming procedure allows us to build in a "look-ahead" feature that makes the method particularly suitable for comparing control strategies where boomerang effects 1 or 2 years after treatment are important.

We can illustrate such modified dynamic programming by using a demonstration study described earlier (Watt, 1963). In order to construct a recursive algorithm that was a reasonably realistic description of a problem in resource management, it was necessary to collect a long sequence of data describing the behavior of an actual population. One of the best such sequences available is the data on four German insect pests, reported by Schwerdtfeger (1932, 1935a,b, 1941). I selected the data on the moth *Bupalus piniarius* L., because this species showed the most wide-amplitude fluctuations, and hence was most suitable as a source of data for simulation studies on pest population dynamics and control. The number of pupae found per 100 square meters of forest soil in December in a thicket 35 kilometers north-northwest of Magdeburg was recorded every year from 1881 to 1940. Schwerdtfeger (1935, 1941) has published the pupal counts for the entire period, and the West German Government Weather Service kindly sent me photographic copies of the Magdeburg weather records for 1881 to 1940. After a great deal of graphing to select an appropriate model, regression analysis showed that temporal trends in *Bupalus* pupal counts could be described by the linear multiple regression equation

$$Y = b_0 + \sum_{i=1}^{4} b_i X_i$$

where $Y = \ln (N_{t+1}/N_t)$

N_t = number of pupae found per square meter of ground in year t

$$X_1 = (\ln N_t)^2$$
$$X_2 = (\ln N_{t-1})^2$$
X_3 = mean monthly Magdeburg October temperature in year $t + 1$
X_4 = mean monthly Magdeburg November temperature in year $t + 1$

It is interesting and very significant that X_2 was shown by analysis of variance to be the most important of the four variables in determining Y. This unusual finding is credible because Morris (1963) found in a large-scale population study on spruce budworm that large larval survival ratios in a given year were more closely correlated with larval densities 1 year previously than with larval densities at the beginning of the year. This discovery of a lag effect in natural biological systems, if it occurs universally, has the most profound significance for resource management. If an action taken in year t is more highly correlated with effects in year $t + 2$ than with effects in year $t + 1$, we need a 2-year "look-ahead" built into any program for comparative assessment of policy. That is, a policy might be optimal in terms of its effects 1 year hence, but suboptimal in terms of its effects in 2 years. If we only look 1 year ahead instead of 2, we could well arrive at an incorrect answer.

The reason for this time lag may be the nature of homeostasis: it can be interspecies as well as intraspecies, and can also occur in reciprocal relations between a species and its environment or food. Consider two entities, A and B. A at time t affects B at $t + 1$, which in turn affects A at $t + 2$. Hence, A at t will affect A at $t + 2$. The time lag occurs because of the time it takes for any such process to occur. For example, the density of a phytophagous animal in year t produces a numerical response in the density of parasites and predators of that species in $t + 1$, since the more food there is for the parasites and predators, the greater their generation survival will be. However, the more parasites and predators there are at $t + 1$, the less of the phytophagous species will survive to $t + 2$. Therefore animal population densities may constitute autocorrelated time series with lags.

Computer experiments

If a computer simulation study to explore the consequences of various pest-control strategies is to be completely realistic, it must handle the following factors:

1. The population level for the first and second year, P1 and P2. (All algebraic designations used here are compatible with Fortran specifications.) P1 and P2 are treated as variables in the program, so that we can determine the effect of starting with, say, very high or very low population levels.

2. Tables of weather records for critical months in the life of the insects.

In the present program TNOV (*I*) and TOCT (*I*) for $I = 1, \ldots, 60$ were read in, standing for mean of daily maximum and minimum temperatures in both November and October in each of 60 years at Magdeburg.

3. A set of parameters for the functions used to compute population levels in *I* from population levels in I − 1 and I − 2, and from other relevant factors affecting pest survival. These parameters are labeled B0, B1, B2, B3, B4, D0, D1, D2 in the present program. They measure the insects' physiological responses to weather and ecological responses to crowding. The parameters are all read in as variables, so we can see the consequences for strategies changing them. (Parameters B0, B1, etc. are regression coefficients obtained from analysis of Magdeburg *Bupalus* data.)

4. The variable THRES, standing for the threshold above which we attempt control. Clearly, if we are considering only use of direct control methods, it is only economically feasible and otherwise desirable to attempt control of a pest species when it is dense enough to be, in fact, a pest.

5. A set of parameters to describe the control policy we adopt at each I. These parameters are labeled Q(1), Q(2), Q(3), . . . , Q(6), and give the proportion of the population surviving control. Since the optimal strategy under certain circumstances will be no control at all, Q(1) = 1.00; Q(6) is set at some rather low level such as .03, meaning that 97 percent of the insects were annihilated by control.

6. R values, which are multipliers to correct survival ratios for the effect of genetic selection, since some types of pest control change the genetic constitution of the population by selectively eliminating the least vigorous members of the population. Where pest control has been very severe [that is, Q(6) = .03], selection pressure is intense, and the group surviving control may consist only of very vigorous individuals indeed. This high mean vigor will be reflected in a sharp upward rise in population survival in the postcontrol generation (Morris, 1963). Where Q is very low, survival may be 10 to 35 times higher than normal. Where Q(1) = 1.00, R(1) = 1.00; R(6) values vary from 10 to 35, and the intermediate R values are calculated as follows. Suppose we assume that vigor is normally distributed in a pest population, and when Q = 0.03, R = 10.0 because the most vigorous 3 percent of the population has an average survival coefficient 10 times that of the entire population. R(J) for any Q(J) can be calculated by using tables of ordinates and areas of the normal distribution as follows:

$$10\mu = \mu + \frac{\sigma}{0.03} \int_{t_{0.97}}^{\infty} z\psi(z)dz = \mu + \frac{\sigma}{0.03}\psi(t_{0.97}) \qquad (13.2)$$

where
$$\psi(z) = \frac{1}{\sqrt{2\pi}}e^{-z^2}$$

and
$$0.03 = \int_{t_{0.97}}^{\infty} \psi(z)dz \tag{13.3}$$

Thus, for $\mu = 1$,

$$\sigma = \frac{9(0.03)}{\psi(t_{0.97})} = \frac{0.27}{0.0681} = 3.96$$

and then

$$R_{0.40} = 1 + \frac{\sigma}{0.4}\psi(t_{0.6}) = 1 + 9.9(0.3863) = 4.27 \tag{13.4}$$

7. Another variable is the length of the look ahead. For each year of each case simulated, three entirely separate sets of computations were performed. Input and output for each of the three were kept separate throughout the entire 60 years.

The first set of calculations produced the defoliation that would result if, at every year, survival were reduced to Q(6) whenever the pest population exceeded THRES. (Fortran designation: CODAM, for "control damage.")

The second set of calculations produced the defoliation that would result if, at every year when the pest population exceeded THRES, dynamic programming was used to choose appropriate control strategy. (Fortran designation: DAMAJ.)

The third set of calculations produced the defoliation that would result if, at every year when the pest population exceeded THRES, look-ahead programming was used to choose appropriate control strategy. (DAM2, for "damage with 2-step look-ahead.")

The criterion functions, or return functions, or payoff functions used to comparatively evaluate the three strategies throughout the 60 years for each case, were merely

$$\sum_{I=1}^{60} \text{CODAM} \qquad \sum_{I=1}^{60} \text{DAMAJ} \qquad \text{and} \qquad \sum_{I=1}^{60} \text{DAM2}.$$

All three of these sums will be referred to as "cumulative defoliation over 60 years."

In order to explain the computer program in more detail, we must define the following Fortran symbols:

COP1, COP2, COP3   Control population sizes at times $I - 1$, I, $I = 1$, respectively. These are population levels for the population if treatment Q(6) is used whenever COP2 > THRES. At the end of each loop, COP3 becomes COP2 for the next loop (when I = I + 1).

COPQ   If COP2 > THRES, COP2 is multiplied by Q(6) to become COPQ. Otherwise, COPQ is set equal to COP2. At the end of each loop, COPQ becomes COP1 for the next loop.

P1, P2, P3(J)   Population sizes at times I − 1, I, I + 1, respectively, if dynamic programming is used whenever P2 > THRES. Six P3(J) values are computed, one for each Q(J). The optimal (smallest) P3(J) becomes P2 for the next loop. If P2 ⩽ THRES, P3(J) = P3(1).

PQ2(J)   Vector, corresponding to COPQ, for dynamic programming case.

P12, P22, P32(J), P4(J,K)   P12 and P22 are scalars, P32(J) is a vector, and P4(J,K) is a matrix representing population sizes at times I − 1, I, I + 1, for look-ahead programming case.

PQ22(J), PQ3(J,K)   Vectors and matrices corresponding to COPQ at times I and I + 1 for look-ahead programming case.

Of course, many other symbols are found in the Fortran listing, but it is not necessary to mention these when describing the logic of the program in a general way.

The computational procedure may be indicated by the highly condensed flow chart given in Figs. 13.1 and 13.2. In the flow diagram, abbreviations C and D are used. The following expressions explain the use of these symbols:

$$COP3 = COPQ \cdot EXP(C)$$

$$= COPQ \cdot EXP\{BO − B1[TOCT(I)] + B2[TNOV(I)] − B3(X)^2 − B4(Z^2)\}$$

where $X = LOG (COP1)$

and $Z = LOG (COPQ)$

$$P4(J,K) = PQ3(J,K)\{EXP[DO − D1(Z^2) − D2(Y^2)]\}$$

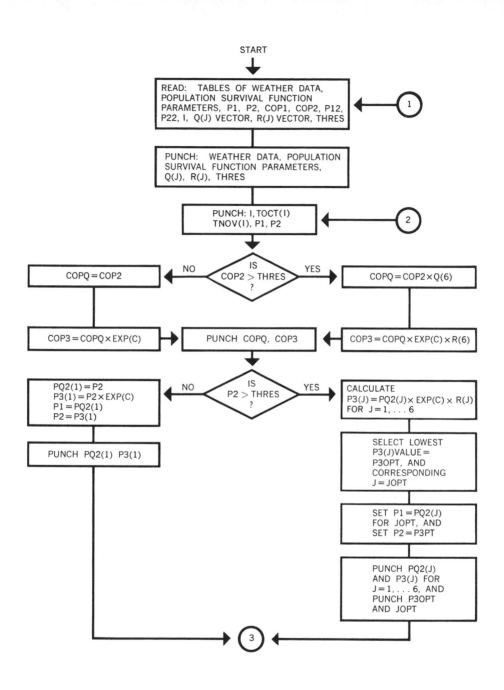

FIGURE 13.1
First half of flow diagram for computation (Watt, 1963).

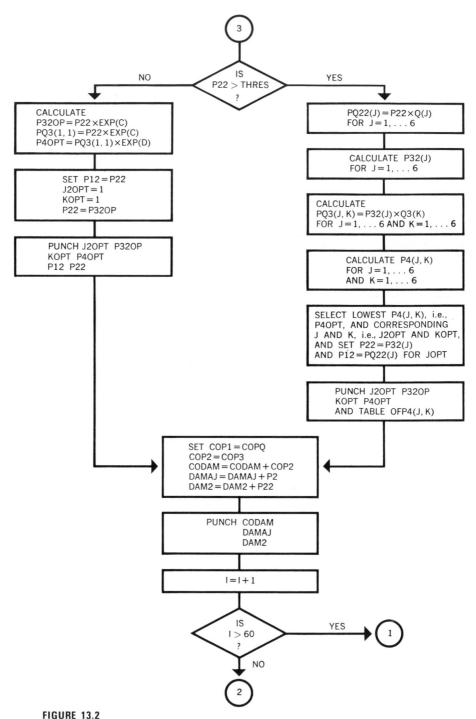

FIGURE 13.2
Second half of flow diagram for computation (Watt, 1963).

where $Y = \text{LOG PQ3}(J,K)$

and $Z = \text{LOG PQ22}(J)$

The coefficients for both the above equations were derived by using the Canada Forestry Department regression routine on the Magdeburg data.

Findings

Some of the most interesting findings from the program are gathered in Table 13.1. (Running time to obtain these results was 20 to 40 minutes for the 60 years of data, depending on the nature of the case, on an IBM 1620.)

Table 13.1 shows that any change in the parameters of a situation produces changes in the relative attractiveness of the three approaches to control policy; however, the control strategy in which we try to kill as many insects as possible whenever the pest density exceeds a threshold level is invariably the worst of the three. Why? The explanation is that when we find an insect existing at pest densities, we are likely dealing with a biological situation temporarily out of balance. Normal biological mechanisms will restore the balance (that is, lower the density of the pest to an equilibrium level). When man uses insecticides to artificially drive down the pest density, he merely elicits a variety of homeostatic mechanisms that force the pest to remain at pest densities most of the time. Clearly, there is powerful motivation for exploring alternative control procedures that might have the short-term power of insecticides without the long-term deleterious side effects.

Two-year look-ahead programming is rarely any better than, and usually not so good as, one-year look-ahead programming (the usual dynamic programming approach). The reason is that we can obtain lowest possible populations of a pest 2 years from now only at the cost of leaving quite dense populations next year. The cost more than compensates for the gain.

What these findings mean is that from one point of view, that of minimizing long-term total defoliation, no control at all may actually sometimes produce better control than some control. Even so, no control at all may not be that attractive because in some industries, the threshold level of pest density that can be tolerated is essentially zero. This will be the case, for example, when the economically important part of the plant is attacked directly by the pest, and the consumer demands that this part of the plant be perfect. The computer findings suggest that the price for this insistence on essentially perfect products may be very high indeed.

The relative merits of various control philosophies appear to depend on the threshold level of pest density at which we apply control. The threshold level in turn depends on the amount of damage plants, animals, and humans

TABLE 13.1
Results of computer experiments (Watt, 1963)

Input data (all populations started with P1 = 536.0 and P2 = 2,549.0)	Threshold	Maximum possible control	Cumulative defoliation	
			Dynamic programming	Look-ahead programming*
Actual Magdeburg Bupalus data				
B0 = 3.066, Q1 = 1.00, R1 = 1.00	10.0	5,723.1	2,549.4	3,505.9
B1 = 0.3952, Q2 = 0.40, R2 = 4.49	20.0	6,661.2	3,348.3	3,505.9
B2 = 0.3662, Q3 = 0.20, R3 = 6.31	30.0	7,170.6	3,518.3	3,505.9
B3 = 0.01537, Q4 = 0.10, R4 = 7.85	40.0	7,514.6	3,645.5	3,505.9
B4 = 0.06505, Q5 = 0.05, R5 = 9.13	50.0	7,586.0	3,638.4	3,505.9
D0 = 0.9683, Q6 = 0.03, R6 = 10.00	60.0	7,556.6	3,662.4	3,505.9
D1 = 0.0114	70.0	7,474.7	3,662.4	3,505.9
D2 = 0.0639	80.0	7,110.4	3,505.9	3,505.9
Greater degree of density-dependence than in Magdeburg Bupalus data				
B0 = 7.087, Q1 = 1.00, R1 = 1.00	100.0	354,007.5	20,682.6	120,882.6
B1 = 0.3952, Q2 = 0.40, R2 = 4.49	200.0	354,007.5	20,682.6	120,882.6
B2 = 0.3662, Q3 = 0.20, R3 = 6.31	300.0	354,007.5	20,682.6	120,882.6
B3 = 0.03074, Q4 = 0.10, R4 = 7.85	400.0	354,007.5	20,682.6	103,035.5
B4 = 0.13010, Q5 = 0.05, R5 = 9.13	500.0	340,013.7	20,682.6	103,035.5
D0 = 4.7333, Q6 = 0.03, R6 = 10.00	600.0	306,983.5	20,682.6	103,035.5
D1 = 0.0228	700.0	257,793.8	20,682.6	103,035.5
D2 = 0.1278	800.0	257,793.8	20,682.6	103,035.5

TABLE 13.1—continued

Input data (all populations started with $P1 = 536.0$ and $P2 = 2,549.0$)	Threshold	Cumulative defoliation		
		Maximum possible control	Dynamic programming	Look-ahead programming*
Greater sensitivity to weather than in Magdeburg Bupalus data				
B0 = 8.087, Q1 = 1.00, R1 = 1.00	100.0	2,634,116.0	292,225.4	292,225.4
B1 = 0.7904, Q2 = 0.40, R2 = 4.49	200.0	2,630,079.0	292,225.4	292,225.4
B2 = 0.7322, Q3 = 0.20, R3 = 6.31	400.0	2,611,855.6	292,225.4	292,225.4
B3 = 0.01537, Q4 = 0.10, R4 = 7.85	800.0	2,578,019.3	292,225.4	292,225.4
B4 = 0.06505, Q5 = 0.05, R5 = 9.13	1,600.0	2,273,680.4	292,225.4	292,225.4
D0 = 0.9683, Q6 = 0.03, R6 = 10.00	3,200.0	993,572.2	292,225.4	292,225.4
D1 = 0.0114	6,400.0	754,417.6	292,225.4	292,225.4
D2 = 0.0639	12,800.0	422,726.3	292,225.4	292,225.4
Lower equilibrium density than in Magdeburg Bupalus data				
B0 = 4.021, Q1 = 1.00, R1 = 1.00	10.0	23,055.5	2,561.4	2,519.4
B1 = 0.3952, Q2 = 0.40, R2 = 4.49	20.0	22,977.8	2,583.7	2,519.4
B2 = 0.3662, Q3 = 0.20, R3 = 6.31	30.0	22,977.8	2,519.4	2,519.4
B3 = 0.03074, Q4 = 0.10, R4 = 7.85	40.0	22,964.8	2,519.4	2,519.4
B4 = 0.13010, Q5 = 0.05, R5 = 9.13	50.0	22,474.7	2,519.4	2,519.4
D0 = 3.7650, Q6 = 0.03, R6 = 10.00	60.0	21,866.7	2,519.4	2,519.4
D1 = 0.0228	70.0	18,250.7	2,519.4	2,519.4
D2 = 0.1278	80.0	15,761.4	2,519.4	2,519.4

*Where all numbers in a column are identical, the strategies are identical regardless of THRES.

can tolerate, which in turn depends on previous damage they have suffered, and on economic factors. The whole question of how best to control an insect pest is very complex indeed, particularly when we consider all possible modes of control.

The fundamental explanation for our difficulties, of course, lies in homeostasis. Every plant and animal species extant has only persisted because it has evolved homeostatic mechanisms allowing it to adjust rapidly to catastrophes. All that insecticides constitute, to the pest species, is the most recent in a long line of catastrophes the pest is well preadapted to withstand.

A great deal of exploratory research is required on the application of dynamic programming to resource management strategy evaluation problems. In particular, we need research on the most appropriate way to define criterion functions (sum of losses this year and next, or lowest loss this year followed by lowest next year, for example). Interested readers can acquire more of the philosophical, mathematical, and computational details of dynamic programming in the books by Bellman (1957, 1961) and Bellman and Dreyfus (1962).

13.3 ADAPTIVE CONTROL PROCESSES

In dynamic programming, we discover how to manage a system using a systematic process of trial and error on the computer, by means of recurrence relations giving us the input for step t as a function of the input for step $t - 1$, and the decision variables (control policies) at $t - 1$. In other words, the management decision process is based on the fact that we have insight into the dynamics of the system we are attempting to manage, and this insight can be incorporated into difference equations. A logical extension of the concept of dynamic programming is to those situations where we do *not* have insight into the dynamic nature of the system, but gain this insight in the course of attempting to manage the system. Bellman (1961, chaps. 13 to 18) discusses the logic and philosophy of adaptive control processes, and outlines a mathematical approach to the description and solution of problems of this type.

In essence, he formulates such problems as consisting, at any stage, of a system state, and a set of information about the system, on the basis of which we decide to transform the system into a new state. After making the transformation, we decide how next to proceed on the basis of a criterion function. The complications that arise with this procedure, where we have an adaptive control process, stem from various types of uncertainty that may exist about the situation at any stage.

1. We may not actually know the state of the system. It may not always be possible to observe the state of the system, which results from actions on it at a prior time, because, for example, it may cost too much to make an adequate set of observations.

2. We may not be aware of the full range of decisions that can be made, or indeed, of the full range of types of decisions that can be made.

3. We may not understand the nature of the dynamic processes that transform the system state from one stage to another.

4. The ultimate purpose of control may not be precisely understood at the beginning of the control operation, but may only become completely explicit and detailed in the course of new information obtained in the course of the control operation.

If we apply these notions of Bellman's to the design of the typical resource management research program, it becomes clear that they imply a rather new way of conducting such research. In the traditional approach, we would continue to collect data on a forest, for example, until we had enough to construct a computer simulation program that could be used to determine how best to manage the forest. However, this completely misses the point that the computer is of great importance as a heuristic device. This suggests a new approach: As soon as we have barely minimum data that can be used to construct even the simplest computer program, we conduct simulation studies, no matter how naive. The output from the computer program, which we use to determine how sensitive the criterion function is to variations of the various independent variables, becomes a check on our research program design. If the computer program shows that the criterion function is most sensitive to perturbations in variable x_7, then obviously this variable must be measured with a high degree of precision and accuracy if the model is to account for the highest possible proportion of the variance in the system. In summary, we can consider the whole experimental program as an adaptive control process involving computer experimentation and actual experimentation in which the various steps in the program occur in the following sequence:

1. The state of an experimental system at time t is specified by a vector $\mathbf{x}_t$. We know that this state is the end result of a prior sequence of states, $\mathbf{x}_0, \mathbf{x}_1, \mathbf{x}_2, \ldots, \mathbf{x}_{t-2}, \mathbf{x}_{t-1}$, and a prior sequence of vectors defining experimental treatments, $\mathbf{y}_0, \mathbf{y}_1, \ldots, \mathbf{y}_{t-2}, \mathbf{y}_{t-1}$. On the basis of an analysis of all this information, we construct a first-approximation mathematical model in the form of a set of difference equations

$$\mathbf{x}_t = f(\mathbf{x}_{t-1}, \mathbf{y}_{t-1})$$

The model is used to create a computer program to simulate the responses of the system to various control policies (or experimental procedures).

2. We use the computer output as a guide to indicate the particular variables to perturbations in which the system is most sensitive, and predict the form of the response to these perturbations. Then we perform new experiments to determine if the experimental responses are as predicted by the computer model.

3. We use the criterion function $\Sigma(Y_{obs} - Y_{cal})^2$ to determine the goodness of fit of the experimental output to that predicted by the computer, and improve the computer model as necessary to make it produce results more like those in the real situation. Then we return to step 2, after obtaining new output.

Thus we have a cyclical sequence of operations, in which each loop consists of performing experiments to test the computer model, refining the computer model, and obtaining new output to suggest further experimental procedures. This cyclical mode of operation exploits the feedback principle to enormously speed up the rate at which we can obtain insight into a complicated system. An immediate implication is that if we wish to hitch the rate of progress of a research program to the tempo of the fastest possible element that can be inserted into such a feedback loop (the computer), it behooves us to begin computer simulation studies as early as possible in the life of a research program.

The important question raised by this whole procedure is, "How does one deduce, in detail, how a model should be improved through analysis of the discrepancy between observed sets of data points and the supposedly corresponding sequences of data points produced by the computer?" Unfortunately there is no completely satisfactory answer; we are forced to fall back on the techniques outlined in Sec. 11.2 for developing models to describe data; only in this case, we need difference equations, not differential equations.

13.4 SYSTEMATIC SEARCH ON COMPUTERS

The object of computer simulation studies on management strategies is to explore the way in which values of the function $Y = f(X_1, X_2, \ldots, X_n)$ change in response to changing values of the independent variables $X_1, X_2, \ldots, X_n$. For large values of n, there are an enormous number of combinations of values of the n independent variables for which Y must be calculated. Rather than computing millions or billions of Y values in order to find that set of X_i values that produce the largest (or smallest) Y, it is desirable to have some systematic procedure that will lead us quickly to the optimal combination of values of the X_i's. Such procedures are discussed in reviews by Wilde (1964) and Spang (1962).

In essence, the problem we encounter in all such methods may be visualized as follows. The function relating Y to the vector $\mathbf{X}$ is an $(n + 1)$-dimensional surface in an $(n + 1)$-dimensional space, where n is the number of independent variables. Our job is that of a blindfolded man climbing in the Himalayas, who wishes to find the highest peak, but knows only the altitude he is at, at any point, and the angle at which his feet point up or down. Clearly, this procedure is vulnerable to a number of mishaps, such as becoming lost in local maxima or minima, which are mistaken for the highest (or lowest) points in the entire space, and wasting time following gradual slopes in troughs, or on ridges, etc. Many methods have been proposed for dealing with these difficulties; rather than reviewing several of them (which are outlined by Wilde and Spang in any case), we will confine our attention to one of the best of the methods, which is illustrative of the whole class of techniques for systematically searching for optima.

Before explaining the method, we shall consider the computational procedure that must be followed in all search methods. Suppose we wish to find the combination of variate values for operating procedures that maximizes the productivity of a natural resource, Y. We must first calculate Y for an arbitrarily chosen array of $\mathbf{X}$, which we believe produced a Y reasonably close to the maximum Y. Then, using some systematic procedure for altering the array of $\mathbf{X}$, we compute a new set of Y values, one for each array of $\mathbf{X}$. If any of the new arrays of $\mathbf{X}$ produces a greater value of Y than the starting array, then the new array becomes the starting point for the next step in the search.

There are two basic questions that must be answered in developing any such search technique. First, at each iteration, how do we determine the direction in which we proceed to obtain the next set of $\mathbf{X}$ values? Second, what will the step size be? (Presumably the new $\mathbf{X}$ values will be obtained by adding to, or subtracting from, our original $\mathbf{X}$ values, a vector $\mathbf{h}$ of step sizes of some predetermined value, or they will be computed according to some predetermined scheme.)

The yardstick by which we evaluate various search techniques is the number of steps taken to find a value of Y that improves at less than a predetermined rate per iteration, as a function of the number of $\mathbf{X}$ variables in the problem. For classical optimization techniques, where there are n independent variables, the number of steps taken to find the optimum Y is proportional to n^3. This is because of the way in which the number of possible search directions increases outward from a point of origin of search, as the

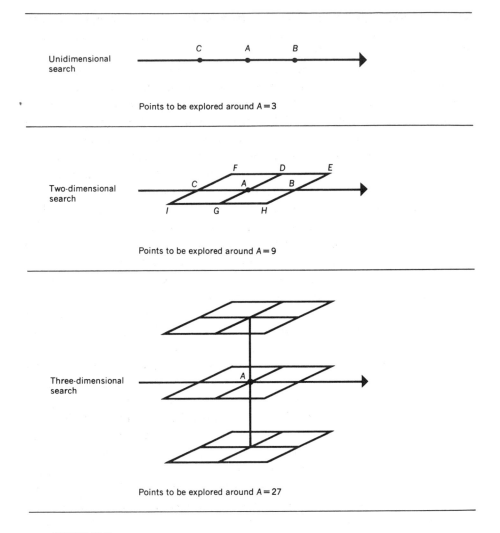

Points to be explored around A = 3

Points to be explored around A = 9

Points to be explored around A = 27

FIGURE 13.3

Increase in computing volume created by a search as the dimensionality of the surface being searched increases.

number of dimensions (that is, the independent variables) increases. This can be understood in terms of Fig. 13.3. Where there is unidimensional search (that is, there is only one independent variable), we progress along the X axis, calculating Y values for given X values in order to find the optimal (maximum or minimum) Y. After we move to A in the direction C to B, if A is the center of a search area, we can only calculate Y for C, A, and B, assuming that Y's are calculated for the center point of the search, and for h units out from the center. Where there are two independent variables, a complete exploration around A now involves, in addition to calculation of Y at C, A, and B, calculation of Y at D, E, F, G, H, and I. For three-dimensional search, we have 27 points at which Y should be calculated. We may form the rule for the relation of number of points to be searched as a function of dimensionality, as follows.

Where there is only one dimension, we calculate Y at the center point, and a point one step-size in front of it and one step-size behind. Increasing the dimensionality by one, we now have 2 points either side of each of the original 3 at which Y must be calculated. The number of points to be searched is now $3 + 2(3) = 9$. Where there are 3 independent variables, we must compute Y at one point above and one below each of our previous 9 points, or a total of $9 + 2(9) = 27$. We are evidently dealing with a series of the following type:

Number of dimensions	Number of points at which Y must be calculated around each search area
1	$3 = 3^1$
2	$3 + 2(3) = 9 = 3^2$
3	$9 + 2(9) = 27 = 3^3$
4	$27 + 2(27) = 81 = 3^4$
5	$81 + 2(81) = 243 = 3^5$
n	3^n

Actually, the number of searches required does not increase quite this rapidly, for two reasons. First, we can make use of tricks in experimental design to cut down the required number of searches. Second, points to be searched around a particular search center may already have been searched around the previous center, and the computer program should be testing to ensure that this duplication of effort is not occurring. Wilde notes that, in classical minimization techniques, the volume of computation grows as n^3, not as 3^n. The reduction of computation is illustrated by the following figures:

Number of dimensions	Number of points at which Y must be calculated around each search area	
	Theoretical	Actual
1	$3^1 = 3$	$1 = 1^3$
2	$3^2 = 9$	$8 = 2^3$
3	$3^3 = 27$	$27 = 3^3$
4	$3^4 = 81$	$64 = 4^3$

5	$3^5 =$	243	$125 = 5^3$
6	$3^6 =$	729	$216 = 6^3$
7	$3^7 =$	2,187	$343 = 7^3$
8	$3^8 =$	6,561	$512 = 8^3$
n	3^n		n^3

In the search procedure of Hooke and Jeeves (1961), the volume of computation required increases not as the cube of the dimensionality, but as the first power of the dimensionality. This extraordinary reduction in computing volume has been achieved by exploiting the following simple, but extremely ingenious, notion. Hypersurfaces describing $Y = f(\mathbf{x})$ geometrically, typically have a ridgelike character. That is, within each of the pairs of dimensions, there will be found long troughs sloping downward toward nadirs, or long ridges sloping upward toward zeniths. If we are hunting for a minimum, there is no point in hunting all over the sides of a valley; the optimum strategy for finding the minimum point in a valley is to stick to the bottom of the valley and travel along it as rapidly as possible toward the nadir. Similarly, if we are hunting for the highest point in a mountain range, we waste time by exploring downward from the crest of a ridge; the optimum strategy is to stick to the crest and follow it along in the direction of most rapid rise until we reach the zenith. In effect, by following troughs of valleys and crests of ridges, we convert search in n space to search along a line, which has one dimension—and achieve enormous reduction in computing volume. As in the case of dynamic programming, where we achieve an enormous reduction in computing volume by analyzing the fundamental nature of problems and devising solution techniques tailored to the nature of the problems, here also reduction in computing volume is attained by dealing with the problem on its own terms.

Hooke and Jeeves' method proceeds as follows. Search begins at a base point $\mathbf{b}_1$ (defined by a vector of values for the n independent variables). For each of the independent variables $x_i (i = 1, 2, \ldots, n)$, we choose a step size, δ_i. Now we define $\boldsymbol{\delta}_i$ as the vector whose ith element is δ_i, all the rest being zero. Y is computed at $\mathbf{b}_1 + \boldsymbol{\delta}_1$. If the latter Y is more optimal than the former, we establish a new search center $\mathbf{t}_{11}$. The subscript 11 indicates that we are developing the first pattern of search, and that we have already altered the variate value of the first independent variable. If, on the other hand, Y at $\mathbf{b} + \boldsymbol{\delta}_1$ is not so good as Y at $\mathbf{b}_1$, we try calculating Y at $\mathbf{b}_1 - \boldsymbol{\delta}_1$. If this Y is better than the original Y, $\mathbf{b}_1 - \boldsymbol{\delta}_1$ becomes the temporary head $\mathbf{t}_{11}$. If we introduce a new type of notation, we can express the preceding routine in symbolic form. The new notation we need is the max symbol. $\text{Max}(X_1, \ldots, X_n)$, represents "the maximum value of the series $X_1, X_2, \ldots, X_n$." Thus, max (17,

3, 25, 39, 4, 22) = 39. Hence we can express the preceding routine in the following concise form, for the case where we are seeking the maximum Y, rather than the minimum Y. The analogous symbolism for the minimization of Y would use min notation.

$$
t_{11} = \begin{cases}
\mathbf{b}_1 + \boldsymbol{\delta}_1 & \text{if } y(\mathbf{b}_1 + \boldsymbol{\delta}_1) > y(\mathbf{b}_1) \\
\mathbf{b}_1 - \boldsymbol{\delta}_1 & \text{if } y(\mathbf{b}_1 - \boldsymbol{\delta}_1) > \max[y(\mathbf{b}_1), y(\mathbf{b}_1 + \boldsymbol{\delta}_1)] \\
\mathbf{b}_1 & \text{if } y(\mathbf{b}_1) > \max[y(\mathbf{b}_1 + \boldsymbol{\delta}_1), y(\mathbf{b}_1 - \boldsymbol{\delta}_1)]
\end{cases}
$$

After we establish the temporary search center, or head, $\mathbf{t}_{11}$, we do the same thing for x_2, the second independent variable, then the third, and so on, until we have established the head $\mathbf{t}_{12}$, where there are n independent variables. This temporary head is now referred to as the second base point, $\mathbf{b}_2$.

The base points $\mathbf{b}_1$ and $\mathbf{b}_2$ now give us the first pattern. We reason that the most probable outcome of a search around $\mathbf{b}_2$ is to discover the same best direction of move as around $\mathbf{b}_1$. Therefore we move in the direction of $\mathbf{b}_2$, but double the distance $\mathbf{b}_2 - \mathbf{b}_1$, to establish the temporary head $\mathbf{t}_{20} = \mathbf{b}_1 + 2(\mathbf{b}_2 - \mathbf{b}_1)$. The subscript 20 means that we are building a second pattern but have not yet altered any of the independent variables. We now repeat the operation conducted around the first base point, altering the n independent variables one at a time. This procedure leads to the establishment of the temporary head $\mathbf{t}_{30}$. If we are having repeated success in the direction along the line $\mathbf{b}_1 - \mathbf{b}_2 - \mathbf{b}_3$, the pattern grows, and

$$\mathbf{b}_3 - \mathbf{b}_2 = 2(\mathbf{t}_{20} - \mathbf{b}_2) = 2(\mathbf{b}_2 - \mathbf{b}_1)$$

If we now find that altering x_2 does not produce any improvement over temporary head $\mathbf{t}_{31}$, but that $\mathbf{t}_{31}$ is better than $\mathbf{b}_3$, then

$$\mathbf{b}_4 = \mathbf{t}_{32} = \mathbf{t}_{31}$$

and the pattern now heads left, still growing in length. If we discover in the fourth pattern that none of the one-at-a-time alterations in the independent variables produces any improvement, but $y(\mathbf{t}_{40}) > y(\mathbf{b}_4)$, then

$$\mathbf{b}_5 = \mathbf{t}_{42} = \mathbf{t}_{41} = \mathbf{t}_{40}$$

and the pattern heads in the same direction as before, but the step size between the base points does not grow. It may develop that none of the explorations about a base point produces any improvement over the last base point. Then we know we are at the peak of the ridge, or crossing over a ridge in another direction, and new tactics are called for. Local exploration in all directions is conducted around the last base point to discover a direction in which we can discover improvement in Y. If this fails, we repeat the operation, but with the step size now cut in half. As soon as a suitable direction is discovered, we begin rapid motion again, doubling the step size between successive pairs of base points. Search is concluded when the improvement in Y falls below a certain rate per movement from one base point to another.

13.5 OPTIMIZING THE ALLOCATION OF EFFORT IN RESEARCH PROGRAMS ON MANAGEMENT OF NATURAL RESOURCES

In the same manner we use operations research methods to increase the efficiency of utilization or management of a natural resource, we can use similar methods to increase efficiency in the research programs designed to obtain data for statistical and simulation studies on the resource.

The underlying concept in this section is of keeping the errors of measurement for different variables in a research program in balance with each other. That is, there is no point in spending a great deal of money on equipment and replication to measure one variable so that the standard error of the mean is 1 percent of the mean, when a slight increase in expenditure on the measurement of another variable included in the systems model would drop the standard error/mean ratio from 65 percent to 10 percent.

The method we shall describe is based on the assumption that we have already collected a great deal of data on a system for, say, 3 years. Now we wish to do trial model-building and statistical analysis to determine if a change in the allocation of sampling or measuring expenditures can produce an increase in the proportion of accounted-for variance in the dependent variable in the systems model.

In essence, our procedure is to determine the relative sensitivity of the model to variability in the different independent variables included in the systems model. We then shift the allocation of sampling and measuring effort in the data-collection program so that the variables that receive an increased share of the overall effort in future modifications of the field program are those variables in which a unit change in value produces the greatest change in the dependent variable.

What are the possible sources of error in a systems model? There are four reasons why the proportion of the variance in the still unaccounted for dependent variable may be too large (Watt, in Morris, 1963).

1. The sampling procedures for some of the input data may produce too high a standard error, due to inadequate or faulty stratification or inadequate replication.

2. There may be one or more major structural defects in the model. For example, suppose we assume that S_L, the survival of large larvae, is expressed as

$$S_L = S_1 S_2 S_3 S_4 S_5 \qquad (13.5)$$

where S_1 = 1.00 (proportion of large larvae killed by parasites)
S_2 = 1.00 (proportion of large larvae killed by other density-dependent factors, such as predators and starvation)
S_3 = 1.00 (proportion of large larvae killed by phenological factors)
S_4 = 1.00 (proportion of large larvae killed by bad weather)
S_5 = 1.00 (proportion of large larvae killed by dispersal due to high degree of stand isolation)

The assumption that the survival of a cohort of individuals of the same age may be expressed as the product of survivorship from a set of mortality factors may be incorrect, and needs to be tested [see Morris (1963) for details, especially pp. 42–63, and 99–115]. We note that (13.5) implies that

$$\ln S_L = \ln b_0 + \sum_{i=1}^{5} b_i \ln S_i \qquad (13.6)$$

where $\qquad b_0 = b_1 = b_2 = b_3 = b_4 = b_5 = 1.00$

Thus we can test the validity of the assumption inherent in (13.5) by subjecting it to a regression analysis as in (13.6), and determine if any of the regression coefficients b_i, $i = 0, \ldots , 5$ departs significantly from 1.00.

There are numerous other tricks for testing the structural validity of models, submodels, and sub-submodels; some of these are considered at the end of Sec. 11.2.

3. Some factors may not have been measured in an appropriate fashion, and some important factors may not have been measured at all. For example, wind may have been measured as miles of wind passing a stationary point per hour, instead of maximum gust velocity within an hour, which would often be much more biologically relevant.

4. Nature is stochastic. That is, a particular vector of variate values for the independent variables in a situation does not produce a particular variate value for the dependent variable, but rather produces a frequency distribution of variate values for the dependent variable.

The first step in determining how to improve a research program is to find out how much of the indeterminancy in the model is due to cause 1 of the four types of causes just mentioned. If cause 1 accounts for the bulk of the indeterminancy, then our efforts to improve the research program clearly should be directed to improving the sampling operation. Otherwise, our trouble originates elsewhere (there is an important variable we are not measuring, or some factor is being measured incorrectly, etc.).

Sophisticated methods of relating variance in a dependent variable to sampling errors for the independent variables are available, in principle, but in practice it turns out that significant biases originate in the approximate methods used to develop the formulas. Therefore a simple ad hoc method is proposed here, which exploits the speed and flexibility of computer simulation.

Suppose we have developed a model of general form

$$Y_t = f(Y_{t-1}, X_1, X_2, \ldots, X_n)$$

Variation in the different independent variables will produce different percentages of variation in the dependent variable for two reasons: some of the independent variables will be more stable than others and the independent variables enter the model differently. The simplest equation with which one can illustrate the second point is

$$Y = aX_1^{X_2}$$

A 10 percent change in X_1 will produce a 10 percent change in Y, but a 10 percent change in X_2 will produce an array of different percent changes in Y depending on the values taken by X_2.

To determine how to allocate sampling error as efficiently as possible to independent variables, assuming we already have a considerable body of measurements on all of them and a trial model with estimates for parameter values, proceed as follows:

1. Compute the variance for each of the measured independent variables.

2. Use computer simulation (or paper and pencil, if the volume of computation is small), to determine the range in Y as a percentage of the mean value for Y produced by varying each of the independent variables in turn from a value equal to the mean less half the variance for that variable, to the mean plus half the variance.

3. Make the sampling effort directed against each independent variable proportional to the percent change in Y produced by that variable (from step 2).

Thus sampling effort is allocated to each factor on the basis of the contribution that factor makes to variation in Y. A high volume of sampling effort will be directed to those factors that vary a lot, and/or enter into the model in such a way and with such parameter values that Y is sensitive to variation in them. A low volume of sampling effort will be expended on independent variables that are relatively stable, and/or enter into the model in such a way and with such parameter values that Y is relatively insensitive to variation in them.

REFERENCES

Bellman, R.: "Dynamic Programming," Princeton University Press, Princeton, N.J., 1957.

————: "Adaptive Control Processes: A Guided Tour," Princeton University Press, Princeton, N.J., 1961.

———— and S. E. Dreyfus: "Applied Dynamic Programming," Princeton University Press, Princeton, N.J., 1962.

Fan, Liang-Tseng: "The Continuous Maximum Principle," John Wiley & Sons, Inc., New York, 1966.

Hald, A.: "Statistical Theory with Engineering Applications," John Wiley & Sons, Inc., New York, 1952.

Hooke, R., and T. A. Jeeves: "Direct Search" Solution of Numerical and Statistical Problems, *J. Assoc. Computing Machinery*, **8**:212–229 (1961).

Morris, R. F. (ed.): The Dynamics of Epidemic Spruce-Budworm Populations, *Mem. Entomol. Soc. Can.*, **31**:1–332 (1963).

Schwerdtfeger, F.: Betrachtungen zur Epidemiologie des Kiefernspanners, *Z. Angew. Entom.*, **19**:104–129 (1932).

————: Studien über den Massenwechsel einiger Forstachädlinge. I. Das Klima der Schadgebiete von *Bupalus piniarius* L., *Panolis flammea* Schiff., *Dendrolimus pini* L., in Deutschland, *Z. Forst-und Jagdwesen*, **67**:15–38, 84–104 (1935a).

————: Studien über den Massenwechsel einiger Forstachädlinge, II. Uber die populationsdichte von *Bupalus piniarius* L., *Panolis flammea* Schiff., *Dendrolimus pini* L., *Sphinx pinastri* L. und ihren zeitlicher Wechsel, *Z. Forst-und Jagdwesen*, **67**:449–482, 513–540 (1935b).

————: Uber die Ursachen des Massenwechsels der Insekten, *Z. Angew. Entom.*, **28**:254–303 (1941).

Spang, H. A.: A Review of Minimization Techniques for Nonlinear Functions, *Soc. Ind. Appl. Math. Rev.*, **4**:343–365 (1962).

Watt, K. E. F.: Dynamic Programming, "Look-Ahead Programming," and the Strategy of Insect Pest Control, *Can. Entomol.*, **95**:525–536 (1963).

Wilde, D. J.: "Optimum Seeking Methods," Prentice-Hall, Inc., Englewood Cliffs, N.J., 1964.

Wouk, A.: Recursive Methods, Dynamic Programming, and Invariant Imbedding, *SIAM Rev.*, **4**:384–393 (1962).

ADDITIONAL READING

A principal theme of this book has been that management of natural resources is a subject that should be viewed from a comparative standpoint, with principles, techniques, and theories drawn from a wide variety of types of resources, and a number of tangential fields. This approach commits the student to perusal of a large body of scientific literature, scattered in an enormous variety of types of publications. Therefore certain guidelines for this vast body of knowledge seem worthwhile and in order.

Anyone concerned with natural resource management on a national or international scale will find himself very dependent on reliable and comprehensive statistics for a wide variety of industries, resources, and human populations. One of the most important sources of such information is the publications of the Food and Agriculture Organization of the United Nations in Rome. The following series are particularly useful:

Yearbook of Fishing Statistics (includes data on whales and whaling).
Food Supply, Time Series.
The State of Food and Agriculture.
Yearbook of Forest Products Statistics.
Commodity Reports.
Food Balance Sheets.
Freedom from Hunger Campaign, Basic Studies. (Some volumes are published by the World Meteorological Organization in Geneva, or the World Health Organization in Geneva, instead of FAO. These bulletins are very useful for obtaining a comprehensive view of the situation in natural resources.)

The following two annual publications of the United Nations Statistical Office are also very useful: *Statistical Yearbook*, and *Demographic Yearbook*.

Another important point for students to note is that reports of research on natural resource systems are too long to lend themselves to the 20-printed-pages-or-less publication format of scientific journals. Therefore the most important publications of this type appear as books, as bulletins or monographs published by government agencies, or as monographs published irregularly by scientific societies that also publish a regularly issued journal. The following are particularly noteworthy:

Bulletins of the Fisheries Research Board of Canada.
Bulletins of the Inter-American Tropical Tuna Commission, La Jolla, California.
Canadian Wildlife Series, Dept. of Northern Affairs and National Resources, National Parks Branch, Canadian Wildlife Service, Ottawa.
Fisheries, U.S. Bureau of Commercial Fisheries, U.S. Dept. of the Interior Special Scientific Reports.
Fisheries Investigations (London), ser. II, Her Majesty's Stationery Office.
Fishery Bulletins, U.S. Fish and Wildlife Service, Dept. of the Interior.
Memoirs of the Entomological Society of Canada.
University of Washington Publications in Fisheries, new ser., Seattle, Wash.
Wildlife Monographs, published by the Wildlife Society.

Some of the most important regularly published journals which contain papers with a point of view similar to this book are as follows:

The American Naturalist.
The Canadian Entomologist.
Ecological Monographs.
Ecology.
The Journal of Animal Ecology.
Journal du Conseil Permanent International Pour L'Exploration de la Mer.
The Journal of the Fisheries Research Board of Canada.
The Journal of Forestry.
The Journal of Theoretical Biology.
Marine Research Series (Scottish Home Department).
Researches on Population Ecology (Japanese Society of Population Ecology).

Another problem is that techniques useful in resource management are often published in journals or books that almost no biologists customarily read. For example, in addition to the few statistical journals such as *Bio-*

metrika and *Biometrics*, which many biologists peruse, very useful mathematical techniques and concepts are presented in:

> *The Journal of the Society of Industrial and Applied Mathematics.*
> *The Society of Industrial and Applied Mathematics Review.*
> *Technometrics.*

General Systems, the yearbook of the Society for General Systems Research, contains numerous papers with very stimulating ideas about systems research.

Finally, the resource management biologist interested in a systems approach should be alert to the fact that people in quite different fields, such as aerospace, chemical engineering, and operations research, are also concerned with systems, and hence are producing books useful to the resource manager. The following are examples:

Bellman, Richard: "Adaptive Control Processes: A Guided Tour," Princeton University Press, Princeton, N.J., 1961.

―――― and S. E. Dreyfus: "Applied Dynamic Programming," Princeton University Press, Princeton, N.J., 1962.

Fan, Liang-Tseng: "The Continuous Maximum Principle," John Wiley & Sons, Inc., New York, 1966.

Roberts, S. M.: "Dynamic Programming in Chemical Engineering and Process Control," Academic Press Inc., New York, 1964.

Wilde, D. J.: "Optimum Seeking Methods," Prentice-Hall, Inc., Englewood Cliffs, N.J., 1964.

A compendium of computer techniques that will be particularly useful to readers of this book is

Ledley, R. S.: "Use of Computers in Biology and Medicine," McGraw-Hill Book Company, New York, 1965.

EPILOGUE

It would be tragic if readers of this book conclude that the problem of expanding human populations versus finite resources can be solved merely by increasing the efficiency of utilization of the resources. Ultimately, such a one-sided approach would reduce men everywhere to the role of pitiful scavengers, constantly combing the litter of a ravaged biosphere in search of scraps overlooked in prior searches by vast hordes of fellow scavengers. Increased efficiency of resource management, unaccompanied by internationally practiced birth control, can only lead our species rapidly down a one-way street to oblivion. Unless a massive worldwide program of birth control is begun *now*, no amount of efficiency in resource management will suffice for the needs of humanity.

AUTHOR INDEX

Radovich, J., 35, 36, 83, 86
Rand Corporation, 351
Rich, E. R., 298, 301, 304
Richards, O. W., 298
Richdale, L. E., 24
Richman, S., 55, 89, 92
Ricker, W. E., 35, 36, 253
Riffenburgh, R. H., 219
Robertson, F. W., 293, 295, 296, 298
Roe, F. G., 120–122, 359
Rosen, M. N., 164
Royce, W. F., 254, 255, 361–364
Runnstrom, S., 177

Salt, R. W., 284
Salvadori, M. G., 247
Sang, J. H., 293, 295, 296, 298
Sargent, F., II, 164, 332
Satomura, H., 275
Schaefer, M. B., 87, 89, 254
Schnabel, Z. E., 204–206
Schuck, H. A., 254, 255
Schwerdtfeger, F., 61, 414
Selleck, D. M., 208
Seton, E. T., 70, 120
Sette, O. E., 30, 84
Shannon, C., 341
Silliman, R. P., 55, 89–91
Siniff, D. B., 200
Skellam, J. G., 326, 348
Skoog, R. O., 200
Slobodkin, L. B., 35, 55, 89–92
Smirnov, E., 294, 303, 304
Smith, Audrey U., 284
Smith, B. R., 111
Smith, E. V., 115, 116
Smith, F. E., 106
Smith, H. S., 36
Smith, R. F., 275
Smith, R. W., 366, 371
Solomon, M. E., 42, 136
Somme, S., 178
Spang, H. A., III, 249, 427
Stallybrass, C. O., 157, 160, 161
Stanley, J., 299
Steven, G. A., 29
Stevenson, J. C., 30
Swingle, H. S., 115–117

Taber, R. D., 129
Takahashi, F., 35, 66
Talbot, L. M., 71, 72

Thompson, W. R., 36
Tibbo, S. W., 28
Trimmer, C. D., 135
Tromp, S. W., 157, 164, 165
Tuck, L. M., 27
Turnbull, A. L., 43, 59
Turner, K. B., 150
Turner, M. E., 268

Uda, M., 85
Ullyett, G. C., 293, 303–305, 309
United Nations Statistical Office, 70
Utida, S., 35, 293, 294, 297, 298, 300, 302,
    303, 308
Uvarov, B. P., 68, 149

Van Etten, R. C., 210
Varga, R. S., 248
Varley, G. C., 36
Vincent, A. B., 379
Volterra, V., 295, 315, 348
Von Bertalanffy, L., 257

Wagner, F. H., 119
Watt, K. E. F., 25, 29, 30, 34–36, 44, 49, 55,
    61, 69, 89, 90, 92, 113, 114, 137, 154,
    156, 213, 216–220, 223, 224, 259, 260,
    263, 265, 270, 285, 293, 312, 313, 315,
    317, 325, 345, 349, 363, 365, 367, 380,
    387, 388, 414, 419, 420, 422, 423
Weaver, W., 341
Wellington, W. G., 150–154, 285–287, 365
White, K., 277, 279
Widrig, T. M., 211
Wigglesworth, V. B., 274, 276
Wigley, R. L., 102, 103, 107–109
Wilde, D. J., 249, 250, 427
Wilson, D. A., 381
Wiolovitsh, N., 294, 303, 304
Wood, H., 79
Wood, R., 178
Wooster, H. A., 16
Wouk, A., 413
Wright, B. S., 134–136
Wright, S., 326, 371

Yearbook of Fishing Statistics, 78, 79, 97–99
Young, H., 220, 221
Young, S. P., 143
Young, W. E., 141

Zimmerman, J. H., 103
Zinsser, H., 157
Zwolfer, H., 43, 44

SUBJECT INDEX

Epidemic, migration factor, 163
  terminating, 337
  wave, 158
Epidemiology, 157
Epizootic wave, 162
Equal sign, 185
Equation, general, underlying, 268
  normal, 238
  solving, 247
European spruce sawfly, 61
Eutrophic, 117
Experimental components approach, 271
Exploitation, maximum possible rate, 88
  rates of, 55, 213
  ruinously hard, 90
Exponential growth law, 192
Extremum, methods, 407
  problem, 3, 249
Eyring model, 281

F tests, interpretation of, 195
Factorial design, 295
Fast-access memory, 339
Feedback-governing mechanism, 80
Feeding frenzy, 141
Fire, 128, 399
Fish, availability of, 224
  populations of, North Sea,
    catch per unit effort, 4
    stock, change in composition of, 4
Fisheries, Japanese, 77
  North Sea, 77
  Peru, 78
  sardine and anchovy, 82
Fishing, cushion, 86
  equilibrium catch, 87
  optimal strategy, 86
  sport, smallmouth bass, 29
Fitted regression line, variation about, 231
Food, cycles, 36
  human, animals' preference for, 142
  populations, 28
  pyramid, 310
  web and information theory, 42
Forest, cutting policy, 403
  economics, 389
Fortran, 338, 345
  programs, 349
  statement, order of occurrence, 360
Function, criterion, 410
  response, 42
Fujita model, 298

Games, 203
Gear, characteristics of, 214
Genetic, heterozygosity, speciation,
    relationship with, 49
  selection, 416
Gradient methods, 249
Great Lakes Fisheries Commission, 103
Grid, 328
Growth rate, individual, animal, 22
Gypsy moth control program, 62

Habitat, African brittle soil characteristics, 73
Hatchery-raised animals, release of, 112
Heat experience, accumulated, 282
Homeostasis, 366, 424
Homoscedasticity, 232
Host-parasite, 35
  system, mathematical model, 137
Human, 8
  birth rate, 12
  death rate, 12
  diets, 15
  environmental resistance, 9
  growth equation, 8
Humidity, effect of, 274
Hunger level, 320
Hunter densities, 174
Hunting, pressure, 173
Hydroelectric system, 396
Hypersurfaces, 430

IF statements, 186
Immunity, 159
Impounded water, 177
Incident radiation, efficiency of utilization
    of, 37
Indian Plague Research Commission, 157
Inefficiency, 190
Inequalities, 405
Infection, 160, 333
Information content, 340
  measurement for, 49
Insect-pest control strategies, 372
Insecticide, 60
Instantaneous, meaning of, 297
Instantaneous fishing-mortality rate, 213
Instantaneous infectivity rate, 337
Instantaneous mortality, 213
Instantaneous removal rate, 337
Interest rate, per annum compound, 125
Interference, 291, 295
International Whaling Commission, 96

Introduced hares, Anticosti Island, 148
Investment policies, alternative, 125
Irrigation diversion and distribution
    facilities, 396
Iterative process, 213

Japanese fishery, 77

Laboratory populations, 89
Lag, 366
    observed, 173
Lagrange multiplier, 250, 408
Lamprey, in Finger Lakes, 107
    in Great Lakes, 4, 26
    and lake trout, 106
    spawning migrations, 108
    spread in Great Lakes, 104
Land-management policies, 69
    antelope, 70
    buffalo, 70
Lea Act, 176
Legislation, 404
Life expectancy, median, 92
Local maxima or minima, 426
Log-log graph paper, 138, 266
Logical branched tree, 264
Logistic, 276
Look ahead, 417
    feature, 414
Loop, inner, 359, 373
    intermediate, 373
    outer, 359, 373
Low repeatability, 203

Magnetic tape, 360
Main effects, 191
Malaria, 157
Malnutrition, 129
Man-eating leopard of Rudraprayag, 142
Management, deer, 125
Marine communities, 46
Mark-recapture method, 204
Markov property, 410
Marquardt algorithm, 250
Mathematical models, predator-prey and
    host-parasite systems, 137
Mathematics, branches of, 184
    extremum problem, 6
    finite, 184
    infinitesimal, 184
    new, 178
Matrix, 233, 329

Matrix, adjugate, 236
    algebra, 232
    inverse, 234
Matter cycles, 36
Maturity index, 221
MAX and MIN, 187
Maximum annual sustained yield, 404
Maximum principle, 412
Maximum yield, 21
Memory requirements, 378
Mesh size, 404
Metabolism, active and basal, 275
Microclimate and population dynamics,
    tent caterpillar, 285
Microtus arvalis, 155
Microtus montanus, 156
Migration, 163
Mink-muskrat system, 136
Model, nonlinear, 243
    structural validity of, 433
    types of, 252
Monoculture, 41
Monte Carlo techniques, 340, 349
Mortality, 213
    natural, 21
    selective, 208
Mountain ranges, intervening, 288
Multiple regression analysis, 232
Multiple use, 390
Multistage decision process, 405
Murres, 27
Myxomatosis, 157

Natality, animal, 22
Negentropy, 345
Nesting sites, animal, 24
Net earnings, 363
Net profit, 125
Net return, 363
Nonlinear model, 243
North Sea fish, 4
North Sea fisheries, 77
Null hypotheses, 230
Nutritional plane, 291

Oligotrophic, 117
One-shot solutions, 406
Operators, restricting the number of, 363
Optimality principle, 410, 413
Optimum, 413
Optimum sample size, 198
Optimum strategy, 430

Productivity, regulating,
    temperature in, 113
  of various terrestrial plant associations, 38
Program structure, 360
  basic, 361
Programming, dynamic, 409, 411
Pseudoalgebraic codes, 184
Pursuit success, 321
Pursuit velocity, 323

Radial dispersal, 365
Random normal deviates, table of, 351
Ranking function, 391
Rate of exploitation, 22
Rats, psychological history of, 221
Reaction-rate kinetics, 277
Recruit, 80
Rectangular coordinate system, 328
Recursive, definition, 413
Regression, iterative, 242
Regression analysis, mixed-mode, 240
Regression line, slope of, 228
Regulation of numbers and average weight, 21
Relative gear efficiency, 363
Repetitive operations, 187
Representativeness, 190
Reproductive potential, 123
Reproductive rate, animal, 22
Reservoirs, 396
  midwestern, 117
Resource, food from the ocean, 15
  multiple-use, 178
Resource management, game and
    simulation, 5
Respiration loss, 37
Response, functional, 136
  numerical, 136
Rice acreage, 176
Ridges, 430
Royce-Schuck formulation, 254

Salmon, 361
  fishery, profitability of, 364
  gear-limitation policy, 361
Sampling effort, 434
  allocation of, 199
Sampling error, 230
Sampling procedures, 432
Sampling schemes, stratified, 190
Scabies, 124
Schooling, 219
Scientific sampling, 3

Search, efficiency, 375
  procedure, 430
  techniques, 427
Semilog graph paper, 266
Sensitivity analysis, 425
Sheep, 123, 124
Signs, 210
Simulation, 338
  studies, 349, 425
    parasite, 386
  technological and economic factors, 361
Slack variable, 250
Small mammals captured on trap line, 222
Smallpox, 157
Spacing, of individual plants, 47
Spatial distribution, of organisms, 46
  pattern, of the organism in the
      community, 40
Speciation, relationship with genetic
    heterozygosity, 49
Spraying, reaction to, 375
Spreading waves of upper-respiratory
    diseases, 339
Spruce budworm, 285
Standard error of estimate, 227
Standing crop, 21, 117
Steepest gradients up, 412
Stepwise approach, 239
Sterility in females, 294
Stimulus, unfamiliar, 319
Stochastic, 340
Stochastic models, 348
  in biology, 350
Stock-recruitment curves, 253
Stocking, pheasant cocks, 119
  ponds in Alabama, 115
  program, 404
Storms, 365
Stratification, 221
  sampling, 198
  variable, 197
Structural weakness in the model, 270
Submodels, 268
Subscripted variables, 187
Subsidiary component, 272
Sub-submodels, 268
Sum buckets, 339
Supraspecific phylogenetic units,
    organization of species, 40, 48
Survival, animal, 22
  pests, 404
  of sardines, 84